T0076236

MATLAB Recipes

A Problem-Solution Approach

Second Edition

Michael Paluszek

Stephanie Thomas

MATLAB Recipes: A Problem-Solution Approach

Michael Paluszek
Princeton, NJ
USA

Stephanie Thomas
Princeton Junction, NJ
USA

ISBN-13 (pbk): 978-1-4842-6123-1
https://doi.org/10.1007/978-1-4842-6124-8

ISBN-13 (electronic): 978-1-4842-6124-8

Copyright © 2020 by Michael Paluszek and Stephanie Thomas

This work is subject to copyright. All rights are reserved by the Publisher, whether the whole or part of the material is concerned, specifically the rights of translation, reprinting, reuse of illustrations, recitation, broadcasting, reproduction on microfilms or in any other physical way, and transmission or information storage and retrieval, electronic adaptation, computer software, or by similar or dissimilar methodology now known or hereafter developed.

Trademarked names, logos, and images may appear in this book. Rather than use a trademark symbol with every occurrence of a trademarked name, logo, or image we use the names, logos, and images only in an editorial fashion and to the benefit of the trademark owner, with no intention of infringement of the trademark.

The use in this publication of trade names, trademarks, service marks, and similar terms, even if they are not identified as such, is not to be taken as an expression of opinion as to whether or not they are subject to proprietary rights.

While the advice and information in this book are believed to be true and accurate at the date of publication, neither the authors nor the editors nor the publisher can accept any legal responsibility for any errors or omissions that may be made. The publisher makes no warranty, express or implied, with respect to the material contained herein.

Managing Director, Apress Media LLC: Welmoed Spahr
Acquisitions Editor: Steve Anglin
Development Editor: Matthew Moodie
Coordinating Editor: MarkPowers

Cover designed by eStudioCalamar

Cover image by Christopher Burns on Unsplash (www.unsplash.com)

Distributed to the book trade worldwide by Springer Science+Business Media New York, 233 Spring Street, 6th Floor, New York, NY 10013. Phone 1-800-SPRINGER, fax (201) 348-4505, e-mail orders-ny@springer-sbm.com, or visit www.springeronline.com. Apress Media, LLC is a California LLC and the sole member (owner) is Springer Science + Business Media Finance Inc (SSBM Finance Inc). SSBM Finance Inc is a **Delaware** corporation.

For information on translations, please e-mail booktranslations@springernature.com; for reprint, paperback, or audio rights, please e-mail bookpermissions@springernature.com.

Apress titles may be purchased in bulk for academic, corporate, or promotional use. eBook versions and licenses are also available for most titles. For more information, reference our Print and eBook Bulk Sales web page at http://www.apress.com/bulk-sales.

Any source code or other supplementary material referenced by the author in this book is available to readers on GitHub via the book's product page, located at www.apress.com/9781484261231. For more detailed information, please visit http://www.apress.com/source-code.

Contents

About the Authors   **XV**

Acknowledgements   **XVII**

Introduction   **XIX**

I   **Coding in MATLAB**   **1**

1   **Coding Handbook**   **3**

     MATLAB Language Primer   3
         Brief Introduction to MATLAB   3
         Everything Is a Matrix   6
         Strings Are Simple   8
         Use Strict Data Structures   9
         Cell Arrays Hold Anything and Everything 11
         Optimize Your Code with Logical Arrays 12
         Use Persistent and Global Scope to Minimize Data Passing 12
         Understanding Unique MATLAB Operators and Keywords 14
         Harnessing the Power of Multiple Inputs and Outputs 16
         Use Function Handles for Efficiency 17
         Numerics 18
         Images 18
         Datastore 20
         Tall Arrays 22
         Sparse Matrices 22
         Tables and Categoricals 23
         Large MAT-files 23
         Advanced Data Types 25
    1.1   Initializing a Data Structure Using Parameters 25
         Problem 26
         Solution 26
         How It Works 26

1.2   Performing mapreduce on an Image Datastore 29
       Problem 29
       Solution 29
       How It Works 29
1.3   Creating a Table from a File 31
       Problem 32
       Solution 32
       How It Works 32
1.4   Processing Table Data 33
       Problem 33
       Solution 33
       How It Works 34
1.5   String Concatenation 37
       Problem 37
       Solution 37
       How It Works 37
1.6   Arrays of Strings 37
       Problem 37
       Solution 37
       How It Works 37
1.7   Non-English Strings 38
       Problem 38
       Solution 38
       How It Works 38
1.8   Substrings 39
       Problem 39
       Solution 39
       How It Works 39
1.9   Using JSON-Formatted Strings 39
1.10   Creating Function Help 40
       Problem 40
       Solution 40
       How It Works 40
1.11   Locating Directories for Data Storage 42
       Problem 42
       Solution 42
       How It Works 42
1.12   Loading Binary Data from a File 43
       Problem 43
       Solution 43
       How It Works 43

1.13   Command-Line File Interaction 45
       Problem 45
       Solution 45
       How It Works 45
1.14   Using a MEX File to Link to an External Library 47
       Problem 47
       Solution 47
       How It Works 47
1.15   Protect Your IP with Parsed Files 50
       Problem 50
       Solution 50
       How It Works 50
1.16   Writing to a Text File 51
       Problem 51
       Solution 51
       How It Works 51
1.17   Using an Explicit Expansion 55
       Problem 55
       Solution 55
       How It Works 55
1.18   Using a Script Subfunction 56
       Problem 56
       Solution 56
       How It Works 56
1.19   Using Memoize 58
       Problem 58
       Solution 58
       How It Works 58
1.20   Using Java 59
       Problem 59
       Solution 59
       How It Works 59
1.21   Creating Documents 61
       Problem 61
       Solution 61
       How It Works 61
1.22   MATLAB Online 64
       Problem 64
       Solution 64
       How It Works 64

2 MATLAB Style   **69**
2.1   Developing Your Own MATLAB Style Guidelines   69
      Problem   69
      Solution   70
      How It Works   70
2.2   Writing Good Function Help   74
      Problem   74
      Solution   74
      How It Works   74
2.3   Overloading Functions and Utilizing `varargin`   77
      Problem   77
      Solution   77
      How It Works   77
2.4   Adding Built-in Inputs and Outputs to Functions   78
      Problem   78
      Solution   78
      How It Works   78
2.5   Adding Argument Checking to Functions   80
      Problem   80
      Solution   80
      How It Works   80
2.6   Adding Dot Indexing   82
      Problem   82
      Solution   82
      How It Works   82
2.7   Smart Structuring of Scripts   82
      Problem   82
      Solution   82
      How It Works   83
2.8   Implementing MATLAB Command-Line Help for Folders   85
      Problem   85
      Solution   85
      How It Works   85
2.9   Publishing Code into Technical Reports   88
      Problem   88
      Solution   88
      How It Works   89
2.10 Integrating Toolbox Documentation into the MATLAB Help System   93
      Problem   93
      Solution   93
      How It Works   93

2.11   Structuring a Toolbox 98

Problem 98

Solution 98

How It Works 98

3   Visualization   101

3.1   Plotting Data Interactively from the MATLAB Desktop 102

Problem 102

Solution 102

How It Works 102

3.2   Incrementally Annotate a Plot 107

Problem 107

Solution 107

How It Works 107

3.3   Create a Custom Plot Page with Subplot 109

Problem 109

Solution 109

How It Works 109

3.4   Create a Heat Map 112

Problem 112

Solution 112

How It Works 112

3.5   Create a Plot Page with Custom-Sized Axes 113

Problem 113

Solution 113

How It Works 114

3.6   Plotting with Dates 115

Problem 115

Solution 116

How It Works 116

3.7   Generating a Color Distribution 118

Problem 118

Solution 119

How It Works 119

3.8   Visualizing Data over 2D or 3D Grids 120

Problem 120

Solution 120

How It Works 120

3.9   Generate 3D Objects Using Patch 123

Problem 123

Solution 123

How It Works 124

3.10   Working with Light Objects 125
       Problem 125
       Solution 125
       How It Works 125
3.11   Programmatically Setting the Camera Properties 129
       Problem 129
       Solution 129
       How It Works 130
3.12   Display an Image 132
       Problem 132
       Solution 132
       How It Works 132
3.13   Graph and Digraph 135
       Problem 135
       Solution 135
       How It Works 135
3.14   Adding a Watermark 138
       Problem 138
       Solution 138
       How It Works 138

4 **Interactive Graphics**   **143**
4.1   Creating a Simple Animation 143
       Problem 143
       Solution 143
       How It Works 143
4.2   Playing Back an Animation 147
       Problem 147
       Solution 147
       How It Works 147
4.3   Animate Line Objects 148
       Problem 148
       Solution 148
       How It Works 149
4.4   Implementation of a `uicontrol` Button 152
       Problem 152
       Solution 153
       How It Works 153
4.5   Display Status of a Running Simulation or Loop 155
       Problem 155
       Solution 155
       How It Works 155

4.6   Create a Custom GUI with App Designer 158
        Problem 158
        Solution 158
        How It Works 158
4.7   Build a Data Acquisition GUI 165
        Problem 165
        Solution 165
        How It Works 166

5   Testing and Debugging   175
5.1   Creating a Unit Test 176
        Problem 176
        Solution 176
        How It Works 177
5.2   Running a Test Suite 184
        Problem 184
        Solution 184
        How It Works 185
5.3   Setting Verbosity Levels in Tests 186
        Problem 186
        Solution 186
        How It Works 186
5.4   Create a Logging Function to Display Data 189
        Problem 189
        Solution 189
        How It Works 189
5.5   Generating and Tracing MATLAB Errors and Warnings 191
        Problem 191
        Solution 191
        How It Works 191
5.6   Testing Custom Errors and Warnings 192
        Problem 192
        Solution 192
        How It Works 192
5.7   Testing Generation of Figures 196
        Problem 196
        Solution 196
        How It Works 196

6   Classes   **199**

6.1   Object-Oriented Programming 199

6.2   State Space Systems Base Class 201

Problem 201

Solution 201

How It Works 201

6.3   State Space Systems Discrete Class 206

Problem 206

Solution 206

How It Works 206

6.4   Using the State Space Class 207

Problem 207

Solution 208

How It Works 208

6.5   Using a Mocking Framework 209

Problem 209

Solution 209

How It Works 209

II   Applications   **213**

7   The Double Integrator   **215**

7.1   Writing the Equations for the Double Integrator Model 215

Problem 215

Solution 215

How It Works 215

7.2   Creating a Fixed-Step Numerical Integrator 216

Problem 216

Solution 217

How It Works 217

7.3   Implement a Discrete Proportional-Derivative

Controller 219

Problem 219

Solution 219

How It Works 220

7.4   Double Integrator Simulation 225

Problem 225

Solution 225

How It Works 225

7.5   Create Time Axes with Reasonable Time Units 229

Problem 229

Solution 229
How It Works 229
7.6   Create Figures with Multiple Subplots 230
Problem 230
Solution 230
How It Works 230

8   Robotics   235
8.1   Creating a Dynamical Model of the SCARA Robot 235
Problem 235
Solution 235
How It Works 236
8.2   Customize a Visualization Function for the Robot 238
Problem 238
Solution 238
How It Works 239
8.3   Using Numerical Search for Robot Inverse Kinematics 243
Problem 243
Solution 243
How It Works 244
8.4   Developing a Control System for the Robot 246
Problem 246
Solution 246
How It Works 246
8.5   Simulating the Controlled Robot 248
Problem 248
Solution 248
How It Works 248
Summary 251

9   Electric Motors   253
9.1   Motor Model 253
Problem 253
Solution 255
How It Works 255
9.2   Controlling the Motor 258
Problem 258
Solution 258
How It Works 258
9.3   Pulsewidth Modulation of the Switches 262
Problem 262

Solution 262
How It Works 262

9.4 Simulating the Controlled Motor 270
Problem 270
Solution 270
How It Works 270

10 Fault Detection   **277**
10.1 Modeling an Air Turbine 277
Problem 277
Solution 277
How It Works 278
10.2 Building a Detection Filter 280
Problem 280
Solution 280
How It Works 281
10.3 Simulating the Fault Detection System 284
Problem 284
Solution 284
How It Works 284
10.4 Building a GUI for the Detection Filter Simulation 287
Problem 287
Solution 287
How It Works 288

11 Chemical Processes   **299**
11.1 Modeling the Chemical Mixing Process 299
Problem 299
Solution 299
How It Works 300
11.2 Sensing the pH of the Chemical Process 303
Problem 303
Solution 303
How It Works 303
11.3 Controlling the Effluent pH 311
Problem 311
Solution 311
How It Works 311
11.4 Simulating the Controlled pH Process 312
Problem 312
Solution 312
How It Works 312

12 Aircraft   **327**

12.1   Creating a Dynamical Model of an Aircraft 327
     Problem 327
     Solution 327
     How It Works 327

12.2   Equilibrium Control 332
     Problem 332
     Solution 332
     How It Works 332

12.3   Designing a Control System for an Aircraft 340
     Problem 340
     Solution 340
     How It Works 340

12.4   Plotting a 3D Trajectory for an Aircraft 342
     Problem 342
     Solution 342
     How It Works 342

12.5   Simulating the Controlled Aircraft 346
     Problem 346
     Solution 346
     How It Works 346

12.6   Draw an Aircraft 351
     Problem 351
     Solution 351
     How It Works 351

13 Spacecraft Attitude Control   **359**

13.1   Creating a Dynamical Model of the Spacecraft 359
     Problem 359
     Solution 360
     How It Works 360

13.2   Computing Angle Errors from Quaternions 364
     Problem 364
     Solution 364
     How It Works 364

13.3   Simulating the Controlled Spacecraft 366
     Problem 366
     Solution 366
     How It Works 366

13.4 Performing Batch Runs of a Simulation 373
     Problem 373
     Solution 373
     How It Works 373

14 Automobiles   **383**
14.1 Automobile Dynamics 383
     Problem 383
     Solution 383
     How It Works 383
14.2 Modeling the Automobile Radar 389
     Problem 389
     Solution 389
     How It Works 389
14.3 Automobile Autonomous Passing Control 393
     Problem 393
     Solution 393
     How It Works 394
14.4 Automobile Animation 395
     Problem 395
     Solution 395
     How It Works 395
14.5 Modeling an Automobile Suspension 400
     Problem 400
     Solution 400
     How It Works 400

Bibliography   **407**

Index   **409**

About the Authors

Michael Paluszek is President of Princeton Satellite Systems, Inc. (PSS) in Plainsboro, New Jersey. Mr. Paluszek founded PSS in 1992 to provide aerospace consulting services. He used MATLAB to develop the control system and simulations for the IndoStar-1 geosynchronous communications satellite. This led to the launch of Princeton Satellite Systems' first commercial MATLAB toolbox, the Spacecraft Control Toolbox, in 1995. Since then, he has developed toolboxes and software packages for aircraft, submarines, robotics, and nuclear fusion propulsion, resulting in Princeton Satellite Systems' current extensive product line. He is working with the Princeton Plasma Physics Laboratory on a compact nuclear fusion reactor for energy generation and space propulsion.

Prior to founding PSS, Mr. Paluszek was an engineer at GE Astro Space in East Windsor, NJ. At GE, he designed the Global Geospace Sciences Polar despun platform control system and led the design of the GPS IIR attitude control system, the Inmarsat-3 attitude control systems, and the Mars Observer delta-V control system, leveraging MATLAB for control design. Mr. Paluszek also worked on the attitude determination system for the DMSP meteorological satellites. He flew communication satellites on over 12 satellite launches, including the GSTAR III recovery, the first transfer of a satellite to an operational orbit using electric thrusters. At Draper Laboratory, Mr. Paluszek worked on the Space Shuttle, Space Station, and submarine navigation. His Space Station work included designing of Control Moment Gyro-based control systems for attitude control.

Mr. Paluszek received his bachelor's degree in Electrical Engineering and master's and Engineer's degrees in Aeronautics and Astronautics from the Massachusetts Institute of Technology. He is author of numerous papers and has over a dozen US patents. Mr. Paluszek is coauthor of MATLAB Recipes, MATLAB Machine Learning, MATLAB Machine Learning Recipes: A Problem-Solution Approach, and Practical MATLAB Deep Learning, all published by Apress.

Stephanie Thomas is Vice President of Princeton Satellite Systems, Inc. in Plainsboro, New Jersey. She received her bachelor's and master's degrees in Aeronautics and Astronautics from the Massachusetts Institute of Technology in 1999 and 2001. Ms. Thomas was introduced to the PSS Spacecraft Control Toolbox for MATLAB during a summer internship in 1996 and has been using MATLAB for aerospace analysis ever since. In her nearly 20 years of MATLAB experience, she has developed many software tools including the Solar Sail Module for the Spacecraft Control Toolbox, a proximity satellite operations toolbox for the Air Force, collision monitoring Simulink blocks for the Prisma satellite mission, and launch vehicle analysis tools in MATLAB and Java. She has developed novel methods for space situation assessment such as a numeric approach to assessing the general rendezvous problem between any two satellites implemented in both MATLAB and C++. Ms. Thomas has contributed to PSS' Attitude and Orbit Control textbook, featuring examples using the Spacecraft Control Toolbox, and written many software user guides. She has conducted SCT training for engineers from diverse locales such as Australia, Canada, Brazil, and Thailand and has performed MATLAB consulting for NASA, the Air Force, and the European Space Agency. Ms. Thomas is coauthor of *MATLAB Recipes*, *MATLAB Machine Learning, MATLAB Machine Learning Recipes: A Problem-Solution Approach*, and *Practical MATLAB Deep Learning*, published by Apress. In 2016, Ms. Thomas was named a NASA NIAC Fellow for the project "Fusion-Enabled Pluto Orbiter and Lander."

Acknowledgments

We would like to acknowledge Joseph Mueller for his expert editing of this book.

Introduction

Writing software has become part of the job description for nearly every professional engineer and engineering student. While there are many excellent prebuilt software applications for engineers, almost everyone can benefit by writing custom software for their own problems.

MATLAB had its origins for that very reason. Scientists that needed to do operations on matrices used numerical software written in FORTRAN. At the time, using computer languages required the user to go through the write-compile-link-execute process that was time-consuming and error-prone. MATLAB presented the user with a scripting language that allowed the user to solve many problems with a few lines of a script that executed instantaneously. MATLAB had built-in visualization tools that helped the user better understand the results. Writing MATLAB was a lot more productive and fun than writing FORTRAN.

MATLAB has grown greatly since those days. The power of the basic MATLAB software has grown dramatically, and hundreds of MATLAB libraries are now available, both commercially and as open source. MATLAB is so sophisticated that most new users only use a fraction of its power.

The goal of *MATLAB Recipes* is to help all users harness the power of MATLAB. This book has two parts. The first part, Chapters 1 through 5, gives a framework that you can use to write high-quality MATLAB code that you, your colleagues, and possibly your customers can utilize. We cover coding practices, graphics, debugging, and other topics in a problem-solution format. You can read these sections from cover to cover or just look at the recipes that interest you and use them in your latest MATLAB code.

The second part of the book, Chapters 6 through 12, shows complete MATLAB applications revolving around the control of and simulation of dynamical systems. Each chapter provides the technical background for the topic, ideas on how you can write a simple control system, and an example of how you might simulate the system. Each system is implemented in a MATLAB script supported by a number of MATLAB functions. Each chapter also highlights a general MATLAB topic, like graphics or writing graphical user interfaces (GUIs). We have deliberately made the control systems simple so that the reader won't need a course in control theory to get results. Control experts can easily take the script and implement their own ideas. We cover a number of areas, ranging from chemical processes to satellites – and we apologize if we didn't write an example for your area of interest!

The book has something for everyone – from the MATLAB novice to the authors of commercial MATLAB packages. We learned new things writing this book! We hope that you enjoy the book and look forward to seeing your software that it inspires.

Part I

Coding in MATLAB

CHAPTER 1

■ ■ ■

Coding Handbook

The purpose of this chapter is to provide an overview of MATLAB syntax and programming, highlighting features that may be underutilized by many users and noting important differences between MATLAB and other programming languages and IDEs. You should also become familiar with the very detailed documentation that is available from the MathWorks in the help browser. The *Language Fundamentals* section describes entering commands, operators, and data types.

MATLAB has matured a lot in the last two decades from its origins as a linear algebra package. Originally, all variables were double precision matrices. Today, MATLAB provides different variable types such as integers, data structures, object-oriented programming and classes, and integration with Java. The MATLAB application is a full IDE with an integrated editor, debugger, command history, and code analyzer and report capabilities. Engineers who have been working with MATLAB for many years may find that they are not taking advantage of the full range of capabilities now offered, and in this text we hope to highlight the more useful new features.

The first part of this chapter provides an overview of the most commonly used MATLAB types and constructs. We'll then provide some recipes that make use of these constructs to show you some practical applications of modern MATLAB.

MATLAB Language Primer

Brief Introduction to MATLAB

MATLAB is both an application and a programming language. It was developed primarily for numerical computing and is widely used in academia and industry. MATLAB was originally developed by a college professor in the 1970s to provide easy access to linear algebra libraries, and the MathWorks was founded in 1984 to continue the development of the product. The name is derived from *MATrix LABoratory*. Today, MATLAB uses the LAPACK libraries for the underlying matrix manipulations. Many toolboxes are available for different engineering disciplines; in this book, we will focus on features available only in the base MATLAB application.

© Michael Paluszek and Stephanie Thomas 2020
M. Paluszek and S. Thomas, *MATLAB Recipes*,
https://doi.org/10.1007/978-1-4842-6124-8_1

The MATLAB application is a rich development environment for the MATLAB language. It provides an editor, command terminal, debugger, plotting capabilities, creation of graphical user interfaces, and more recently the ability to install third-party apps. MATLAB can interface with other languages including FORTRAN, C, C++, Java, and Python. A code analyzer and profiler are built-in. Extensive online communities provide forums for sharing code and asking questions.

The main components of the MATLAB application are

Command Window – Terminal for entering commands and operating on variables in the base workspace. The MATLAB prompt is $>>$.

Command History – List of previously executed commands.

Workspace display – List of the variables and their values in the current workspace (application memory). Variables remain in the memory once created until you explicitly clear them or close MATLAB.

Current Folder – File browser displaying contents of the current folder and providing file system navigation. Recent versions of MATLAB can also display SVN status on configuration managed files.

File details – Panel displaying information on the file selected in the Current Folder panel.

Editor – Editor for m-files with syntax coloring and a built-in debugger. This can also display any type of text file and will recognize and appropriately color other languages including Java, C/C++, and XML/HTML.

Variables editor – Spreadsheet-like graphical editor for variables in the workspace.

App Designer – Application development window.

Help browser – Searchable help documentation on all MATLAB products and third-party products you have installed.

Profiler – Tool for timing code as it runs.

These components can be docked in various configurations. The default layout of the main application window or *desktop* contains the first five components listed earlier and is shown in Figure 1.1. The Command Window is in the center. The upper-left panel shows a file browser with the contents of the Current Folder. Under this is a file information display. On the right-hand side is the Workspace display and the Command History panel. The *base workspace* is all the variables currently in the application memory. Commands from the history can be dragged onto the command line to be executed, or double-clicked. The extensive toolbar includes buttons for running the code analyzer and opening the code profiler and the help window, as well as typical file and data operations. Note the PLOTS and APPS tabs above the toolbar. The PLOTS

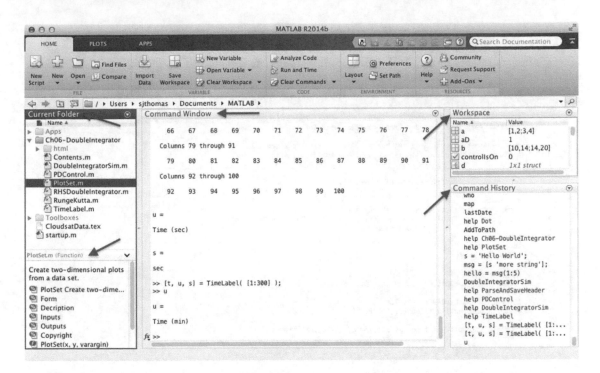

Figure 1.1: *MATLAB desktop with the Command Window.*

tab allows the graphical creation and management of plots from data selected in the workspace browser. The APPS tab allows you to access and manage third-party apps that you install.

You can rearrange the components in the application window, moving, resizing, or hiding them, and save your own layouts. You can "undock" any component, moving it to its own window. You can also revert back to the default layout at any time or choose from several other available configurations. You can also hide the toolstrip to get more real estate for your windows. There are new capabilities to customize your interface with each version, so explore what's new!

The editor with the default syntax coloring is shown in Figure 1.2, with a file from this chapter shown. The horizontal lines show the division of the code into "cells" using a double-percent sign, which can be used for sequential execution of code and for creating sections of text when publishing. The cell titles are bolded in the editor. MATLAB keywords are highlighted in blue, comments in green, and strings in pink. The toolbar includes buttons for commenting code, indenting, and running or debugging the code. The "Go To" pop-up menu gives access to subfunctions within a large file (see Section 1.10). Note the PUBLISH and VIEW tabs with additional features on publishing, covered in the next chapter, and options for the editor view.

The last window we will show is the help browser in Figure 1.3. MATLAB has extensive help including examples and links to online videos and tutorials. Third-party toolboxes can also install help into this browser. Like any browser, you can have open multiple tabs, there is

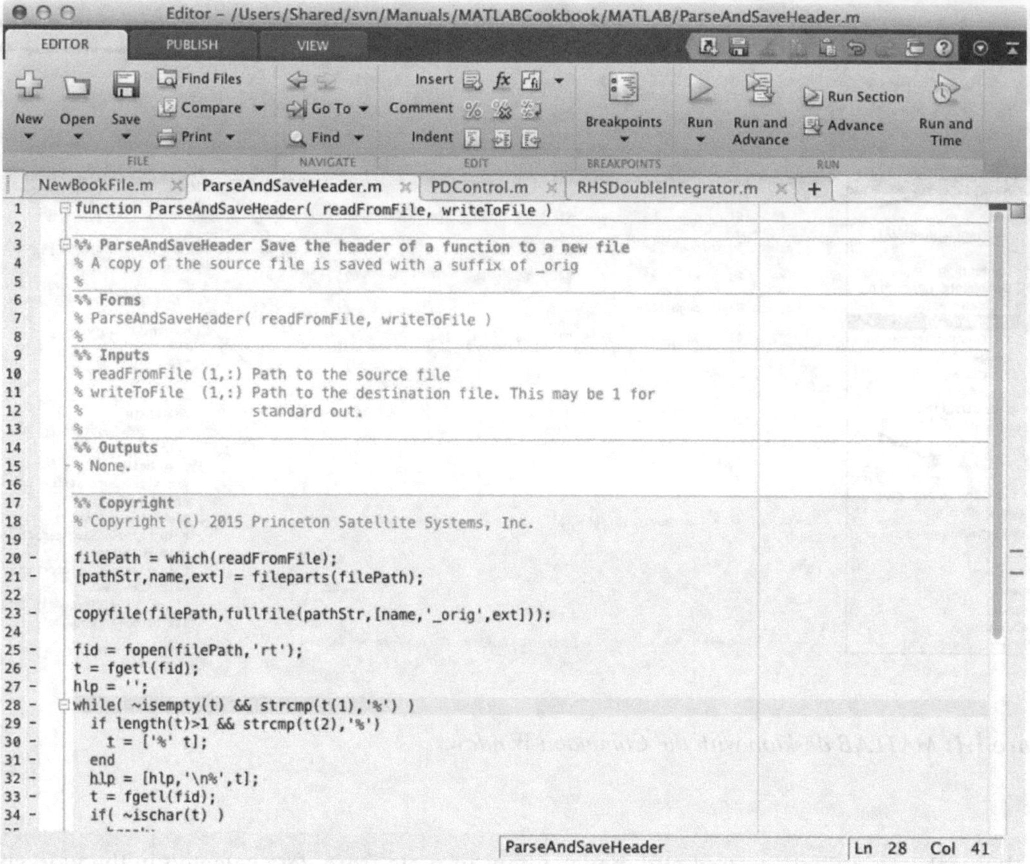

Figure 1.2: *MATLAB file editor.*

a search utility, and you can mark favorite topics. We will refer to topics available in the help browser throughout this book.

Everything Is a Matrix

By default, all variables in MATLAB are double precision matrices. You do not need to declare a type for these variables. Matrices can be multidimensional and are accessed using one-based indices via parentheses. You can address elements of a matrix using a single index, taken column-wise, or one index per dimension. Use square brackets to enclose the matrix data and semicolons to mark the end of rows. Use a final semicolon to end the line, or leave it off to print the result to the command line. To create a matrix variable, simply assign a value to it, like this 2x2 matrix a and 2x1 matrix b:

```
>> a = [1 2; 3 4];
>> a(1,1)
```

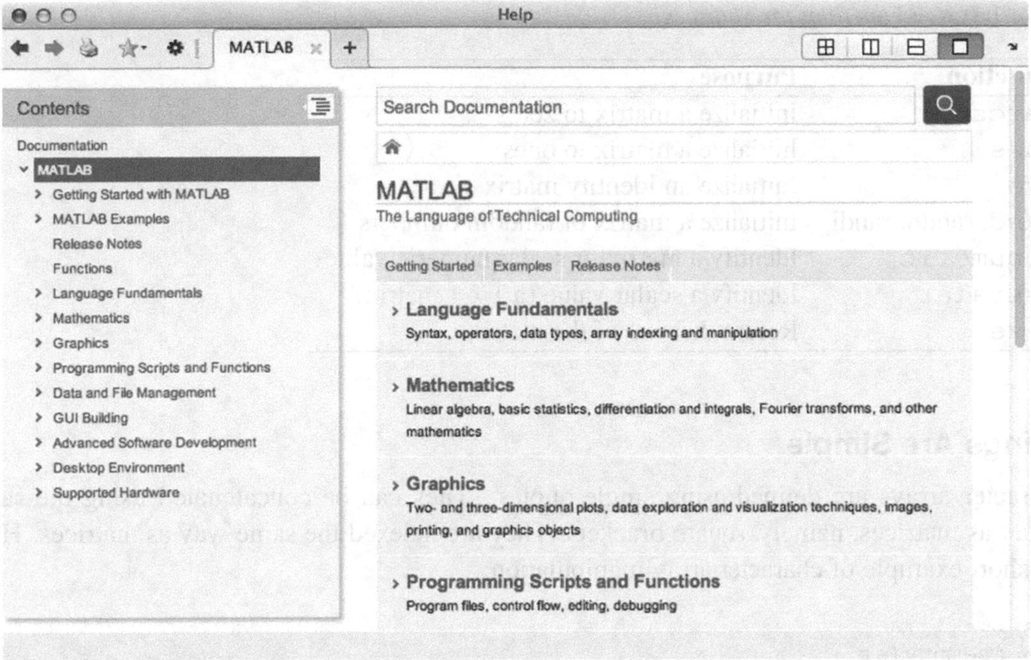

Figure 1.3: *MATLAB help window.*

```
      1
>> a(3)
      2
>> b = [5; 6]
b =

      5
      6
```

You can simply add, subtract, multiply, and divide matrices with no special syntax. The matrices must be the correct size for the linear algebra operation requested. A transpose is indicated using a single quote suffix, A', and the matrix power uses the operator ^.

```
>> b = a'*a;
>> c = a^2;
>> d = b + c;
```

By default, every variable is a numerical variable. You can initialize matrices to a given size using the `zeros`, `ones`, `eye`, or `rand` functions, which produce zeros, ones, identity matrices (ones on the diagonal), and random numbers, respectively. Use `isnumeric` to identify numeric variables. Table 1.1 shows key matrix functions.

7

Table 1.1: *Key Functions for Matrices*

Function	Purpose
`zeros`	Initialize a matrix to zeros
`ones`	Initialize a matrix to ones
`eye`	Initialize an identity matrix
`rand, randn, randi`	Initialize a matrix of random numbers
`isnumeric`	Identify a matrix or scalar numeric value
`isscalar`	Identify a scalar value (a 1 x 1 matrix)
`size`	Return the size of the matrix

Strings Are Simple

Character arrays are defined using single quotes. They can be concatenated using the same syntax as matrices, namely, square brackets. They are indexed the same way as matrices. Here is a short example of character array manipulation:

```
>> s = '';
>> isempty(s)
ans =
  logical
   1
>> s = 'Hello World';
>> msg = [s ' more chars']
msg =
    'Hello World more chars'
>> hello = msg(1:5)
hello =
    'Hello'
```

Use `ischar` to identify character variables. Also note that `isempty` returns TRUE for an empty array, that is, `''`.

Since R2016b, MATLAB has also provided a string type defined using regular quotes. Some newer functions are designed to operate specifically on strings, but most work on both text types. If you concatenate strings using square brackets, they are maintained as separate elements in an array rather than combined as character arrays are. To append strings, use the "+" operator (see Recipe 1.5). `isempty` returns FALSE for an empty string, that is, `` `' ' ``; this creates a 1-by-1 string with no characters rather than an empty string.

```
>> str = "";
>> isempty(str)
ans =
  logical
   0
>> s = "Hello World";
>>  msg = [s "additional string"]
msg =
```

```
1x2 string array
  "Hello World"    "additional string"
```

For a description of string syntax, type `help strings` at the MATLAB command line, and for a comprehensive list of string and character functions, type `help strfun`. Table 1.2 shows a selection of key string functions.

Table 1.2: *Key Functions for Strings*

Function	Purpose
ischar	Identify a character array
isstring	Identify a string
char	Convert integer codes or cell array to character array
sprintf	Write formatted data to a string
strcmp, strncmp	Compare strings
strfind	Find one string within another
num2str, mat2str	Convert a number or matrix to a string
lower	Convert a string to lowercase
contains	Search for patterns in string arrays
split	Split strings at whitespace

Use Strict Data Structures

Data structures in MATLAB are highly flexible, leaving it up to the user to enforce consistency in fields and types. You are not required to initialize a data structure before assigning fields to it, but it is a good idea to do so, especially in scripts, to avoid variable conflicts.

Replace

```
d.fieldName = 0;
```

with

```
d = struct;
d.fieldName = 0;
```

In fact, we have found it is generally a good idea to create a special function to initialize larger structures that are used throughout a set of functions. This is similar to creating a class definition. Generating your data structure from a function, instead of typing out the fields in a script, means you always start with the correct fields. Having an initialization function also allows you to specify the types of variables and provide sample or default data. Remember, since MATLAB does not require you to declare variable types, doing so yourself with default data makes your code that much clearer.

■ **TIP**   Create an initialization function for data structures.

You make a data structure into an array simply by assigning an additional copy. The fields must be in the same order, which is yet another reason to use a function to initialize your structure. You can nest data structures with no limit on depth.

```
1  d = MyStruct;
2  d(2) = MyStruct;
3
4  function d = MyStruct
5
6  d = struct;
7  d.a = 1.0;
8  d.b = 'string';
```

MATLAB now allows for *dynamic field names* using variables, that is, `structName.`(`dynamicExpression`). This provides improved performance over `getfield`, where the field name is passed as a string. This allows for all sorts of inventive structure programming. Take our data structure array in the previous code snippet, and let's get the values of field a using a dynamic field name; the values are returned in a cell array.

```
>> field = 'a';
>> values = {d.(field)}

values =

    [1]     [1]
```

Use `isstruct` to identify structure variables and `isfield` to check for the existence of fields. Note that `isempty` will return *false* for a struct initialized with `struct`, even if it has no fields. Table 1.3 lists some key functions for interacting with structs.

Table 1.3: *Key Functions for Structs*

Function	Purpose
struct	Initialize a structure with or without fields
isstruct	Identify a structure
isfield	Determine if a field exists in a structure
fieldnames	Get the fields of a structure in a cell array
rmfield	Remove a field from a structure
deal	Set fields in a structure to a value

Cell Arrays Hold Anything and Everything

One variable type unique to MATLAB is the cell array. This is really a list container, and you can store variables of any type in elements of a cell array. Cell arrays can be multidimensional, just like matrices, and are useful in many contexts.

Cell arrays are indicated by curly braces, {}. They can be of any dimension and contain any data, including string, structures, and objects. You can initialize them using the `cell` function, recursively display the contents using `celldisp`, and access subsets using parentheses just like for a matrix. The following is a short example.

```
>> c = cell(3,1);
>> c{1} = 'some text';
>> c{2} = false;
>> c{3} = [1 2; 3 4];
>> b = c(1:2);
>> celldisp(b)
b{1} =
string

b{2} =
    0
```

Using curly braces for access gives you the element data as the underlying type. When you access elements of a cell array using parentheses, the contents are returned as another cell array, rather than the cell contents. MATLAB help has a special section called *Comma-Separated Lists* which highlights the use of cell arrays as lists. The code analyzer will also suggest more efficient ways to use cell arrays, for instance:

Replace

```
    a = {b{:} c};
```

with

```
    a = [b {c}];
```

Cell arrays are especially useful for sets of strings, with many of MATLAB's string search functions optimized for cell arrays, such as `strcmp`.

Use `iscell` to identify cell array variables. Use `deal` to manipulate structure array and cell array contents. Table 1.4 shows a selection of key cell array functions.

Table 1.4: Key Functions for Cell Arrays

Function	Purpose
cell	Initialize a cell array
cellstr	Create cell array from a character array
iscell	Identify a cell array
iscellstr	Identify a cell array containing only strings
celldisp	Recursively display the contents of a cell array

Optimize Your Code with Logical Arrays

A *logical array* is composed of only ones and zeros. You can initialize logical matrices using the `true` and `false` functions, and there is an `islogical` function to test if a matrix is logical. Logical arrays are outputs of numerous built-in functions, like `isnan`, and are often recommended by the code analyzer as a faster alternative to manipulating array indices. For example, you may need to set any negative values in your array to zero.

Replace

```
1  k = find(x<0);
2  x(k) = 0;
```

with

```
1  x(x<0) = 0;
```

where `x<0` produces a logical array with 1 where the values of x are negative and 0 elsewhere.

MATLAB provides both traditional relational operators, that is, && for AND and || for OR, as well as unique element-wise operators. These element-wise operators, that is, single & and |, compare matrices of the same size and return logical arrays. Table 1.5 shows some key functions for logical operations.

Table 1.5: *Key Functions for Logical Operations*

Function	Purpose
logical	Convert numeric values to logical
islogical	Identify a logical array (composed of 1s and 0s)
true	Return a true value (1) or array (M,N)
false	Return a false value (0) or array (M,N)
any	Return true if any value in the array is a nonzero number
all	Return true if none of the values in the array is 0
and, or	Functional forms of element-wise operators & and	
isnan, isinf, isfinite	Values testing functions returning logical arrays

Use Persistent and Global Scope to Minimize Data Passing

In general, variables defined in a function have a local scope and are only available within that function. Variables defined in a script are available in the workspace and, therefore, from the command line.

MATLAB has a *global* scope which is the same as any other language, applying to the base workspace and maintaining the variable's value throughout the MATLAB session. Global variables are empty once declared, until initialized. The `clear` and `clearvars` functions each have flags for removing only the global variables. This is shown in the example below.

```
>> global MY_GLOBAL_VAR; % variable is empty
>> MY_GLOBAL_VAR = 1.0;
>> whos
  Name                    Size            Bytes  Class      Attributes

  MY_GLOBAL_VAR           1x1                 8  double     global

>> clearvars -GLOBAL
```

MATLAB has a unique scope that pertains to a single function, `persistent`. This is useful for initializing a function that requires a lot of data or computation and then saving that data for use in later calls. The variable can be reset using the `clear` command on the function, that is, `clear functionName`. This can also be a source of bugs so it is important to note the use of persistent variables in a function's help comments, so you don't get unexpected results when you switch models.

■ **TIP**   Use a persistent variable to store initialization data for subsequent function calls.

Variables can also be in scope for multiple functions defined in a single file, if the `end` keyword is used appropriately. In general, you can omit a final `end` for functions, but if you use it to wrap the inner functions, the functions become *nested* and can access variables defined in the parent function. This allows subroutines to share data without passing large numbers of arguments. The editor will highlight the variables that are so defined.

In the following example, the `constant` variable is available to the nested function inside the parent function.

NESTED FUNCTION

```
1   function y = parentFunction( x )
2
3   constant = 3.0;
4   y = nestedFunction( x );
5
6       function z = nestedFunction( x )
7
8       z = constant*x;
9
10      end
11
12  end
```

Table 1.6 shows a selection of scope functions.

Table 1.6: *Key Functions for Scope Operations*

Function	Purpose
`persistent`	Specify persistent scope for a variable in a function
`global`	Specify global scope for a variable
`clear`	Clear a function or variable
`who, whos`	List variables in a workspace
`mlock, munlock`	Lock (and unlock) a function or MEX-file which prevents it from being cleared

Understanding Unique MATLAB Operators and Keywords

Some common operators have special features in MATLAB, which we call attention to here.

Colon

The colon operator for creating a list of indices in an array is unique to MATLAB. A single colon used by itself addresses all elements in that given dimension; a colon used between a pair of integers creates a list.

```
>> a(1,1:2)
ans =
    1    2

>> a(:,1)
ans =
    1
    3
```

The colon operator applies to all variable types when accessing elements of an array: cell arrays, strings, data structure arrays.

The colon operator can also be used to create an array using an interval, as a shorthand to `linspace`. The interval and the endpoints can be doubles. Using it for matrix indices is really an edge case using a default interval of 1. For example, `0.1:0.2:0.5` produces `0.1  0.3  0.5`.

```
>> a = 0.1:0.2:0.5
a =
        0.1         0.3         0.5
```

14

Tilde

The tilde ($\sim$) is the logical NOT operator in MATLAB. The output is a logical matrix of the same size as the input, with values of 1 if the input value is 0 and a value of 0 otherwise.

```
>> a = [0 -1; 1 0];
>> ~a
ans =
     1     0
     0     1
```

In newer versions, it also can be used to ignore an input or output to a function, and this is suggested often in the code analyzer as preferable to the use of a dummy variable.

```
1  [~,b] = MyFunction(x,y);
```

Dot

By *dot*, we mean using a period with a standard arithmetic operator, like .* or .\ or .^. This is a special syntax in MATLAB used to apply an operator on an element per element basis over the matrices, instead of performing the linear algebra operation otherwise implied. This is also termed an *array operation* as opposed to a *matrix operation*. Since the matrix and array operations are the same for addition and subtraction, the dot is not required.

```
1  y = a.*b;
```

MATLAB is optimized for array operations. Using this syntax is a key way to reduce `for` loops in your MATLAB code and make it run faster. Consider the traditional alternative code:

```
1  a = rand(1,1000);
2  b = rand(1,1000);
3  y = zeros(1,1000);
4  for k = 1:1000
5    y(k) = a(k)*b(k);
6  end
```

Even this simple example takes two to three times as long to run as the vectorized version shown above.

end

The end keyword serves multiple purposes in MATLAB. It is used to terminate `for`, `while`, `switch`, `try`, and `if` statements, rather than using braces as in other languages. It is also used to serve as the last index of a variable in a given dimension. Using end appropriately can make your code more robust to future changes in the size of your data.

```
>> a = [1 2 3; 4 5 6; 7 8 9];
>> b = a(1:end-1,2:end)
b =
```

```
2       3
5       6
```

Harnessing the Power of Multiple Inputs and Outputs

Uniquely, MATLAB functions can have multiple outputs. They are specified in a comma-separated list just like the inputs. Additionally, you do not need to specify the data types of the inputs or outputs, and you can silently override the output types by assigning any data you want to the variables. Thus, a function can have an infinite number of syntaxes defined within a single file. Outputs must be assigned the names given in the signature; you cannot pass a variable to the `return` keyword.

MATLAB provides helper functions for specifying a variable number of inputs or outputs, namely, `varargin` and `varargout`. These variables are cell arrays, and you access and assign elements using curly braces. Here is an example function definition:

```
1   function [y,varargout] = varargFunction(x,varargin)
2
3   y = varargin{1};
4   varargout{1} = size(x,1);
5   varargout{2} = size(x,2);
```

The following example demonstrates that the outputs were correctly assigned.

USING VARARGOUT AND VARARGIN

```
>> [y,a,b] = varargFunction(rand(3,2),1.0)

y =
     1

a =
     3

b =
     2
```

This allows you to accept unlimited arguments or parameter pairs in your function. It is up to you to create consistent forms for your function and document them clearly in the help comments.

You can also count the input and output arguments for a given call to your function using `nargin` and `nargout` and use this with logical statements or a `switch` statement to handle multiple cases.

If you need very complex input handling, MATLAB now provides an `inputParser` class, which allows you to parse and validate an input scheme. You can define functions to validate the inputs, optional arguments, and predefine parameter pairs.

Use Function Handles for Efficiency

Function handles are pointers to functions. They are closely related to anonymous functions, which allow you to define a short function inline, and return the function handle. When you create a handle, you can change the input scheme and give values for certain inputs, that is, parameters. Using handles as inputs to integrators and similar routines is much faster than passing in a string variable of the function name.

In the following snippet, we create an anonymous function handle to `myFunction` with a different signature and a specific value for a. Note the use of the @, which designates a function handle. The handle can be evaluated with inputs just like a regular function.

```
1  function y = myFunction(a, b, c)
2  ...
3
4  a = 2;
5  h = @(c,b) myFunction(a,b,c);
6  y = h(c,b);
```

The handle `h` can be passed to a function such as an integrator that is expecting a signature with only two variables. You will also commonly use function handles to specify an events function for integrators or similar tools, as well as output functions that are called between major steps. Output functions can print information to the screen or a figure. See, for example, `odeplot` and `odeprint`.

In order to test if a variable is a function handle, you need to use the function handle class name with `isa`, that is:

```
1  isa(f,'function_handle')
```

as `ishandle` works only for graphics handles. For more information, see the help documentation for `function_handle`. Table 1.7 provides the few key functions for dealing with function handles.

Table 1.7: *Key Functions for Handles*

Function	Purpose
feval	Execute a function from a handle or string
func2str	Construct a string from a function handle
str2func	Construct a handle from a function name string
isa	Test for a function handle

Numerics

While MATLAB defaults to doubles for any data entered at the command line or in a script, you can specify a variety of other numeric types, including `single`, `uint8`, `uint16`, `uint32`, `uint64`, `logical` (i.e., an array of booleans). The use of the integer types is especially relevant to using large data sets such as images. Use the minimum data type you need, especially when your data sets are large.

Images

MATLAB supports a variety of formats including GIF, JPG, TIFF, PNG, HDF, FITS, and BMP. You can read in an image directly using `imread`, which can determine the type automatically from the extension, or `fitsread`. (FITS stands for Flexible Image Transport System, and the interface is provided by the CFITSIO library.) `imread` has special syntaxes for some image types, such as handling alpha channels for PNG, so you should review the options for your specific images. `imformats` manages the file format registry and allows you to specify handling of new user-defined types, if you can provide read and write functions.

You can display an image using either `imshow`, `image`, or `imagesc`, which scales the colormap for the range of data in the image.

For example, we use a set of images of cats in Chapter 7, Face Recognition. The following is the image information for a typical image:

```
>> imfinfo('IMG_4901.JPG')
ans =
            Filename: 'MATLAB/Cats/IMG_4901.JPG'
         FileModDate: '28-Sep-2016 12:48:15'
            FileSize: 1963302
              Format: 'jpg'
       FormatVersion: ''
               Width: 3264
              Height: 2448
            BitDepth: 24
           ColorType: 'truecolor'
     FormatSignature: ''
     NumberOfSamples: 3
        CodingMethod: 'Huffman'
       CodingProcess: 'Sequential'
             Comment: {}
                Make: 'Apple'
               Model: 'iPhone 6'
         Orientation: 1
         XResolution: 72
         YResolution: 72
      ResolutionUnit: 'Inch'
            Software: '9.3.5'
```

18

```
         DateTime: '2016:09:17 22:05:08'
 YCbCrPositioning: 'Centered'
   DigitalCamera: [1x1 struct]
         GPSInfo: [1x1 struct]
   ExifThumbnail: [1x1 struct]
```

This is the metadata that tells the camera software, and image databases, where and how the image was generated. This is useful when learning from images as it allows you to correct for resolution (`width` and `height`) bit depth and other factors.

If we view this image using `imshow`, it will publish a warning that the image is too big to fit on the screen and that it is displayed at 33%. If we view it using `image`, there will be a visible set of axes. `image` is useful for displaying other two-dimensional matrix data as individual elements per pixel. Both functions return a handle to an image object; only the axes properties are different. Figure 1.4 shows the use of imshow and image.

```
>> figure; hI = image(imread('IMG_2398_Zoom.png'))
hI =
  Image with properties:

           CData: [680x680x3 uint8]
    CDataMapping: 'direct'

  Show all properties
```

Figure 1.4: *Image display options.*

Table 1.8 shows key image functions.

Table 1.8: *Key Functions for Images*

Function	Purpose
imread	Read an image in a variety of formats
imfinfo	Gather information about an image file
imformats	Determine if a field exists in a structure
imwrite	Write data to an image file
image	Display image from an array
imagesc	Display image data scaled to the current colormap
imshow	Display an image, optimizing figure, axes, and image object properties and taking an array or a filename as an input
rgb2gray	Write data to an image file
ind2rgb	Convert index data to RGB
rgb2ind	Convert RGB data to indexed image data
fitsread	Read a FITS file
fitswrite	Write data to a FITS file
fitsinfo	Information about a FITS file returned in a data structure
fitsdisp	Display FITS file metadata for all HDUs in the file

Datastore

Datastores allow you to interact with files containing data that are too large to fit in memory. There are different types of datastores for tabular data, images, spreadsheets, databases, and custom files. Each datastore provides functions to extract smaller amounts of data that do fit in memory for analysis. For example, you can search a collection of images for those with the brightest pixels or maximum saturation values. We will use the directory of cat images included with the code as an example.

```
>> location = pwd

location =

    '/Users/Mike/svn/MATLABBooks/MATLABCookbook2/MATLAB/Chapter_01/Cats'

>> ds = datastore(location)

ds =

  ImageDatastore with properties:

                     Files: {
                            ' .../svn/MATLABBooks/MATLABCookbook2/
                              MATLAB/Chapter_01/Cats/IMG_0191.png';
```

```
                              ' .../svn/MATLABBooks/MATLABCookbook2/
                                  MATLAB/Chapter_01/Cats/IMG_1603.png';
                              ' .../svn/MATLABBooks/MATLABCookbook2/
                                  MATLAB/Chapter_01/Cats/IMG_1625.png'
                              ... and 8 more
                              }
   AlternateFileSystemRoots: {}
                   ReadSize: 1
                     Labels: {}
                    ReadFcn: @readDatastoreImage
```

Once the datastore is created, you use the applicable class functions to interact with it. Datastores have standard container-style functions like `read`, `partition`, and `reset`. Each type of datastore has different properties. The `DatabaseDatastore` requires the Database Toolbox and allows you to use SQL queries.

MATLAB provides the MapReduce framework for working with out-of-memory data in datastores. The input data can be any of the datastore types, and the output is a key-value datastore. The map function processes the datastore input in chunks, and the reduce function calculates the output values for each key. `mapreduce` can be sped up by using it with the MATLAB Parallel Computing Toolbox, Distributed Computing Server, or Compiler. Table 1.9 shows key datastore functions.

Table 1.9: *Key Functions for Datastore*

Function	Purpose
datastore	Create a datastore
read	Read a subset of data from the datastore
readall	Read all of the data in the datastore
hasdata	Check to see if there is more data in the datastore
reset	Initialize a datastore with the contents of a folder
partition	Excerpt a portion of the datastore
numpartitions	Estimate a reasonable number of partitions
ImageDatastore	Datastore of a list of image files
TabularTextDatastore	A collection of one or more tabular text files
SpreadsheetDatastore	Datastore of spreadsheets
FileDatastore	Datastore for files with a custom format, for which you provide a reader function
KeyValueDatastore	Datastore of key-value pairs
DatabaseDatastore	Database connection, requires the Database Toolbox

Tall Arrays

Tall arrays were introduced in R2016b. They are allowed to have more rows than will fit in memory. You can use them to work with datastores that might have millions of rows. Tall arrays can use almost any MATLAB type as a column variable, including numeric data, cell arrays, strings, datetimes, and categoricals. The MATLAB documentation provides a list of functions that support tall arrays. Results for operations on the array are only evaluated when they are explicitly requested using the gather function. The histogram function can be used with tall arrays and will execute immediately.

The MATLAB Statistics and Machine Learning Toolbox™, Database Toolbox, Parallel Computing Toolbox, Distributed Computing Server, and Compiler all provide additional extensions for working with tall arrays. For more information about this new feature, use the following topics in the documentation:

- Tall Arrays
- Analysis of Big Data with Tall Arrays
- Functions That Support Tall Arrays
- Index and View Tall Array Elements
- Visualization of Tall Arrays
- Extend Tall Arrays with Other Products
- Tall Array Support, Usage Notes, and Limitations

Table 1.10 shows key tall array functions and Table 1.11 shows key sparse matrix functions.

Table 1.10: *Key Functions for Tall Arrays*

Function	Purpose
tall	Initialize a tall array
gather	Execute the requested operations
summary	Display summary information to the command line
head	Access first rows of a tall array
tail	Access last rows of a tall array
istall	Check the type of the array to determine if it is tall
write	Write the tall array to disk

Sparse Matrices

Sparse matrices are a special category of matrix in which most of the elements are zero. They appear commonly in large optimization problems and are used by many such packages. The zeros are "squeezed" out, and MATLAB stores only the nonzero elements along with index data

Table 1.11: *Key Functions for Sparse Matrices*

Function	Purpose
sparse	Create a sparse matrix from a full matrix or from a list of indices and values
issparse	Determine if a matrix is sparse
nnz	Number of nonzero elements in a sparse matrix
spalloc	Allocate nonzero space for a sparse matrix
spy	Visualize a sparsity pattern
spfun	Selectively apply a function to the nonzero elements of a sparse matrix
full	Convert a sparse matrix to full form

such that the full matrix can be recreated. Many regular MATLAB functions, such as `chol` or `diag`, preserve the sparseness of an input matrix.

Tables and Categoricals

Tables were introduced in release R2013 of MATLAB and allow tabular data to be stored with metadata in one workspace variable. It is an effective way to store and interact with data that one might put in, or import from, a spreadsheet. The table columns can be named, assigned units and descriptions, and accessed as one would fields in a data structure, that is, `T.DataName`. See `readtable` on creating a table from a file, or try out the Import Data button from the Command Window. Table 1.12 shows key Tables functions.

Categorical arrays allow for storage of discrete nonnumeric data, and they are often used within a table to define groups of rows. For example, time data may have the day of the week, or geographic data may be organized by state or county. They can be leveraged to rearrange data in a table using `unstack`. This is more efficient searching than elements of a cell array. See `categorical` and `categories`.

You can also combine multiple data sets into single tables using `join`, `innerjoin`, and `outerjoin`, which will be familiar to you if you have worked with databases.

Large MAT-files

You can access parts of a large MAT-file without loading the entire file into memory by using the `matfile` function. This creates an object that is connected to the requested MAT-file without loading it. Data is only loaded when you request a particular variable or part of a variable. You can also dynamically add new data to the MAT-file.

Table 1.12: Key Functions for Tables and Categoricals

Function	Purpose
table	Create a table with data in the workspace
readtable	Create a table from a file
join	Merge tables by matching up variables
innerjoin	Join tables A and B retaining only the rows that match
outerjoin	Join tables including all rows
stack	Stack data from multiple table variables into one variable
unstack	Unstack data from a single variable into multiple variables
summary	Calculate and display summary data for the table
categorical	Arrays of discrete categorical data
iscategorical	Create a categorical array
categories	List of categories in the array
iscategory	Test for a particular category
addcats	Add categories to an array
removecats	Remove categories from an array
mergecats	Merge categories

For example, we can load a MAT-file of neural net weights.

```
>> m = matfile('PitchNNWeights','Writable',true)

m =

  matlab.io.MatFile

  Properties:
      Properties.Source: '/Users/Mike/svn/MATLABBooks/MATLABCookbook2/
          MATLAB/Chapter_01/PitchNNWeights.mat'
    Properties.Writable: true
                      w: [1x8 double]

  Methods
```

We can access a portion of the previously unloaded w variable or add a new variable name, all using this object m.

```
>> y = m.w(1:4)
y =
      1      1      1      1
>> m.name = 'Pitch Weights'
m =
  matlab.io.MatFile

  Properties:
      Properties.Source: '/Users/Shared/svn/Manuals/MATLABMachineLearning
```

```
          /MATLAB/PitchNNWeights.mat'
    Properties.Writable: true
                   name: [1x13 char]
                      w: [1x8  double]
>> d = load('PitchNNWeights')
d =
      w: [1 1 1 1 1 1 1 1]
    name: 'Pitch Weights'
```

There are some limits to the indexing into unloaded data, such as struct arrays and sparse arrays. Also, `matfile` requires MAT-files using version 7.3, which is not the default for a generic `save` operation as of R2016b. You must either create the MAT-file using `matfile` to take advantage of these features or use the `-v7.3'` flag when saving the file.

Advanced Data Types

The data types discussed so far are all that are needed for most engineering programming. However, for specialized applications, there are additional options for data types, including:

Classes – Classes, with properties and methods, can be defined using the `classdef` keyword in an m-file similar to writing a function. See also the `properties`, `methods`, and `events` keywords. See Chapter 6 for recipes using classes.

Time series – The `timeseries` object and the related `tscollection` object provide methods for associating data samples with timestamps. Plotting a `timeseries` object will use the stored time vector automatically.

Map containers – The map container allows you to store and look up data using a key which may be nonnumeric. This is an object instantiated via `containers.Map`.

Primer Recipes

The next part of this chapter provides recipes for some common tasks in modern MATLAB, like using different data types, adding help to your functions, loading binary data, writing to a text file, creating a MEX file, and parsing functions into "pcode."

1.1   Initializing a Data Structure Using Parameters

It's always a good idea to use a special function to define a data structure you are using as a type in your codebase, similar to writing a class but with less overhead. Users can then overload individual fields in their code, but there is an alternative way to set many fields at once: an initialization function which can handle a parameter pair input list. This allows you to do additional processing in your initialization function. Also, your parameter string names can be more descriptive than you would choose to make your field names.

Problem

We want to initialize a data structure so that the user clearly knows what they are entering.

Solution

The simplest way to implement the parameter pairs is using `varargin` and a switch statement. Alternatively, you could write an `inputParser`, which allows you to specify required and optional inputs as well as named parameters. In that case, you have to write separate or anonymous functions for validation that can be passed to the `inputParser`, rather than just write out the validation in your code.

How It Works

We will use the data structure developed for the automobile simulation as an example. The header lists the input parameters along with the input dimensions and units, if applicable.

AutomobileInitialize.m

```
1  %% AUTOMOBILEINITIALIZE Initialize the automobile data structure.
2  %
3  %% Form
4  %   d = AutomobileInitialize( varargin )
5  %
6  %% Description
7  % Initializes the data structure using parameter pairs.
8  %
9  %% Inputs
10 % varargin:   ('parameter',value,...)
11 %
12 % 'mass'                              (1,1)  (kg)
13 % 'steering angle'                    (1,1)  (rad)
14 % 'position tires'                    (2,4)  (m)
15 % 'frontal drag coefficient'          (1,1)
16 % 'side drag coefficient'             (1,1)
17 % 'tire friction coefficient'         (1,1)
18 % 'tire radius'                       (1,1)  (m)
19 % 'engine torque'                     (1,1)  (Nm)
20 % 'rotational inertia'                (1,1)  (kg-m^2)
21 % 'state'                             (6,1)  [m;m;m/s;m/s;rad;rad/s
       ]
```

The function first creates the data structure using a set of defaults and then handles the parameter pairs entered by a user. After the parameters have been processed, two areas are calculated using the dimensions and the height.

```
30 function d = AutomobileInitialize( varargin )
31
32 % Defaults
33 d.mass       = 1513;
34 d.delta      = 0;
```

```
35  d.r            = [   1.17 1.17 -1.68 -1.68;...
36                      -0.77 0.77 -0.77  0.77];
37  d.cDF          = 0.25;
38  d.cDS          = 0.5;
39  d.cF           = 0.01; % Ordinary car tires on concrete
40  d.radiusTire   = 0.4572; % m
41  d.torque       = d.radiusTire*200.0; % N
42  d.inr          = 2443.26;
43  d.x            = [0;0;0;0;0;0];
44  d.fRR          = [0.013 6.5e-6];
45  d.dim          = [1.17+1.68 2*0.77];
46  d.h            = 2/0.77;
47  d.errOld       = 0;
48  d.passState    = 0;
49  d.model        = 'MyCar.obj';
50  d.scale        = 4.7981;
51
52  fNames = fieldnames(d);
53  for k = 1:2:length(varargin)
54    if isfield(d,varargin{k})
55      d.(varargin{k}) = varargin{k+1};
56    else
57      warning('Parameter %s is not a valid field name',varargin{k});
58    end
59  end
60
61  names = {'mass','mass',1513;...
62           'steering angle','delta',0;...
63           'position tires','r',[   1.17 1.17 -1.68 -1.68;-0.77 0.77 -0.77
                  0.77];...
64           'frontal drag coefficient','cDF',0.25;...
65           'side drag coefficient','cDS',0.5;...
66           'tire friction coefficient','cF',0.01;...
67           'tire radius','radiusTire',0.4572;...
68           'engine torque','torque',0;...
69           'rotational inertia','inr',2443.26;...
70           'state','x',[0;0;0;0;0;0];...
71           'rolling resistance coefficients','fRR',[0.013 6.5e-6];...
72           'side and frontal automobile dimensions','dim',[1.17+1.68
                  2*0.77];...
73           'height automobile','h',2/0.77;...
74           'errOld','errOld',0;...
75           'passState','passState',0;...
76           'car model','model','MyCar.obj';...
77           'car scale','scale',4.7981}
78
79  d = cell2struct(names(:,3),names(:,2),1);
80  d.torque = d.radiusTire*200.0; % N
81
82  missed = {};
83  for k = 1:2:length(varargin)
84    % match to a descriptive parameter name
```

27

```
85    match = strcmpi(varargin{k},names(:,1));
86    if ~any(match)
87      % match to a field name
88      match = strcmp(varargin{k},names(:,2));
89    end
90    if ~any(match)
91      warning('No match for the parameter %s',varargin{k});
92      missed{end+1} = varargin{k};
93      continue;
94    end
95    d.(names{match,2}) = varargin{k+1};
96  end
97
98  if ~isempty(missed)
99    error('Unprocessed parameters.')
100 end
101
102 % Processing
103 d.areaF = d.dim(2)*d.h;
104 d.areaS = d.dim(1)*d.h;
105 d.g     = LoadOBJ(d.model,[],d.scale);
```

To perform the same tasks with inputParser, you add either an addRequired, addOptional, or addParameter call for every item in the switch statement. The named parameters require default values. You can optionally specify a validation function; in the following example, we use isNumeric to limit the values to numeric data.

```
>> p = inputParser
p.addParameter('mass',0.25);
p.addParameter('cDF',1513);
p.parse('cDF',2000);
d = p.Results


p =

  inputParser with properties:

        FunctionName: ''
       CaseSensitive: 0
       KeepUnmatched: 0
      PartialMatching: 1
         StructExpand: 1
          Parameters: {1?0 cell}
              Results: [1?1 struct]
            Unmatched: [1?1 struct]
       UsingDefaults: {1?0 cell}

d =

  struct with fields:
```

```
cDF: 2000
mass: 0.2500
```

In this case, the results of the parsed parameters are stored in a Results substructure.

1.2   Performing mapreduce on an Image Datastore

Problem

We discussed the `datastore` class in the introduction to the chapter. Now let's use it to perform analysis on the full set of cat images using `mapreduce`, which is scalable to very large numbers of images. This involves two steps, first a *map* step that operates on the datastore and creates intermediate values and then a *reduce* step which operates on the intermediate values to produce a final output.

Solution

We create the `datastore` by passing in the path to the folder of cat images. We also need to create a map function and a reduce function, to pass into `mapreduce`. If you are using additional toolboxes like the Parallel Computing Toolbox, you would specify the reduce environment using `mapreducer`.

How It Works

First, create the `datastore` using the path to the images.

```
>>  imds = imageDatastore('Cats')

imds =

  ImageDatastore with properties:

                    Files: {
                            ' .../svn/MATLABBooks/MATLABCookbook2/
                              MATLAB/Chapter_01/Cats/IMG_0191.png';
                            ' .../svn/MATLABBooks/MATLABCookbook2/
                              MATLAB/Chapter_01/Cats/IMG_1603.png';
                            ' .../svn/MATLABBooks/MATLABCookbook2/
                              MATLAB/Chapter_01/Cats/IMG_1625.png'
                            ... and 8 more
                            }
     AlternateFileSystemRoots: {}
                 ReadSize: 1
                   Labels: {}
                  ReadFcn: @readDatastoreImage
```

Second, we write the map function. This must generate and store a set of intermediate values that will be processed by the reduce function. Each intermediate value must be stored as a key in the intermediate key-value datastore using add. In this case, the map function will receive one image each time it is called. We call it catColorMapper, since it processed the red, green, and blue values for each image using a simple average.

```
1  function catColorMapper(data, info, intermediateStore)
2
3  % Calculate the average (R,G,B) values
4  avgRed = mean(mean(data(:,:,1)));
5  avgGreen = mean(mean(data(:,:,2)));
6  avgBlue = mean(mean(data(:,:,3)));
7
8  % Store the calculated values with text keys
9  add(intermediateStore, 'Avg Red', struct('Filename',info.Filename,'Val'
       , avgRed));
10 add(intermediateStore, 'Avg Green', struct('Filename',info.Filename,'
       Val', avgGreen));
11 add(intermediateStore, 'Avg Blue', struct('Filename',info.Filename,'Val
       ', avgBlue));
```

The reduce function will then receive the list of the image files from the datastore once for each key in the intermediate data. It receives an iterator to the intermediate datastore as well as an output datastore. Again, each output must be a key-value pair. The hasnext and getnext functions used are part of the mapreduce ValueIterator class. In this case, we find the minimum value for each key across the set of images.

```
1  function catColorReducer(key, intermediateIter, outputStore)
2
3  % Iterate over values for each key
4  minVal = 255;
5  minImageFilename = '';
6  while hasnext(intermediateIter)
7    value = getnext(intermediateIter);
8
9    % Compare values to find the minimum
10   if value.Val < minVal
11       minVal = value.Val;
12       minImageFilename = value.Filename;
13   end
14 end
15
16 % Add final key-value pair
17 add(outputStore, ['Minimum -  ' key], minImageFilename);
```

Finally, we call `mapreduce` using function handles to our two helper functions. Progress updates are printed to the command line, first for the mapping step and then for the reduce step (once the mapping progress reaches 100%).

```
>> minRGB = mapreduce(imds, @catColorMapper, @catColorMapper);

********************************
*      MAPREDUCE PROGRESS      *
********************************
Map   0% Reduce    0%
Map  13% Reduce    0%
Map  27% Reduce    0%
Map  40% Reduce    0%
Map  50% Reduce    0%
Map  63% Reduce    0%
Map  77% Reduce    0%
Map  90% Reduce    0%
Map 100% Reduce    0%
Map 100% Reduce   33%
Map 100% Reduce   67%
Map 100% Reduce  100%
```

The results are stored in a MAT-file, for example, `results_1-28-Sep-2016_16-28-38_347`. The store returned is a key-value store to this MAT-file, which in turn contains the store with the final key-value results.

```
>> output = readall(minRGB)
output =
        Key                                Value

    _____

    ''Minimum - Avg Red'      '/MATLAB/Cats/IMG_1625.png'
    ''Minimum - Avg Blue'     '/MATLAB/Cats/IMG_4866.jpg'
    ''Minimum - Avg Green'    '/MATLAB/Cats/IMG_4866.jpg'
```

You'll notice that the image files are different file types. This is because they came from different sources. MATLAB can handle most image types quite well.

1.3   Creating a Table from a File

Often, with big data, we have complex data in many files. MATLAB provides functions to make it easier to handle massive sets of data. In this section, we will collect data from a set of weather files and perform a Fast Fourier Transform (FFT) on data from two years. First, we will write the FFT function.

Problem

We want to do Fast Fourier Transforms.

Solution

Write a function using `fft` and compute the energy from the FFT. The energy is just the real part of the product of the FFT output and its transpose.

How It Works

The following functions take in data `y` with a sample time `tSamp` and perform an FFT:

FFTEnergy.m

```
26   n = size( y, 2 );
27
28   % If the input vector is odd drop one sample
29   if(2*floor(n/2) ~= n )
30     n = n - 1;
31     y = y(1:n,:);
32   end
33
34   x  = fft(y);
35   e  = real(x.*conj(x))/n;
36
37   hN = n/2;
38   e  = e(1:hN,:);
39   r  = 2*pi/(n*tSamp);
40   w  = r*(0:(hN-1));
41
42   if( nargout == 0 )
43     tL = sprintf('FFT Energy Plot: Resolution = %10.2e rad/sec',r);
44           PlotSet(w,e','x label','Frequency (rad/sec)','y label', ...
45       'Energy','plot title', tL,'plot type', 'xlog', 'figure title', 'FFT
                ');
46     clear e
47   end
```

We get the energy using these two lines:

```
34   x  = fft(y);
35   e  = real(x.*conj(x))/n;
```

Taking the real part just accounts for numerical errors. The product of a number and its complex conjugate should be real.

The function computes the resolution. Notice it is a function of the sampling period and number of points.

```
39   r  = 2*pi/(n*tSamp);
```

32

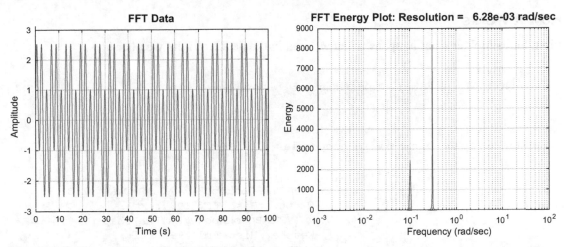

Figure 1.5: *The input data for the FFT and the results.*

The built-in demo creates a series with a frequency at 1 rad/sec and a second at 2 rad/sec. The higher frequency one, with an amplitude of 2, has more energy as expected.

```
49  function Demo
50  %% Demo
51  tSamp    = 0.1;
52  omega1   = 1;
53  omega2   = 3;
54  t        = linspace(0,1000,10000)*tSamp;
55  y        = sin(omega1*t) + 2*sin(omega2*t);
56
57  PlotSet(t,y,'x label', 'Time (s)', 'y label','Amplitude',...
58      'plot title','FFT Data', 'figure title', 'FFT Data');
59  FFTEnergy( y, tSamp );
```

Figure 1.5 shows the data and the FFT. Note the clearly visible frequencies in the FFT plot that match the oscillations in the time plot.

1.4   Processing Table Data

Problem

We want to compare temperature frequencies in 1999 and 2015 using data from a table.

Solution

Use `tabularTextDatastore` to load the data and perform a Fast Fourier Transform on the data.

How It Works

First, let us look at what happens when we read in the data from the weather files.

```
>> tds        = tabularTextDatastore('./Weather')

tds =

  TabularTextDatastore with properties:

                      Files: {
                             ' .../MATLABCookbook2/MATLAB/Chapter_01/
                               Weather/HistKTTN_1990.txt';
                             ' .../MATLABCookbook2/MATLAB/Chapter_01/
                               Weather/HistKTTN_1993.txt';
                             ' .../MATLABCookbook2/MATLAB/Chapter_01/
                               Weather/HistKTTN_1999.txt'
                             ... and 5 more
                             }
               FileEncoding: 'UTF-8'
   AlternateFileSystemRoots: {}
      PreserveVariableNames: false
         ReadVariableNames: true
              VariableNames: {'EST', 'MaxTemperatureF', 'MeanTemperatureF
                             ' ... and 20 more}
             DatetimeLocale: en_US

  Text Format Properties:
              NumHeaderLines: 0
                   Delimiter: ','
                RowDelimiter: '\r\n'
               TreatAsMissing: ''
                MissingValue: NaN

  Advanced Text Format Properties:
             TextscanFormats: {'%{uuuu-MM-dd}D', '%f', '%f' ... and 20
                    more}
                    TextType: 'char'
          ExponentCharacters: 'eEdD'
                CommentStyle: ''
                  Whitespace: ' \b\t'
    MultipleDelimitersAsOne: false

  Properties that control the table returned by preview, read, readall:
       SelectedVariableNames: {'EST', 'MaxTemperatureF', 'MeanTemperatureF
          ' ... and 20 more}
            SelectedFormats: {'%{uuuu-MM-dd}D', '%f', '%f' ... and 20
                    more}
                   ReadSize: 20000 rows
```

WeatherFFT selects the data to use. It finds all the data in the mess of data in the files. When running the script, you need to be in the same folder as WeatherFFT.

WeatherFFT.m

```
6   c0 = cd;
7   p = mfilename('fullpath');
8   cd(fileparts(p));
9   secInDay = 86400;
11
12  %% Create the datastore from the directory of files
13  tDS                        = tabularTextDatastore('./Weather/');
14  tDS.SelectedVariableNames = {'EST','MaxTemperatureF'};
15
16  preview(tDS)
17  z = readall(tDS);
18
19  % The first column in the cell array is the date. year extracts the
        year
20  y      = year(z{:,1});
21  k1993 = find(y == 1993);
22  k2015 = find(y == 2015);
23  tSamp = secInDay;
24  t      = (1:365)*tSamp;
25  j      = {[1 2]};
26
27  %% Plot the FFT
28
29  % Get 1993 data
30  d1993      = z{k1993,2}';
31  m1993      = mean(d1993);
32  d1993      = d1993 - m1993;
```

If the data does not exist, TabularTextDatastore puts NaN in the data points place. We happen to pick two years without any missing data. We use preview to see what we are getting.

```
>> WeatherFFT
Warning: Variable names were modified to make them valid MATLAB
    identifiers.

ans =

  8x2 table

      EST           MaxTemperatureF

    _____      _____


    1990-01-01          39
    1990-01-02          39
    1990-01-03          48
```

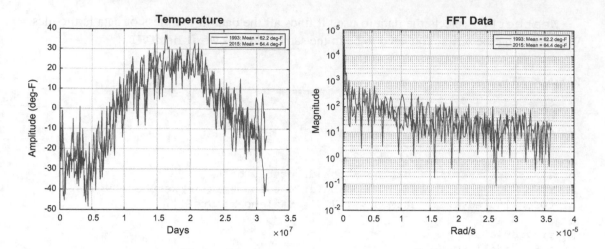

Figure 1.6: *1993 and 2015 data.*

1990-01-04	51
1990-01-05	46
1990-01-06	43
1990-01-07	42
1990-01-08	37

In this script, we get an output from `FFTEnergy` so that we can combine the plots. We chose to put the data on the same axes. Figure 1.6 shows the temperature data and the FFT.

We get a little fancy with `plotset`. Our legend entries are computed to include the mean temperatures.

```
35
36   % Get 2015 data
37   d2015      = z{k2015,2}';
38   m2015      = mean(d2015);
39   d2015      = d2015 - m2015;
40   [e2015,f] = FFTEnergy( d2015, tSamp );
41
42   lG = {{sprintf('1993: Mean = %4.1f deg-F',m1993) sprintf('2015: Mean =
         %4.1f deg-F',m2015)}};
43
44   PlotSet(t,[d1993;d2015],  'x label', 'Days', 'y label','Amplitude (deg-
         F)',...
45     'plot title','Temperature', 'figure title', 'Temperature','legend',lG
         ,'plot set',j);
46
47   PlotSet(f,[e1993';e2015'],'x label', 'Rad/s','y label','Magnitude',...
48     'plot title','FFT Data', 'figure title', 'FFT','plot type','ylog','
         legend',lG,'plot set',j);
49
50   cd(c0);
```

36

1.5   String Concatenation

In this next set of recipes, we will give examples of operations that work with strings but not with character arrays. Strings are a fairly new data type in MATLAB (since R2016b).

Problem

We want to concatenate two strings.

Solution

Create the two strings and use the "+" operator.

How It Works

You can use the + operator to concatenate strings. The result is the second string after the first.

```
>> a = "12345";
>> b = "67";
>> c = a + b

c =

    "1234567"
```

1.6   Arrays of Strings

Problem

We want any array of strings.

Solution

Create the two strings and put them in a matrix.

How It Works

We create the same two strings as shown earlier and use the matrix operator. If they were character arrays, we would need to pad the shorter with blanks to be the same size as the longer.

```
>> a = "12345";
>> b = "67";
>> c = [a;b]

c =

  2 x 1 string array
```

```
      "12345"
      "67"
>> c = [a b]

c =

  1 x 2 string array

    "12345"    "67"
```

You could have used a cell array for this, but strings are more convenient.

1.7   Non-English Strings

Problem

We want to write a string in Japanese.

Solution

Copy the characters into a string array.

How It Works

Strings do not have to be in English. Copy any unicode characters into a MATLAB string.

```
    >> str = [
```
”恋に悩み、苦しむ。”
”恋の悩みで苦しむ。”
”空に星が輝き、瞬いている。”
”空の星が輝きを増している。”]

The resulting string is
```
>>str
```

```
str =
```

4×1 string array
    ”恋に悩み、苦しむ。”
”恋の悩みで苦しむ。”
”空に星が輝き、瞬いている。”
)”空の星が輝きを増している。”

1.8 Substrings

Problem

We want to get the portion of strings after a fixed prefix.

Solution

Create a string array and use `extractAfter`.

How It Works

Create a string array of strings to search and use `extractAfter`.

```
>> a = ["1234";"12456";"12890"];
f = extractAfter(a,"12")


f =

  3 x 1 string array

    "34"
    "456"
    "890"
```

Most of the string functions work with `char`, but strings are a little cleaner. Here is the preceding example with cell arrays.

```
>> a = {'1234';'12456';'12890'};
>> f = extractAfter(a,"12")

f =

  3 x 1 cell array

    {'34' }
    {'456'}
    {'890'}
```

1.9 Using JSON-Formatted Strings

JSON (JavaScript Object Notation) is a lightweight data-interchange format used in JavaScript. MATLAB has functions for going to and from JSON. JSON covers all types of data. Encoding and decoding works with both cell arrays and data structures. This example code will get you started.

```
1  m = {rand(3,3), 'text', 3};
2  t = jsonencode(m);
3  n = jsondecode(t)
```

```
1  n =
2
3    3x1 cell array
4
5      {3x3 double}
6      {'text'    }
7      {[        3]}
```

1.10   Creating Function Help

Problem

You need to document your functions so that others may use them, and you remember how they work in the future.

Solution

MATLAB provides a mechanism for providing command-line access to documentation about your function or script using the `help` command provided you put the documentation in the right place.

How It Works

The comments you provide at the top of your function file, called a *header*, become the function help. The help can be printed at the command line by typing `help MyFunction`. While we will cover the style and format of these comments in the next chapter, we draw your attention to the functionality here.

The help comments can go either above or below the declarative line of your function. If you include the words "see also" in your comments followed by the names of additional functions, MATLAB will helpfully supply links to those functions' help. All comments are printed until the first blank line is reached.

Consider the help for a function that calculates a dot product. The first line should be a single sentence description of the function, which is utilized by `lookfor`. If you insert your function name in all capital letters, MATLAB will automatically replace it with the true case version when printing the help. Your comments might look like this:

```
1  function d = Dot( w, y )
2  %% DOT Dot product of two arrays.
3  %% Forms
4  %   d = Dot( w, y )
5  %   d = Dot( w )
6  %
7  %% Description
8  % Dot product with support for arrays. The number of columns of w and y
        can be:
9  %
10 % * Both > 1 and equal
```

```
11  %  * One can have one column and the other any number of columns
12  %
13  % If there is only one input the dot product will be taken with itself.
14  %
15  %% See also
16  % Cross
```

When printed to the command line, MATLAB will remove the percent signs and just display the text, like this:

```
>> help Dot
  Dot Dot product of two arrays.
  % Forms
    d = Dot( w, y )
    d = Dot( w )

  % Description
  Dot product with support for arrays. The number of columns of w and y
      can be:

  * Both > 1 and equal
  * One can have one column and the other any number of columns

  If there is only one input the dot product will be taken with itself.

  % See also
  Cross
```

You can link to additional help documentation attached to subfunctions in your file. (Subfunctions are visible to other functions in the same file, but not outside the file in which they are defined. However, you can output a handle to a subfunction.) This can be handy for providing more detailed examples or descriptions of algorithms. In order to do so, you have to embed an HTML link in your help comments, for example:

```
1  % More detailed help is in the <a href="matlab: help foo>extended_help
       ">extended help</a>.
2
3  function extended_help
4  %EXTENDED_HELP Additional technical details and examples
5  %
6  % Describe additional details of your algorithms or provide examples.
7
8  error('This is a placeholder function just for helptext');
```

This typesets in the Command Window as

```
More detailed help is in the extended help.
```

MATLAB also provides the capability for you to create HTML help for your functions that will appear in the help browser. This requires the creation of XML files to provide the content hierarchy. See the MATLAB help topic *Display Custom Documentation* and the related recipe in the next chapter.

You can also run help reports to identify functions which are missing help or missing certain sections of help such as a copyright notice. To learn how to launch this report on your operating system, see the help topic *Check Which Programs Have Help*.

1.11   Locating Directories for Data Storage

Problem

A variety of demos and functions in your toolbox generate data files, and they end up all over your file system. You can't use an absolute path on your computer because the code is shared among multiple engineers.

Solution

Use `mfilename` to save files in the same location as the generating file or to locate a dedicated data directory that is relative to your file location.

How It Works

It's easy to sprinkle `save` commands throughout your scripts, or print figures to image files, and end up with files spread all over your file system. MATLAB provides a handy function, `mfilename`, which can provide the path to the folder of the executing m-file. You can use this to locate a data folder dedicated to either input files or output files for your routine. This uses the MATLAB functions `fileparts` and `fullfile`.

For example, to save an output MAT-file in the same location as your function or script:

```
1  thisPath = mfilename('fullpath'); % path to your file
2  thisDir = fileparts(thisPath);     % path to your file's parent
      directory
3  save(fullfile(thisDir,'fileName'),x,y);
```

To save an output to a dedicated directory, you only need an additional call to `fileparts`. In this case, the directory is called DataDir. Say, for example, that your function is located in ToolsDir at the same level as DataDir in MyToolbox:

```
MyToolbox/DataDir
MyToolbox/ToolsDir
```

```
1  thisPath = mfilename('fullpath');              % file in ToolsDir
2  prevDir = fileparts(fileparts(thisPath));  % path to MyToolbox
3  dataDir = fullfile(prevDir,'DataDir','fileName');
4  save(fullfile(dataDir,'fileName'),x,y);
```

If you are printing images, you can either use the functional form of `print` as with `save` or change the path to the directory you want. You should save the current directory and return there when your script is complete.

```
1  thisPath = mfilename('fullpath');
2  cd0 = cd;
3  cd(fileparts(thisPath));
4  print -dpng MyFigure
5  cd(cd0)
```

Table 1.13 shows key functions for path operations.

Table 1.13: *Key Functions for Path Operations*

Function	Purpose
mfilename	Name and, optionally, full path to the current executing m-file
fileparts	Divide a path into parts (directory, filename, extension)
fullfile	Create a system-dependent filename from parts
cd	The current directory
path	The current MATLAB path

1.12   Loading Binary Data from a File

Problem

You need to store data in a binary file, perhaps for input to another software program.

Solution

MATLAB provides low-level utilities for creating and writing to binary files including specifying the endianness.

How It Works

Reading and writing binary data do introduce some complexities beyond text files. Let's start with MATLAB's example of creating a binary file of a magic square. This demonstrates `fopen`, `fwrite`, and `fread`. The options for precision are specified in the help for `fread`. For example, a 32-bit integer can be specified with the MATLAB-style string `'int32'` or the C-style string `'integer*4'`.

```
>> magic(4)

ans =

    16     2     3    13
     5    11    10     8
     9     7     6    12
     4    14    15     1
```

```
>> fid = fopen('magic4.bin','wb');
>> fwrite(fid,magic(4),'integer*4');
>> fclose(fid);
```

Now, let's try to read this data file back in. Since the data was stored as 32-bit integers, we have to specify this precision to get the data back.

```
>> fid = fopen('magic4.bin','rb');
>> c = fread(fid,inf,'integer*4')

c =

    16
     5
     9
     4
     2
    11
     7
    14
     3
    10
     6
    15
    13
     8
    12
     1
```

The shape of our matrix was not preserved, but we can see that the data was printed to the file in column-wise order. To fully recreate our data, we need to reshape the matrix.

```
>> data = reshape(c,5,5)

data =

    16     2     3    13
     5    11    10     8
     9     7     6    12
     4    14    15     1
```

If you need to specify the endianness of the data, you can do so in both fopen and fread. The local machine format is used by default, but you can specify the IEEE floating point with little endian byte ordering, the same with big ending ordering, and both with 64-bit long data type. This may be important if you are using binary data from an online source or using data on embedded processors.

For example, to write the same data in a big endian format, simply add the 'ieee-be' parameter.

44

```
>> fid = fopen('magic5.bin','wb','ieee-be');
>> fwrite(fid,magic(5),`integer*4');
>> fclose(fid);
```

Table 1.14 lists key functions for interacting with binary data.

Table 1.14: *Key Functions for Binary Data*

Function	Purpose
fopen	Open a file in text or binary mode
fwrite	Write to a file
fread	Read the contents of a file
fclose	Close the file

1.13   Command-Line File Interaction

Problem

You have some unexpected behavior when you try to run a script MATLAB, and you suspect a function conflict among different toolboxes.

Solution

MATLAB provides functions for locating and managing files and paths from the command line.

How It Works

MATLAB has a file browser built-in to the Command Window, but it is still helpful to be familiar with the commands for locating and managing files from the command line. In particular, if you have a lot of toolboxes and files in your path, you may need to identify name conflicts.

For example, if you get the wrong behavior or a strange error from a function and you recently changed your path, you may have a file shadowing it in your path. To check for duplicate copies of a function name, use which with the -all switch. Shadowed versions of the function will be marked. which can take a partial pathname.

```
>> which DemoPSS -all
/Users/.../Toolboxes/Missions/Demos/DemoPSS.m
/Users/.../Toolboxes/Math/Demos/DemoPSS.m           % Shadowed
/Users/.../Toolboxes/Imaging/Demos/DemoPSS.m        % Shadowed
/Users/.../Toolboxes/Common/Demos/DemoPSS.m         % Shadowed
```

To display the contents of a file at the command line, which is helpful if you need to see something in the file but don't need to open the file for editing, use type, as in Unix.

To list the contents of a directory, use `what`. A partial path can be used if there are multiple directories with the same name on your path. Specifying an output returns the results in a data structure array. MATLAB identifies which files are code, MAT-file, p-files, and so on. `what` is recursive and will return all directories with the given name anywhere in the path – useful if you use the same name of a directory for functions and demos, as follows:

```
>> what Database

MATLAB Code files in folder /Users/Shared/svn/Toolboxes/SourceCode/Core/
    Common/Demos/Database

Contents     TConstant

MATLAB Code files in folder /Users/Shared/svn/Toolboxes/SourceCode/Core/
    Common/Database

BuildConstant     Contents           MergeConstantDB
Constant          Database
```

Use `exist` to determine if a function or variable exists in the path or workspace. The code analyzer will prompt you to use the syntax with a second argument specifying the desired type, that is, 'var', 'file', 'dir'. The output is a numerical code indicating the type of the file or variable found.

Open a file in the editor from the command line using `edit`.

Load a MAT-file or ascii file using `load`. Give an output to store the data in a variable, or else it will be loaded directly into the workspace. For MAT-files, you can also specify particular variables to load.

```
d = load('MyMatFile','var1','var2');
```

The final command-line function we will introduce is `lookfor`. This function searches through all help available on the MATLAB path for a keyword. The keyword must appear in the first line of the help, that is, the one-line help comment or "H1" line. The printed result looks like a Contents file and includes links to the help of the found functions. Here is an example for the keyword *integration*.

```
>> lookfor integration
IntegrationAccuracyDemo        - Integration Accuracy for
    PropagateOrbitPlugin.
PropagatorComparison           - Integration accuracy study comparing RK4
    , RK45, and ode113.
lotkademo                      - Numerical Integration of Differential
    Equations
cumtrapz                       - Cumulative trapezoidal numerical
    integration.
trapz                          - Trapezoidal numerical integration.
```

Table 1.15 lists key functions to use at the command line.

Table 1.15: *Key Functions for Command-Line Interaction*

Function	Purpose
which	Location of a function in the path
what	List the MATLAB-specific files in the directory
type	Display the contents of a file
dbtype	Display the contents of a file with line numbers
exist	Determine if a function or variable exists
edit	Open a file in the editor
load	Load a MAT-file into the workspace
lookfor	Search help comments in the path for a keyword

1.14 Using a MEX File to Link to an External Library

Problem

There is an external C++ library that you need to use for an application, and you would like to perform the analysis in MATLAB.

Solution

You can write and compile a special function in MATLAB using the C/C++ matrix API that will allow you to call the external library functions via a MATLAB function. This is called a *MEX file*.

How It Works

A MEX function is actually a shared library compiled from C/C++ or FORTRAN source code, and is callable from MATLAB. This can be used to link to external libraries such as GLPK, BLAS, and LAPACK. When writing a MEX function, you provide a gateway routine `mexFunction` in your code and use MATLAB's C/C++ Matrix Library API. You must have a MATLAB-supported compiler installed on your machine.

```
1  #include "mex.h"
2
3  void mexFunction( int nlhs, mxArray *plhs[], int nrhs, const mxArray *
     prhs[])
```

You can see that, as with regular MATLAB functions, you can provide multiple inputs and multiple outputs. `mxArray` is a C language type, actually the fundamental data type for all matrices in MATLAB, provided by the MATLAB API.

You use the `mex` function to compile your C, C++, or FORTRAN function into a binary. Passing the verbose flag, `-v`, provides verbose output familiar to C programmers. An extension such as "mexmaci64", as determined on your system by `mexext`, is appended, and you can then call the function from MATLAB like any other m-file. For example, on Mac, MATLAB detects and uses Xcode automatically when compiling one of the built-in examples, `yprime.c`. This function "Solves simple 3 body orbit problem." First, you need to copy the example into a local working directory.

```
>> copyfile(fullfile(matlabroot,'extern','examples','mex','yprime.c')
   ,'.','f');
```

Some excerpts from the verbose compile are shown as follows:

```
>> mex -v -compatibleArrayDims yprime.c
Verbose mode is on.
No MEX options file identified; looking for an implicit selection.
... Looking for compiler 'Xcode with Clang' ...
... Looking for environment variable 'DEVELOPER_DIR' ...No.
... Executing command 'xcode-select -print-path' ...Yes ('/Applications/
    Xcode.app/Contents/Developer').
... Looking for folder '/Applications/Xcode.app/Contents/Developer' ...
    Yes.
... Executing command 'which xcrun' ...Yes ('/usr/bin/xcrun').
... Looking for folder '/usr/bin' ...Yes.
...
Found installed compiler 'Xcode with Clang'.
------------------------------------------------------------------
        Compiler location: /Applications/Xcode.app/Contents/Developer
        Options file: /Applications/MATLAB_R2014b.app/bin/maci64/mexopts/
            clang_maci64.xml
        CC : /usr/bin/xcrun -sdk macosx10.9 clang
        DEFINES : -DMX_COMPAT_32   -DMATLAB_MEX_FILE
        MATLABMEX : -DMATLAB_MEX_FILE
        CFLAGS : -fno-common -arch x86_64 -mmacosx-version-min=10.9 -
            fexceptions -isysroot /Applications/Xcode.app/Contents/
            Developer/Platforms/MacOSX.platform/Developer/SDKs/MacOSX10
            .9.sdk
        INCLUDE : -I"/Applications/MATLAB_R2014b.app/extern/include" -I"/
            Applications/MATLAB_R2014b.app/simulink/include"
        COPTIMFLAGS : -O2 -DNDEBUG
        LD : /usr/bin/xcrun -sdk macosx10.9 clang
        LDFLAGS : -Wl,-twolevel_namespace -undefined error -arch x86_64 -
            mmacosx-version-min=10.9 -Wl,-syslibroot,/Applications/Xcode.
            app/Contents/Developer/Platforms/MacOSX.platform/Developer/
            SDKs/MacOSX10.9.sdk -bundle  -Wl,-exported_symbols_list,"/
            Applications/MATLAB_R2014b.app/extern/lib/maci64/mexFunction.
            map"
        LDBUNDLE : -bundle
        LINKEXPORT : -Wl,-exported_symbols_list,"/Applications/
            MATLAB_R2014b.app/extern/lib/maci64/mexFunction.map"
        LINKLIBS : -L"/Applications/MATLAB_R2014b.app/bin/maci64" -lmx -
            lmex -lmat -lstdc++
        OBJEXT : .o
        LDEXT : .mexmaci64
```

```
-----------------------------------------------------------------
Building with 'Xcode with Clang'.
/usr/bin/xcrun -sdk macosx10.9 clang -c -DMX_COMPAT_32    -
    DMATLAB_MEX_FILE -I"/Applications/MATLAB_R2014b.app/extern/include" -
    I"/Applications/MATLAB_R2014b.app/simulink/include" -fno-common -arch
     x86_64 -mmacosx-version-min=10.9 -fexceptions -isysroot /
    Applications/Xcode.app/Contents/Developer/Platforms/MacOSX.platform/
    Developer/SDKs/MacOSX10.9.sdk -O2 -DNDEBUG /Users/Shared/svn/Manuals/
    MATLABCookbook/MATLAB/yprime.c -o /var/folders/22/
    l715021s5rnghdtkxsy_cbk40000gp/T//mex_47653762085718_983/yprime.o
/usr/bin/xcrun -sdk macosx10.9 clang -Wl,-twolevel_namespace -undefined
    error -arch x86_64 -mmacosx-version-min=10.9 -Wl,-syslibroot,/
    Applications/Xcode.app/Contents/Developer/Platforms/MacOSX.platform/
    Developer/SDKs/MacOSX10.9.sdk -bundle  -Wl,-exported_symbols_list,"/
    Applications/MATLAB_R2014b.app/extern/lib/maci64/mexFunction.map" /
    var/folders/22/l715021s5rnghdtkxsy_cbk40000gp/T//
    mex_47653762085718_983/yprime.o  -O -Wl,-exported_symbols_list,"/
    Applications/MATLAB_R2014b.app/extern/lib/maci64/mexFunction.map"  -L
    "/Applications/MATLAB_R2014b.app/bin/maci64" -lmx -lmex -lmat -lstdc
    ++ -o yprime.mexmaci64
rm -f /var/folders/22/l715021s5rnghdtkxsy_cbk40000gp/T//
    mex_47653762085718_983/yprime.o
MEX completed successfully.
```

Now, assuming you copied the source into an empty directory, if you now print the contents, you will see something like the following:

```
>> dir

.               ..              yprime.c        yprime.mexmaci64
```

and you can run a test of the compiled library.

```
>> T = 0;
>> Y = rand(1,4);
>> yprime(T,Y)
```

Writing MEX files is not for the faint of heart and requires substantial programming knowledge in the base language. In the preceding printout, you can see that the standard C++ library is included, but you need to provide links and include explicitly to other libraries you want to use. Note that your MEX-file will not have any function help, so it is a good idea to provide a companion m-file that supplies the help comments and calls your MEX function internally.

■ **TIP**   Provide a separate m-file with your MEX-file that contains help comments and, optionally, calls the MEX-file.

See the help articles including *Components of MEX-File* in MATLAB as well as the many included examples for help writing MEX-files. In the case of GLPK (GNU Linear Programming Kit), an excellent MEX file is available under the GNU public license. This was written by

Nicolo Giorgetti and is maintained by Niels Klitgord. This is now available from SourceForge, http://glpkmex.sourceforge.net.

1.15   Protect Your IP with Parsed Files

Problem

You want to share files with customers or collaborators without compromising your intellectual property in the source code.

Solution

Create protected versions of your functions using MATLAB's pcode function. Create a separate file with the help comments so users will have access to the documentation.

How It Works

The pcode function provides a capability to parse m-files into executable files with the content obscured. This can be used to distribute your software while protecting your intellectual property. A pcoded file on your path with a "p" extension, takes precedence over an m-file of the same name. Parsing an m-file is simple:

```
>> pcode Dot
```

The only argument available is the -INPLACE flag to store the p-file in the same directory as the source m-file; otherwise, it will be saved to the current directory.

One difficulty you may encounter is that once you have parsed your functions and moved them into a new folder, you no longer have access to the function help you created. The command-line help is not implemented for pcoded files, and typing "help MyFunction" will no longer work. You have to create a separate .m file with the help comments as for MEX-files. We can write a function to extract the header from an m-file and save it. We will use fprintf for this, so it's important that the header not contain any special characters like backslashes.

ParseAndSaveHeader.m

```
1   %% PARSEANDSAVEHEADER Save the header of a function to a new file.
19  function ParseAndSaveHeader( readFromFile, writeToFile )
20
21  filePath = which(readFromFile);
22  [pathStr,name,ext] = fileparts(filePath);
23
24  copyfile(filePath,fullfile(pathStr,[name,'_orig',ext]));
25
26  fid = fopen(filePath,'rt');
27  t = fgetl(fid);
28  hlp = '';
29  while( ~isempty(t) && strcmp(t(1),'%') )
30    if length(t)>1 && strcmp(t(2),'%')
31      t = ['%' t];
```

```
32    end
33    hlp = [hlp,'\n%',t];
34    t = fgetl(fid);
35    if( ~ischar(t) )
36      break;
37    end
38  end
39  hlp = [hlp,'\n%%\n%% This function was parsed on ',date,'\n\n'];
40  fclose(fid);
41  if ischar(writeToFile)
42    fid = fopen(writeToFile,'wt');
43  else
44    fid = writeToFile;
45  end
46  fprintf(fid,hlp);
47  if ischar(writeToFile)
48    fclose(fid);
49  end
50
51  pcode(filePath);
```

We save a copy of the m-file with a suffix _orig, to prevent unpleasant mistakes with deleted files. Note that we add a final comment at the end with the date the function was parsed.

1.16   Writing to a Text File

Problem

You need to write some information from MATLAB to a text file. One example is creating a template for new functions following a preferred format.

Solution

We will use `fopen`, `fprintf`, and `fclose` to open a new text file, print desired lines to it, and then close it. The `input` function is used to allow the user to enter a one-line summary of the function.

How It Works

MATLAB has a full set of functions for input and output, including writing to files. See `help iofun` for a detailed listing. You can write to text files, spreadsheets, binary files, XML, images, or zip files.

One useful example is creating a template for new functions for your company, following your preferred header format. This requires using `fopen` and `fclose`, and `fprintf` to print the lines to the file. The first input is the desired name of the new function. Note that `fprintf` will print to the command line if given a file ID of 1. We provide an option to do so with the second input, which is a boolean flag. We use the `date` function to get the current year for the copyright notice, which returns a string in the format 'dd-mmm-yyyy'; we use the string function `strsplit` to break the string into tokens. Using string indices would be an alternative. In addition, this demonstrates using `input` to prompt the user for a string, namely, a one-line description of the new function.

NewBookFile.m

```
1  %% NEWBOOKFILE Create a new function with the default header style.
2  % Pass in a file name and the header template will be printed to that
       file
3  % in the current directory. You will be asked to enter a one-line
       summary.
4  %% Form
5  %  NewBookFile( fName, outputIsFile )
6  %% Input
7  %  fName           (1,1)     File name
8  %  outputIsFile    (1,1)     True if a file is created, otherwise header
       is
9  %                            printed to the command line.
10 %% Output
11 % None.
16
17 function NewBookFile( fName, outputIsFile )
18
19 if (nargin < 2)
20   outputIsFile = false;
21 end
22
23 if (nargin == 0 || isempty(fName))
24   fName = input('Function name: ','s');
25 end
26
27 % Check if the filename is valid and if such a function already exists.
28 if (~isvarname(fName))
29     error('Book:error','invalid name');
30 end
31 if (outputIsFile && exist(fName,'file'))
32     error('Book:error','file %s already exists',fName);
33 end
34
35 % Get a one-line description (H1 line) from the user.
36 comment = input('One-line description: ','s');
37
38 % Open the file or specify command line output.
39 if (outputIsFile)
40   fid = fopen([fName '.m'],'wt');
41   c = onCleanup(@() fclose(fid));
42 else
43   fid = 1;
```

```
44    fprintf(fid,'\n');
45  end
46
47  % Write the header to the file. Use the current year for the copyright
48  % notice.
49  fprintf(fid,'%%%% %s %s\n',upper(fName),comment);
50  fprintf(fid,'%% Description.\n');
51  fprintf(fid,'%%%% Form\n');
52  fprintf(fid,'%%   y = %s( x )\n',fName);
53  fprintf(fid,'%%%% Input\n');
54  fprintf(fid,'%%   x    (1,1)    Description\n%%\n');
55  fprintf(fid,'%%%% Output\n');
56  fprintf(fid,'%%   y    (1,1)    Description\n%%\n');
57  fprintf(fid,'%%%% Reference\n');
58  fprintf(fid,'%% Insert the reference.\n');
59  fprintf(fid,'%%%% See also\n');
60  fprintf(fid,'%% List pertinent functions.\n\n');
61
62  today = strsplit(date,'-');
63  year = today{end};
64
65  fprintf(fid,'%%%% Copyright\n');
66  fprintf(fid,'%% Copyright (c) %s Princeton Satellite Systems, Inc.\n%%
          All rights reserved.\n',year);
67  fprintf(fid,'\nfunction y = %s(x)\n',fName);
68
69  if outputIsFile
70    edit(fName);
71  end
```

Note that this function checks for two errors, in the case of a bad function name and if a function with the same name already exists on the path. We use the two-input form of `error` where the first input is a message identifier. The message identifier is useful if an error is returned from a `catch` block. The message identifier can be verified using `lasterr`. For instance, if we fail to enter a valid function name when prompted, we can see the results of the first error.

```
>> NewBookFile([])
Function name:
Error using NewBookFile (line 29)
invalid name

>> [LASTMSG, LASTID] = lasterr

LASTMSG =

Error using NewBookFile (line 29)
```

```
invalid name

LASTID =

Book:error
```

The function includes the ability to print the header to the Command Window, instead of creating a file, which is useful for testing – or if you went ahead and started with a blank file and need to add a header after the fact. This is accomplished by using 1 for the file identifier. Here is what the header will look like:

```
>> NewBookFile('Test')
One-line description: This is a test function.

%% TEST This is a test function.
% Description.
%% Forms
%   y = Test( x )
%% Input
%   x    (1,1)     Description
%
%% Output
%   y    (1,1)     Description
%
%% Reference
% Insert the reference.
%% See also
% List pertinent functions.

%% Copyright
% Copyright (c) 2015 Princeton Satellite Systems, Inc.
% All rights reserved.

function y = Test(x)
```

Table 1.16 summarizes some key functions for interacting with text files.

Table 1.16: *Key Functions for Interacting with Text Files*

Function	Purpose
fprintf	Print formatted text to a file
strsplit	Split a string into tokens using a delimiter
fgetl	Get one line of a file (until a newline character)
input	Get string input from the user via the command line

1.17   Using an Explicit Expansion

Problem

You need to use an explicit expansion.

Solution

Use the dot operator to produce an output with a higher dimension.

How It Works

Explicit expansion expands the dimensionality of the result automatically. In this case, we have an eight-element row and an eight-element column array. Prior to 2016b, these could not be multiplied. Now if you multiply them, it will automatically expand the dimensions of the results.

```
1  %% Explicit expansion
2
3  a = 2:1:9
4  b = (0:7)'
5  a.*b
```

The results are shown in the following. The first clement of b creates the first row of the output by multiplying every element of a. This is continued for the remaining elements of b.

```
>> ImplicitExpansion

a =    2      3      4      5      6      7      8      9
b =    0
       1
       2
       3
       4
       5
       6
       7
ans =  0      0      0      0      0      0      0      0
       2      3      4      5      6      7      8      9
       4      6      8     10     12     14     16     18
       6      9     12     15     18     21     24     27
       8     12     16     20     24     28     32     36
      10     15     20     25     30     35     40     45
      12     18     24     30     36     42     48     54
      14     21     28     35     42     49     56     63
```

1.18   Using a Script Subfunction

Problem

You need to write a script with a subfunction. This is useful when a script needs a function that is not of general utility. You can use this feature to avoid generating many separate function files that are only used in one place. We commonly use it for right-hand sides of simulation loops or for plotting methods needed to visualize data generated by the script.

Solution

We will add a subfunction to a script that provides the dynamics of a model.

How It Works

We model slosh as a mass at the end of a rod attached to the center of mass of the rest of the spacecraft. When the main engine fires, the acceleration on the mass provides a spring-like restoring force on the mass. The damping is due to the interaction of the fluid with the walls of the fuel tank. Figure 1.7 shows the model.

The function `Slosh.m` provides a simple numerical model for fuel slosh on a spacecraft. The subfunction `RHSSlosh` starts with `function` and finishes with `end`. The script does not have an end before the `function` statement.

Slosh.m

```
1  %% Slosh Model
2  % A slosh model. This also demonstrates a script with a sub function.
3  % You can run this with a step input or a doublet
4
5  n          = 1000; % Number of steps
6  dT         = 0.1; % Time step
7  doublet        = true; % Logical
8
9  % Create the data structure for the right hand side
```

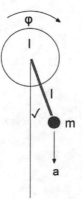

Figure 1.7: *Simple slosh model.*

```
10  d.i0        = 100; % Inertia of the spacecraft
11  m0          = 1000; % Mass of the spacecraft used only for the
        acceleration
12
13  % Control system
14  zeta        = 0.7071; % Damping ratio of the controller
15  omega       = 0.01; % Undamped natural frequency of the controller
16  d.pD              = d.i0*omega*[2*zeta omega];
17
18  % Set up the pendulum model
19  thrust      = 100; % Thrust that produces the slosh pendulum
20  l           = 0.5; % Length of the slosh pendulum
21  m           = 10; % Mass of the fuel
22  a           = thrust/m0; % Acceleration
23  d.i1        = m*l^2; % Inertia of the fuel slosh disk
24  d.k         = a*m; % Spring stiffness
25  d.damp      = 0.05; % Damping
26
27  %% Simulate
28  torque      = 1;
29  xP          = zeros(4,n);
30  x           = [0;0;0;0];
31
32  for k = 1:n
33    xP(:,k)   = x;
34    if( doublet )
35      if( k == 1 )
36        d.torque = torque;
37      elseif( k == 2)
38        d.torque = -torque;
39      else
40        d.torque = 0;
41      end
42    else
43      d.torque = torque;
44    end
45    x         = RK4(@RHSSlosh,x,dT,0,d); % We use a pointer
46  end
47
48  %% Plot the results
49  [t,tL]    = TimeLabel((0:n-1)*dT);
50  yL        = {'\phi (rad)' '\theta (rad)' '\omega_0 (rad/s)' '\omega_1 (
        rad/s)'};
51
52  PlotSet(t,xP,'x label',tL,'y label',yL,'figure title','Slosh');
53
54  %% Right-hand-side is a script sub function
55  function xDot = RHSSlosh(x,~,d)
56
57  phi         = x(1);
58  theta       = x(2);
59  omega0      = x(3);
```

```
60  omega1      = x(4);
61
62  torqueHinge = d.k*(theta-phi) + d.damp*(omega1 - omega0);
63
64  omega0Dot   = (d.torque - d.pD*x([3 1]) + torqueHinge)/d.i0;
65  omega1Dot   = -torqueHinge/d.i1;
66
67  xDot        = [omega0;omega1;omega0Dot;omega1Dot];
68
69  end
```

A few other things are worth noting. The subfunction RHSSlosh has an ∼, in the argument list because time is not used. We pass the names of the y-labels as a cell array using LaTeX symbols for the Greek letters. We set up the data structure, d, by just adding fields.

1.19   Using Memoize

Problem

You have a time-consuming process that you may have to do many times. However, the same inputs always result in the same outputs, so over many inputs, there is a duplication of computation.

Solution

We will use the memoize function. This is an optimization technique to speed up computation by storing the results of function calls in a cache, which can be accessed automatically when the same inputs are seen again. This is equivalent to storing a data table of outputs in a MAT-file, but faster.

How It Works

memoize stores the results in a buffer for later reuse. This is shown in the following script UseMemoize.m:

UseMemoize.m

```
1  %% Use memoize to store the results
2
3  x = rand(2000,2000);
4
5  % Create the memorized function
6  memY = memoize(@schur);
7
8  % Evaluate it
9  tic
10  y = memY(x);
11  toc
12
13  % Now evaluate the memorized version
```

```
14  tic
15  y2 = memY(x);
16  toc
```

The results are

```
>> UseMemoize
Elapsed time is 2.195378 seconds.
Elapsed time is 0.001552 seconds.
```

You can see that the second call is much faster, over 1000 times. This can be handier than storing the output. Note that a function to be memoized should not have any additional side effects, like affecting a global state, based on the inputs. Those side effects would not be repeated on additional calls to the function.

1.20   Using Java

Problem

We want to use MATLAB as a computational engine in Java.

Solution

We will use the MATLAB Engine.

How It Works

This will show how to use Java on Mac OS X. It will be similar on Linux and Windows. MATLAB provides a demo with a Java Swing GUI. First, get the demo function (or write your own Java).

```
>> copyfile(fullfile(matlabroot,'extern','examples','engines','java','
    EngineGUIDemo.java'),'EngineGUIDemo.java')
ls
EngineGUIDemo.java
```

This is a fairly complex function. An example of using the engine in Java to compute roots is shown in the following. You define MATLABEngine.

```
1   import com.mathworks.engine.*;
2   public class JavaDemo
3   {
4       public static void main(String[] args) throws Exception{
5           MatlabEngine eng = MatlabEngine.startMatlab();
6           double[] a = {1.0, 2.0, 3.0};
7           double[] roots = eng.feval("roots", a);
8           eng.close();
9       }
10  }
```

You will need the MATLAB root to build your Java from Terminal.

```
>> matlabroot

ans =

    '/Applications/MATLAB_R2019b.app'
```

Open Terminal on Mac. `cd` (change directory) to the directory containing your Java. Compile using

```
1  javac -classpath /Applications/MATLAB_R2019b.app/extern/engines/java/
       jar/engine.jar EngineGUIDemo.java
```

Notice where we have the matlabroot. Run using

```
1  java -Djava.library.path=/Applications/MATLAB_R2019b.app/bin/maci64 -
       classpath .:/Applications/MATLAB_R2019b.app/extern/engines/java/jar
       /engine.jar EngineGUIDemo
```

You will see the GUI shown in Figure 1.8. Start MATLAB and then enter the number for which you want the factorial.

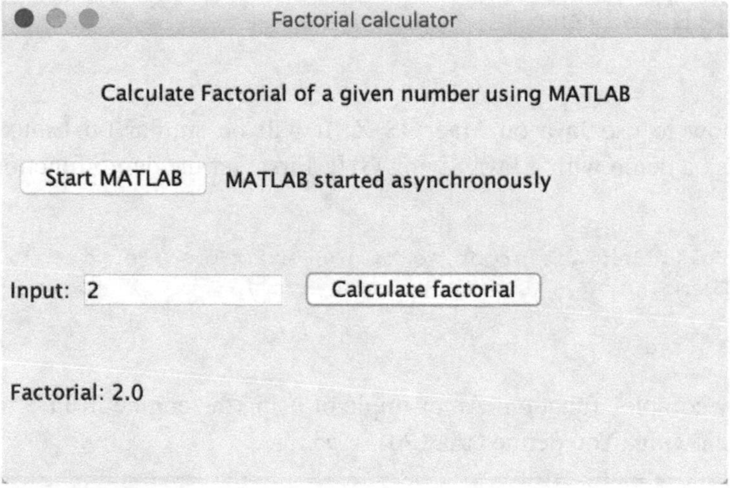

Figure 1.8: *EngineGUIDemo.*

1.21   Creating Documents

Problem

We want to use MATLAB to create a document, combining text, code fragments, and plots. This can be a way to create a technical memo or document an analysis to be handed off to another engineer.

Solution

Use live scripts.

How It Works

Live scripts allow you to create documents with active MATLAB code, text, equations, and images. Click New Live Script and you will see

Add text and code in the Live Editor tab.

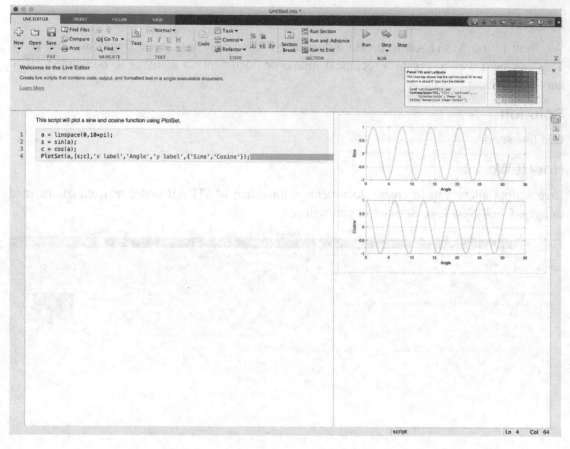

To add equations and graphics, click the Insert tab. The equation editor is like the one in Microsoft Word.

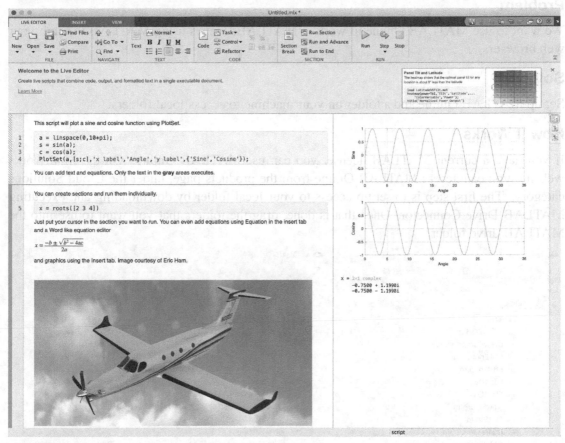

Type

```
>> open ID.mlx
```

to open this document. If having trouble with running live scripts, you may need to disable any VPNs you have installed. Live scripts have many uses. For example, you can create interactive technical memos in which the reader can execute the MATLAB code needed to produce the results. This way, the reader can try different cases easily. You no longer need to list the MATLAB code used. Another use is interactive training.

1.22   MATLAB Online

Problem

We want to use MATLAB Online to work in MATLAB. This is using MATLAB through your web browser.

Solution

Set up MATLAB Online and a folder on your machine to access local folders.

How It Works

If you have a current MATLAB license, you can use MATLAB Online. On the MathWorks website, you can access MATLAB Online from the products page, under the Cloud Solutions category. The first step is to set up access to your local folder by downloading and executing MATLAB Drive Connector. Once that is done, drag the folders that you want to use into the MATLAB drive folder.

Name		Date Modified	Size	Kind
▼ Published		Today at 11:52 AM	--	Folder
▼ General		Yesterday at 5:09 PM	--	Folder
	CToDZOH.m	Dec 6, 2019 at 2:11 PM	1 KB	Objective-C
	DrawComponents.m	Dec 6, 2019 at 2:11 PM	2 KB	Objective-C
	FFTEnergy.m	Dec 6, 2019 at 2:11 PM	1 KB	Objective-C
	Frustrum.m	Mar 15, 2020 at 4:35 PM	2 KB	Objective-C
	LoadOBJ.m	Dec 6, 2019 at 2:11 PM	6 KB	Objective-C
	Mat2Q.m	Yesterday at 5:09 PM	2 KB	Objective-C
	NewFigure.m	May 2, 2020 at 9:57 AM	505 bytes	Objective-C
	PlotSet.m	Dec 6, 2019 at 2:11 PM	4 KB	Objective-C
	Q2Mat.m	Yesterday at 5:09 PM	2 KB	Objective-C
	QMult.m	Yesterday at 5:09 PM	2 KB	Objective-C
	RungeKutta.m	Dec 6, 2019 at 2:11 PM	1 KB	Objective-C
	SaveStructure.m	Dec 6, 2019 at 2:11 PM	1 KB	Objective-C
	TimeLabel.m	Dec 6, 2019 at 2:11 PM	1 KB	Objective-C
	Unit.m	Yesterday at 5:09 PM	1 KB	Objective-C

In this case, we dragged in the General folder from the code for this book. Go to the MATLAB Online web page and start the session.

We already started the session. We created q and tried to run Q2Mat, which didn't work.

You first need to click the folder with the down arrow in the toolbar to select the General folder.

```
                    FILE
  📤 📥   ☁  /  >  MATLAB Drive  >
RENT FOLDER
```

This pop-up window will appear.

```
Select a New Folder                                           ✕

  Name
  ▸ 🗀 MATLAB Add-Ons
  ▴ 🗀 MATLAB Drive
     ▴ 🗀 Published (my site)
        ▸ 🗀 General




                                            [ OK ] [ Cancel ]
```

Fortunately, you don't have to do this every time! Here is how it works when we return to the MATLAB Online.

Summary

This chapter reviewed the basic syntax for MATLAB programming. We highlighted differences between MATLAB and similar languages, like C and C++, in the language primer. Recipes give tips for efficient usage of key features, including writing to binary and text files. Tables at the end of each section highlight key functions you should have at your fingertips.

This chapter did not provide any information on using MATLAB's computational tools, like integration and numerical search, as those will be left to subsequent chapters. Interacting with MATLAB graphics is also left to a later chapter.

CHAPTER 2

■ ■ ■

MATLAB Style

This chapter provides guidelines and recipes for suggested style elements to make your code and tools more understandable and easier to use.

When structuring a function, we have specific guidelines. The comments should be clear and descriptive and follow the formatting guidelines set by your institution. The same goes for naming conventions. In addition, we recommend supplying "built-in" inputs and outputs, that is, example parameters, so the function can be completely executed without any input from the user. These additional demo forms should be listed with the different syntaxes you create for the function.

Documenting your code goes beyond adding a header and some comments to your code. MATLAB now allows you to integrate HTML help into your toolboxes that displays in the browser along with MATLAB's documentation. You can also use the publishing utility to create comprehensive technical reports. Incorporating these features into your style guidelines from the beginning will save you a lot of work when you want to release your toolbox to others.

2.1   Developing Your Own MATLAB Style Guidelines

Problem

Each engineer in your group has their own favorite naming and whitespace styles. Whitespace is the blank spaces and lines that are left between code elements. Some writers like the code very compact; others add a lot of whitespace to set things off, which may make the code longer. When people work on each other's code, you end up with a mishmash that makes the code more difficult to read.

© Michael Paluszek and Stephanie Thomas 2020
M. Paluszek and S. Thomas, *MATLAB Recipes*,
https://doi.org/10.1007/978-1-4842-6124-8_2

Solution

Develop and publish your own style guidelines. MATLAB can help enforce some guidelines, such as tab sizes, in the preferences.

How It Works

We recommend the classic book *The Elements of MATLAB Style* by Richard K. Johnson as a starting point for developing your own style guidelines. Many of the recommendations are generic to good coding practice across programming languages, and others are specific to MATLAB, such as using publishing markup syntax in your comments. The book addresses formatting, naming, documentation, and programming.

We deviate from the book's recommendations in a few ways. For one, we prefer to capitalize the names of functions. This distinguishes your custom functions from built-in MATLAB functions in scripts. In another, we like to use single-letter variables for structures, rather than long camel case names such as MyFancyDataStructure. However, the key to clear MATLAB code, which is also emphasized in Johnson's text, is to treat MATLAB code like compiled code. Be mindful of variable types, use parentheses even when MATLAB doesn't explicitly require them, and write plentiful comments. This means, generally, at least one comment with each code block and loop explaining its purpose. If you've left a blank line between code bits or indented a section of code, there should be a comment.

For instance, when a variable value is a double, indicate this with a decimal point. This avoids confusion that the parameter may be an integer.

Replace

```
value = 1;
```

with

```
value = 1.0;
```

In `if` statements, always use parentheses. If you ever want to port the code to another language in the future, this saves you time, and it makes the code clearer and easier to read for programmers versed in multiple languages.

Replace

```
if thisIsTrue && thatIsTrue
```

with

```
if (thisIsTrue && thatIsTrue)
```

You should always avoid "magic numbers" in your code, which are easy to use when quickly typing out a function to test a concept. This is a number value that is typed in, such as to a logical statement, instead of assigned to a variable. Take the time to define a properly named variable and add a comment with the source of the number.

Replace

```
if (value > 2.0 || value < 0.0)
```

with

```
if (value > minValue || value < maxValue)
```

With the advent of color-coding IDEs, such as MATLAB's editor, adding a lot of whitespace to delineate code sections has fallen out of favor in style guidelines. Generally, one line of blank space is enough between blocks of code. We suggest adding an additional line of whitespace between the end of a function and the start of a subfunction. You shouldn't put whitespace between lines of code that are closely related.

Some programmers prefer to align blocks of code on their equal signs. This can be helpful, especially when coding sets of equations from a reference. However, it can also be tedious to maintain when code is under active development. If you like this style, you may prefer to wait on adding the aligning space until the function has passed internal code review and is ready for release. In our code, we generally align on equals signs only within smaller blocks as delineated by comments or whitespace.

Consider:

```
1  % Initialization
2  myVar1 = linspace(0,1);
3  b = 1.0;
4
5  % Calculation
6  [result1, result2] = MyFunction(myVar1,b);
7  plotH = plot(myVar1,result2);
```

This could be aligned in multiple ways, such as

```
1  % Initialization
2  myVar1 = linspace(0,1);
3  b      = 1.0;
4
5  % Calculation
6  [result1, result2] = MyFunction(myVar1,b);
7  plotH              = plot(myVar1,result2);
```

or, if aligning across the blocks, as

```
1  % Initialization
2  myVar1             = linspace(0,1);
3  b                  = 1.0;
4
5  % Calculation
6  [result1, result2] = MyFunction(myVar1,b);
7  plotH              = plot(myVar1,result2);
```

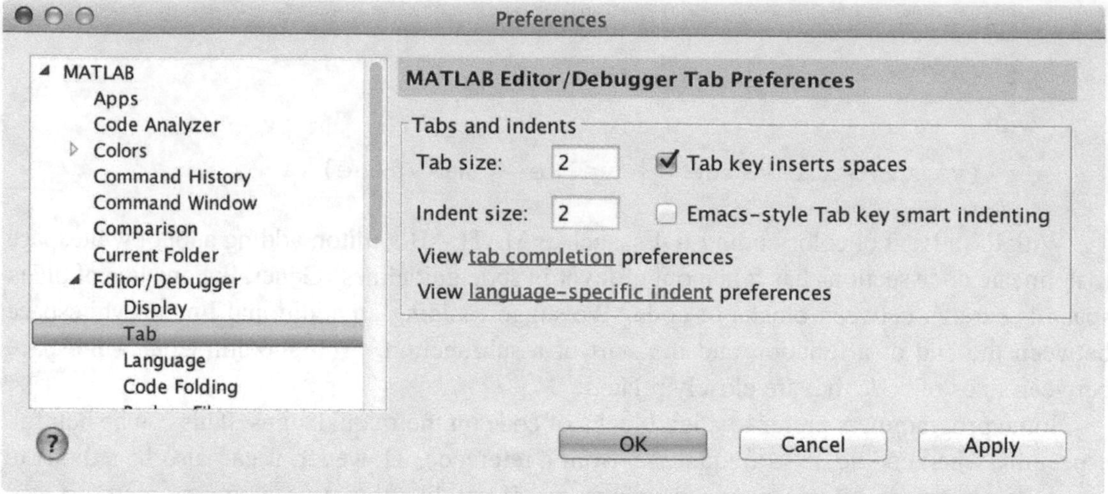

Figure 2.1: *Tab preferences with size of two and spaces option checked.*

In our code for this book, you will see the former, per-block style of alignment.

Another consideration with whitespace is tab sizes. Some guidelines recommend larger tabs of four or eight spaces, arguing that MATLAB code is rarely deeply nested. We routinely write a lot of deeply nested code so our internal guideline is for two spaces. When you set the tab size in the MATLAB preferences, and set it to insert spaces for tabs, you can use the Smart Indent feature to easily highlight and update code blocks. Figure 2.1 shows the tab preferences pane in MATLAB R2014b, on a Mac.

We prefer to use uppercase for function names (`MyFunction`) specifically to distinguish them from the lowercase function names of the built-in MATLAB functions. Otherwise, we use camel case (`myVariableName`) for variables and often single-letter or very short names for structures. For index variables, we tend to use k to avoid confusion with the variables i and j and their association with imaginary numbers. We follow the standard convention of capitalizing constant names, that is, for the Earth's gravitational constant, `MU_EARTH`.

In other words, you need to establish naming conventions for the following:

- Function names
- Variable names
- Structure names
- Index variables
- Constants

Additional naming conventions might include standard prefixes or suffixes for certain types of files or variables. One example is using the letters "RHS" in the name of a function that provides dynamics for integration, that is, the right-hand side of the equations when the derivatives are on the left. The word "Demo" is helpful in the name of a script that demonstrates a particular function or feature. You should be consistent about the order of variable name elements. For example, if r means radius and the second element is the name of the planet, then use R_EARTH and R_MOON. Don't make the second MOON_R. The order should be consistent throughout your code base.

Further rules could address the names of boolean variables or the use of verbs in function names. The most important step is to create a set of conventions for your organization and write them up, or create some function templates, so that your engineers write consistent code.

The guidelines we have described here and use throughout this book are summarized as follows:

Naming – Naming guidelines

> **Function names** – Use camel case for function names with the first letter capitalized. The first word is ideally an action verb or "RHS."
>
> **Script names** – If the script is a demo of a particular function or set of functions, append "Demo" to the name.
>
> **Variable names** – Use camel case for variable names with a lowercase first letter.
>
> **Constants** – Use all uppercase with underscores to identify constants.
>
> **Variable name length** – Most variable names should be at least three characters. This helps enforce uniqueness and makes the variables more readily distinguishable. Exceptions include commonly used data structures, index variables, and when replicating equations from a text where single-letter variable names are standard and easily recognizable to someone in the field.
>
> **Index variables** – When using a single index variable, use k; when using two, j and k; for additional variables, use l, m.

Doubles – Always use a decimal point when typing out a double value.

Magic numbers – Avoid magic numbers in your code; prefer the use of a variable to specify a number.

Comments – Always add a comment describing the source or rationale for a hard-coded number in your code.

If statements – Always use parentheses around the conditional portion of IF statements.

Tabs – Use a tab size of two spaces and set MATLAB to insert spaces for tabs. Use Smart Indent to enforce consistent tabs before committing files.

Blank lines – Use one blank line between most code sections and two blank lines between subfunctions.

Alignment – Align code on the equals sign only within the code block, as separated by blank lines.

Guidelines for function headers are addressed in the next recipe.

2.2 Writing Good Function Help

Problem

You look at a function a couple months (or years) after you wrote it, or a colleague wrote it, and find it has only one cryptic comment at the top. You no longer remember how the function works, what it was supposed to do, or exactly what your comment means.

Solution

Establish a format for your function headers and stick to it. Use the publishing markup to enable you to generate good-looking documentation from the m-file.

How It Works

Write the header for your function at the top, using the publishing markup. This means that the very first line should start with a section break, %%, and include the name of your function, as that will be the title of the published page. This line should also include a one-sentence summary of the function; this must be in the first non-empty comment line of the file and is also termed the "H1" line. This summary can be used automatically by MATLAB when generating `Contents.m` files for your folders and by the `lookfor` function, which searches files on the path for keywords.

Document inputs and outputs separately using section titles. Indicate the type or size of the variable and provide a description. Using two spaces between the comment sign % and line's text will generate monospaced text, which we use for the input and output lists. We have developed the following keys to indicate the variable type and size:

- {} – Cell array
- (1,1) – Scalar value
- (:) or ' ' – String
- (:,:) – Matrix of variable size
- (1,:) – Row of variable length
- (:,1) – Column of variable length
- (m,n) – Matrix with row and column sizes (m,n) that must match other inputs or outputs
- (.) – Data structure

- (:) – Data structure array
- (*) – Function handle

Always include a copyright or authorship notice. Take credit for authoring your code! The standard is to start with the initial year the function is created, then add a range of years when you update the function, that is, Copyright (c) 2012, 2014–2015. The "c" in parentheses approximates the actual copyright symbol. After copyright, the next line should state "All Rights Reserved." Add a contract number, project number, or distribution statement if pertinent. Add a blank line between the main header and the copyright notice to suppress it from the command-line help display.

We show an example in the following for a function which computes a dot product column-wise for two matrices. Note that this will still be legible in the command window output of `help Dot`, with the first % of the cell breaks suppressed. We use the * markup for a bulleted list. The output will always be one row which is indicated in the size key.

FUNCTION HEADER EXAMPLE

```
%% DOT Dot product of two arrays.
%% Forms
%   d = Dot( w, y )
%   d = Dot( w )
%% Description
% Dot product with support for arrays. The number of columns of w and y can be:
%
% * Both > 1 and equal
% * One can have one column and the other any number of columns
%
% If there is only one input the dot product will be taken with itself.
%% Inputs
%   w  (:,:)  Array of vectors
%   y  (:,:)  Array of vectors
%% Outputs
%   d  (1,:)  Dot product of w and y
%% See also
% Cross
```

When published to HTML, this will appear as follows, ignoring the generated Contents section:

DOT Dot product of two arrays.

Forms

```
d = Dot( w, y )
d = Dot( w )
```

Description

Dot product with support for arrays. The number of columns of w and y can be:

- Both > 1 and equal
- One can have one column and the other any number of columns

If there is only one input the dot product will be taken with itself.

Inputs

```
w  (:,:)   vector
y  (:,:)   vector
```

Outputs

```
d  (1,:)   Dot product of w and y
```

See also

Cross

Finally, remember to describe any plots created or files generated, that is, "side effects." It's also a good idea to identify if a function uses persistent or global variables, which may require a `clear` command to reset. The following list summarizes the parts of the header, in order:

1. **H1 line** – Start with a single line description of the function.

2. **Syntax** – List the syntax supporter.

3. **Description** – Provide a more detailed description. Describe any built-in demos, default values for parameters, persistent or global variables users need to be aware of, and any "side effects" including plots or files saved. Indicate if a function will request input from the user.

4. **Inputs** – List the inputs with a size/format key. Include units, if applicable.

5. **Outputs** – List the outputs as for inputs.

6. **See also** – List any related functions.

7. **Reference** – If applicable, list any references.

8. **Copyright** – Include a copyright notice. There should be a blank line between the rest of the header and the copyright notice.

2.3   Overloading Functions and Utilizing `varargin`

Problem

You want to reuse a section of code you have written, but you may have to use it in different situations or extract additional data for it in some circumstances but not others.

Solution

You can overload functions in MATLAB easily and implicitly. This means calling the same function with different inputs and outputs. `varargin` and `varargout` make it simple to manage variable-length input and output lists.

How It Works

MATLAB allows you to overload a function in any way you would like, inside the file that defines it. This applies to the inputs and the outputs. There is generally a trade-off between writing the clearest code you can, with a single calling syntax, and avoiding duplication of code. Perhaps there are intermediate variables which may be useful as outputs in some cases, or you want to provide backward compatibility with an older syntax. When creating libraries for numerical computations, there always seem to be additional syntaxes that are useful. We recommend the following when overloading functions:

- Use `varargin` and `varargout` when possible and rename the variables with descriptive names as close to the top of the function as you can.
- Be sure to document all input and output variants clearly in the header. Adding another optional input or output and neglecting to document it is the #1 reason for out-of-date headers.
- Use comments to clearly identify what the outputs are when you are renaming them to match the function's syntax, or use `varargout`.
- Clear the function outputs if you are creating a plot and they are not needed, to avoid unnecessary printing to the command line.

The following example highlights the use of these guidelines. We often use functions with a string "action" defining multiple input variations by name. This provides additional clarity beyond depending on the input number or type to select an overloaded method.

FUNCTION OVERLOADING

```
1  % d = OverloadedFunction('default data');
2  % OverloadedFunction('demo');
3  % [y,d] = OverloadedFunction('update',x,d);
4
5  function varargout = OverloadedFunction( action, varargin )
6
```

```
7   switch action
8     case 'default data'
9       d = DefaultData;
10      varargout{1} = d;
11
12    case 'demo'
13      d = DefaultData;
14      x = linspace(0,1);
15      y = OverloadedFunction('update',x,d);
16      figure('name','OverloadedFunction Demo');
17      plot(x,y);
18
19    case 'update'
20      x = varargin{1};
21      d = varargin{2};
22      y = Update(x,d);
23      varargout{1} = y;
24      varargout{2} = d;
25
26  end
```

2.4   Adding Built-in Inputs and Outputs to Functions

Problem

You would like to provide default values for some optional inputs or provide a short demonstration of how a function works.

Solution

Add built-in inputs and outputs to your function using an enumerated input or with `nargin`. This can include a full demo that calls the function and generates plots, as appropriate.

How It Works

Built-in inputs provide an example set of parameters that produce output. In many cases, we provide an input range that can create a plot demonstrating the computation performed in the function. In the case of MATLAB, you must explicitly handle input options in the code, as you can't add a default value in the function definition itself.

One convention that we find useful is to allow for an empty matrix, [], to be entered for an input to use its default value. This allows you to request a default for one input but provide values for subsequent inputs. The following example shows both a demo that creates a plot and a default value for a constant.

78

```
1  function output = MyFunction( variable, constant )
2
3  if (nargin == 0)
4    % perform demo
5    variable = linspace(0,100);
6    output = MyFunction( variable );
7    return;
8  end
9  if (nargin < 2 || isempty(constant))
10   % default value of constant
11   constant = 2.05;
12 end
```

Notice that the built-in demo, which is performed when there are no inputs at all, calls the function itself and then returns. This makes the demo also a built-in test. The code to generate the built-in outputs, which could be a text report to the command line or a plot, generally comes at the end of the function. This enables you to create the built-in outputs with inputs the user specifies and not just the built-in inputs. For instance, there might be alternative values of the constant. Note that in the following output generation example, the name of the figure is specified including the name of the function, which is exceedingly helpful if you routinely generate dozens of plots during your work.

```
1  ... body of function with calculations ...
2
3  if (nargout==0)
4    % Default output is a plot
5    figure('Name',sprintf('Demo of %s',mfilename))
6    plot(variable, output)
7    clear output
8  end
```

■ **TIP**   Assign a name to figures that you create. Include the name of the function or demo for clarity. The name will be displayed in the title bar of the figure and in MATLAB's Windows menu.

Writing all your functions this way has several advantages. For one, you are showing valid ranges of the variables up front, without requiring a reader to refer to a separate test function or demo in another folder. Having this hard data available every time you open the function helps keep your code and your comments consistent. Also, you have a test of the function, which you can easily rerun at any time right from the editor. You can publish the function with execution turned on which will perform the demo and include the command-line output and plots right in the HTML file (or LaTeX or Word, if you so choose.) All of this helps reduce bugs and documents your function for other readers or yourself in the future.

Following this guideline, here is the general format we follow for all functions in this book:

1. Detailed header

2. Copyright

3. Function definition

4. Default inputs

5. Function demo that calls itself

6. Code body with calculations

7. Default output

Note that no final `return` statement is necessary.

In summary, we have enabled the following usages of this function by adding default inputs and outputs:

```
1  output = MyFunction( variable, constant );
2  output = MyFunction( variable );     % uses default value of constant
3  MyFunction;                          % performs built-in demo
4  MyFunction(variable, constant);      % creates a plot for the given input
```

2.5   Adding Argument Checking to Functions

Problem

You would like to check arguments to make sure the user has entered valid values. Perhaps the function only works for a certain range of values or only positive inputs. You would like to give the user specific feedback on why a function has failed due to a known input issue rather than let the function throw an error.

Solution

Add argument validation. The MATLAB document has details about the syntax.

How It Works

We create the following function, `ArgCheckFun`, and use the new `arguments` feature available since R2019b. This allows you to explicitly define the type of each function argument. You can define default values for arguments and supply validation functions. A table of predefined functions exists, including `mustBePositive` and `mustBeNegative`.

```
1   function [x,c] = ArgCheckFun(a,b,c)
2
3   arguments
4       a (2,1) double
5       b (2,1) double
6       c (1,:) char
7   end
8
9   x = a'*b;
```

Now let's run a few tests.

```
>> ArgCheckFun
Error using ArgCheckFun
Invalid input argument list. Not enough input arguments.
Function requires 3 input(s).
```

This first test fails as expected.

```
>> ArgCheckFun(1,1,1)

ans =
     2
```

The second does not fail, even though the arguments are not what is specified in the argument block. Note that the third input is unused except for being passed back as an output, so the wrong form does not cause an error.

```
>> ArgCheckFun(rand(3,1),1,1)
Error using ArgCheckFun
Invalid input argument at position 1. Value must be vector with 2
    elements.
```

The third does fail, as it should, as the first input has too many elements.

```
>> [x,c] = ArgCheckFun(rand(2,1),rand(2,1),'test1')

x =
    0.2652

c =
    'test1'
```

The last text works.

There may be cases where this is useful. However, if you properly document your arguments in the header, you won't have much need for this feature.

2.6 Adding Dot Indexing

Problem

You would like to use dot indexing with a function that outputs a data structure.

Solution

Create a function with a single data structure output. Dot indexing is a new feature available by default.

How It Works

Create the following function:

```
1  function d = StructFun(v)
2  d = struct('x',v,'y',0);
```

Now run two tests, passing 2 for the value v, and in the first test, accessing x, and in the second, accessing y.

```
>> StructFun(2).x

ans =
     2

>> StructFun(2).y

ans =
     0
```

You can see that you can now access a field directly. It is not necessary to use a data structure variable as an intermediate step if you only need a single field's value.

2.7 Smart Structuring of Scripts

Problem

You write a few lines of code in a script to test some idea. Can you figure out what it does a year later?

Solution

Treat your scripts like functions, and structure them well. Take the time to follow a template.

How It Works

A script is any collection of commands that you put in a file and execute together. In our toolboxes, we treat scripts as demos of our toolbox functions, and therefore as instructional. Here are some guidelines we recommend when creating scripts:

Create help – Help headers are not just for functions; write them for your scripts too. Will you remember what this script does in a year? Will someone else in your company be able to understand it? Write a detailed description including a list of any files required or generated.

Use publishing markup – Create cells in your scripts (using %%) to delineate sections. Write detailed comments after the section headings. Publish your script to HTML and see how it looks. You can even add equations using LaTeX markup or insert images.

Initialize your variables – Take care to fully initialize your variables or you could have conflicts when you run multiple scripts in a row. This especially applies to data structures and cell arrays. See the recipes for data types in Chapter 1 for the correct way to initialize different variables.

Specify a directory for saved files – Make sure you are saving any data into a particular location and not just wherever the current directory happens to be.

Our scripts follow the following pattern. Cell breaks are used between the sections.

1. Detailed header using publishing markup.

2. Copyright notice.

3. User parameters, meant to be changed between runs, are grouped at the top.

4. Constants are defined.

5. Initialize plotting arrays before loops.

6. Perform calculations.

7. Create plots.

8. Store outputs in files, if desired.

A complete example, which can be executed, is shown as follows:

ScriptDemo.m

```matlab
1  %% This is a template for a script layout.
2  % A detailed description of the script includes any files loaded or
       generated
3  % and an idea of what data and plots will be created. We will calculate
       a sine
4  % or cosine with or without scaling of the input. The script creates
       one plot
5  % and saves the workspace to a file called Demo.mat.
6  %% See also
7  % sin, cos
8
9  %% User parameters
10 param1  = 0.5;
11 nPoints = 50;
12 useSine = false;
13
14 %% Constants
15 MY_CONSTANT = 0.25;
16
17 %% Calculation loop
18 yPlot = zeros(2,nPoints);
19 x       = linspace(0,4*pi,nPoints);
20 for k = 1:nPoints
21   if (useSine)
22     y = sin( [1.0;param1]*x(k) + MY_CONSTANT );
23   else
24     y = cos( [1.0;param1]*x(k) + MY_CONSTANT );
25   end
26   yPlot(:,k) = y;
27 end
28
29 %% Plotting
30 figure('Name','DEMO');
31 plot(x,yPlot);
32
33 %% Save workspace to a file
34 saveDir = fileparts(mfilename('fullpath'));
35 save(fullfile(saveDir,'Demo'))
```

We can verify that the data is stored by clearing the workspace and loading the mat-file after the demo has been run.

```
>> clear all
>> ScriptDemo
>> clear all
>> load Demo.mat
>> who

Your variables are:

MY_CONSTANT  nPoints      useSine       y
k            param1       x             yPlot
```

2.8 Implementing MATLAB Command-Line Help for Folders

Problem

You have a set of folders in your code base and you would like users to easily navigate them as they can the built-in MATLAB library.

Solution

Placing Contents.m files in each folder can provide metadata for the contents of the folders, and this can be displayed on the command line.

How It Works

Command-line help isn't just for functions and scripts. Folders can also have help in the form of a contents listing, which includes the function names and a single-line description of each. Toolboxes can also provide documentation in response to a ver command with a toolbox-level contents listing. This information is provided in a Contents.m file that consists entirely of comments.

The built-in *Contents Report* can generate Contents.m files for you. It can also check and fix existing Contents.m files. This will automatically use the "H1" line, or the first line of the header in the function or script. In Recipe 2.2, we provide an example of a function header that includes this line. This report is accessed, along with other reports, in the Current Folder window, using the pop-up menu. To read more and learn how to run the report on your operating system, see the MATLAB help topic "Create Help Summary Files." It's typically accessed from the context menu in the Current Folder window.

Version information isn't limited to a single Contents file per toolbox; it is generated by special lines inserted into the top of any Contents.m file:

```
% Version xxx dd-mmm-yyyy
```

You can also add a descriptive line above the version information and add subheadings to groups of files. For example, consider the output from the codetools directory included in MATLAB:

```
>> help codetools
  Commands for creating and debugging code
  MATLAB Version 8.4 (R2014b) 08-Sep-2014

  Editing and publishing
    edit                    - Edit or create a file
    grabcode                - Copy MATLAB code from published HTML
    mlint                   - Check files for possible problems
    notebook                - Open MATLAB Notebook in Microsoft Word (on
       Microsoft Windows platforms)
    publish                 - Publish file containing cells to output file
    snapnow                 - Force snapshot of image for published
       document

  Directory tools
    mlintrpt                - Run mlint for file or folder, reporting
       results in browser
    visdiff                 - Compare two files (text, MAT, or binary) or
       folders
```

As with the header of a function, there can be no blank lines in the Contents file, only comments. This is shown in an example from Chapter 7 of this book, the double integrator, where we have added letters of the alphabet as section breaks.

```
1  % MATLAB/Ch06-DoubleIntegrator
2  %
3  % D
4  %     DoubleIntegratorSim - Double Integrator Demo
5  %
6  % P
7  %     PDControl            - PDCONTROL Design and implement a PD
      Controller in sampled time.
8  %     PlotSet              - PlotSet Create two-dimensional plots from a
      data set.
9  %
10 % R
11 %     RHSDoubleIntegrator - RHSDoubleIntegrator Right hand side of a
      double integrator.
12 %     RungeKutta          - RungeKutta Fourth order Runge-Kutta
      numerical integrator.
```

```
13  %
14  %  T
15  %     TimeLabel              - TimeLabel Produce time labels and scaled
       time vectors
```

Figure 2.2 shows how to access the Contents Report for this folder from the command window. If a `Contents.m` file doesn't already exist, the report will create one for you, which will then be available from the command line via `help FolderName`.

The actual report is shown in Figure 2.3. You can see that there are links to edit the `Contents.m` file, such as for adding version information, fixing the spacing, or fixing all problems. The report will detect if you have changed the H1 description line of the function, and it conflicts with the text in the Contents file. It allows you to update those descriptions with a single click, avoiding any copy/paste.

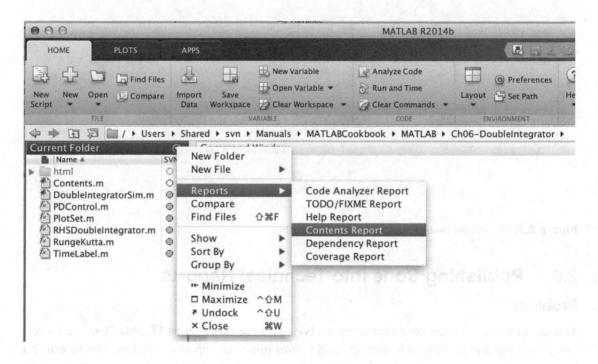

Figure 2.2: *Access the Contents Report on double integrator.*

```
● ○ ○                    Web Browser – Contents File Report
  Contents File Report  ✕  +                              ⊞ ▯ ◨ ▢  ⬎

 ←  ➡  ↻  🖶  🔍
```

Contents File Report

The Contents Report displays information about the integrity of the Contents.m file for the folder (Learn More).

[Rerun This Report] [Run Report on Current Folder]

[edit Contents.m | fix spacing | fix all]

Report for folder /Users/Shared/svn/Manuals/MATLABCookbook/MATLAB/Ch06-DoubleIntegrator

```
MATLAB/Ch06-DoubleIntegrator

D

    DoubleIntegratorSim - Double Integrator Demo

P

    PDControl            - PDCONTROL Design and implement a PD Controller in sampled time.
    PlotSet             - Create two-dimensional plots from a data set.

R

    RHSDoubleIntegrator - Right hand side of a double integrator.
    RungeKutta          - Fourth order Runge-Kutta numerical integrator.

T

    TimeLabel           - Produce time labels and scaled time vectors
```

Figure 2.3: *Completed double integrator contents report.*

2.9   Publishing Code into Technical Reports

Problem

You are creating a report based on some analysis you are doing in MATLAB. You are laboriously copying and pasting code snippets and figures into your report document. You discover a bug in your code, and you have to do it all over again.

Solution

The publishing feature in MATLAB allows you to run a script and automatically generate a document containing the results, including images of each figure generated and the code itself, with text and equations that you insert. These reports can be easily regenerated when you change your code.

How It Works

The publishing features allow you to generate HTML, LaTeX, Word, and PowerPoint documents from your code. These documents can display the code itself as well as command-line output and plots. You can even capture snapshots of your figures during loops and include equations using LaTeX markup. Every programmer should become familiar with these features. We highlight the main features in the following.

The very first section at the top of your file gives a title to the published document. The comments which follow in your header will be published as discussed in Recipe 2.2. Having a good header is important since this can be displayed at the command line, up until the first blank line of your function. However, you can also add more sections, text, equations, and images throughout your code. MATLAB will automatically generate a table of contents of all the sections and will insert the generated plots and command-line output in each section.

You need to be careful about putting section breaks inside loops, since this will produce a snapshot of any figures at every iteration. This could be a desired behavior if you want to capture the evolution of a figure, but could also accidentally produce hundreds of unwanted images. The following is an example script, MemoExample.m, created to demonstrate publishing.

CREATE A TECHNICAL MEMO FROM YOUR CODE

MemoExample.m

```
1   %% Technical Memo Example
2   % Summary of example objective.
3   % Evaluate a function, in this case $\sin(x)$, in a loop. Show how the
4   % equation looks on its own line:
5   %
6   % $$ y = sin(x) $$
7
8   %% Section 1 Title
9   % Description of first code block.
10  % Define a customizable scale factor that is treated as a constant.
11  SCALE_FACTOR = 1.0;
12
13  %% Section 2 Title
14  % Description of second code block.
15  % Perform a for loop that updates a figure.
16  %
17  h = figure('Name','Example Memo Figure');
18  hold on;
19  y = zeros(1,100);
20  x = linspace(0,2*pi);
21  for k = 1:100
22          %%% Evaluate the function. Comments not in a block after the
                  title will
23          %%% not be included in the main text.
24          y(k) = sin(SCALE_FACTOR*x(k));
25          plot(x(k),y(k),'.')
26  end
27
```

```
28  %% Conclusions
29  % You can add additional text throughout your script. You can insert
       lists,
30  % HTML, links, images, etc.
```

Figure 2.4 shows this script in the publishing tab of the MATLAB editor, with the pop-up menu opened to access the publishing options.

There are a number of settings that apply to publishing. You can save a set of settings with a name and easily reuse it for all of your files. The default settings for code are to both evaluate it and include the source code in the published document, but these may be turned off independently. To create a technical memo from a script without including the source code itself, you set the "include code" option to *false*. You can set maximum dimensions on figures and select the format – JPEG, PNG, bitmap, or TIFF. You can even specify a MATLAB

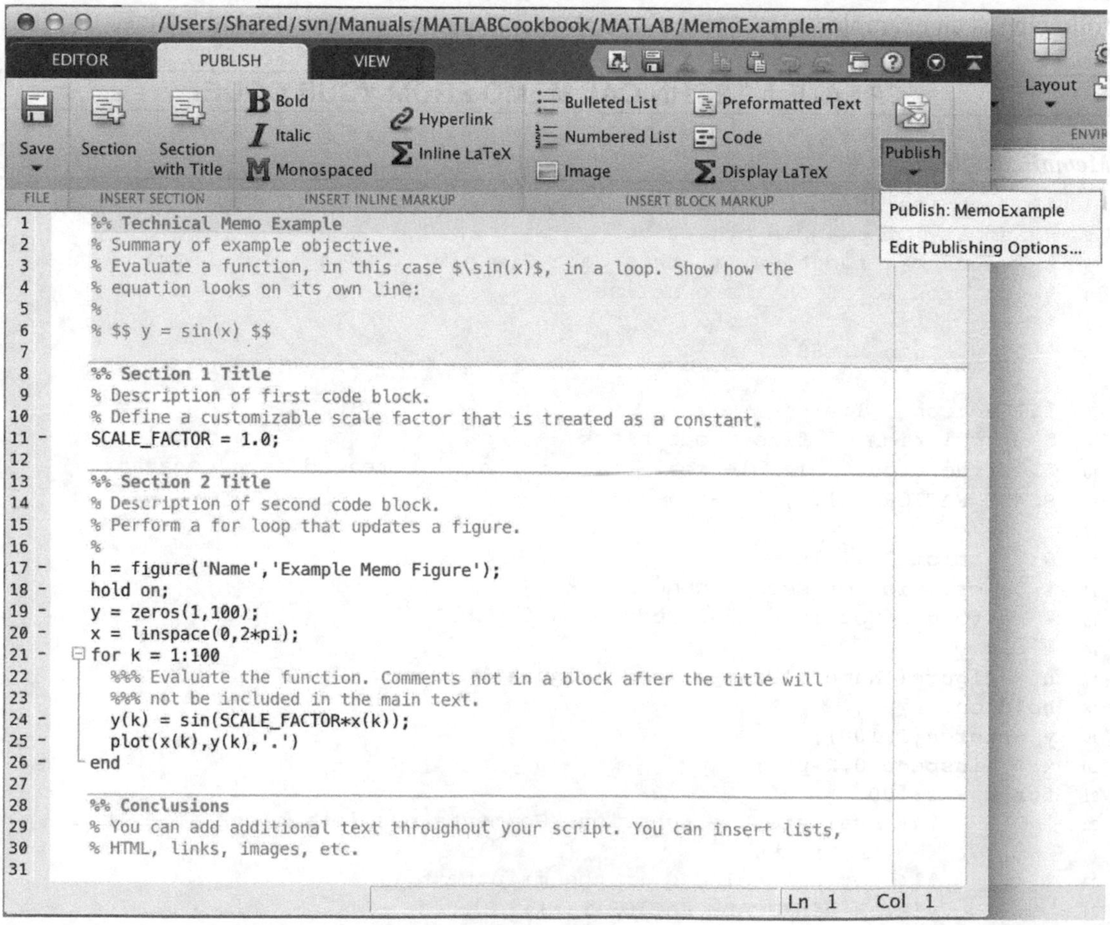

Figure 2.4: *Preparing to publish a script in the editor.*

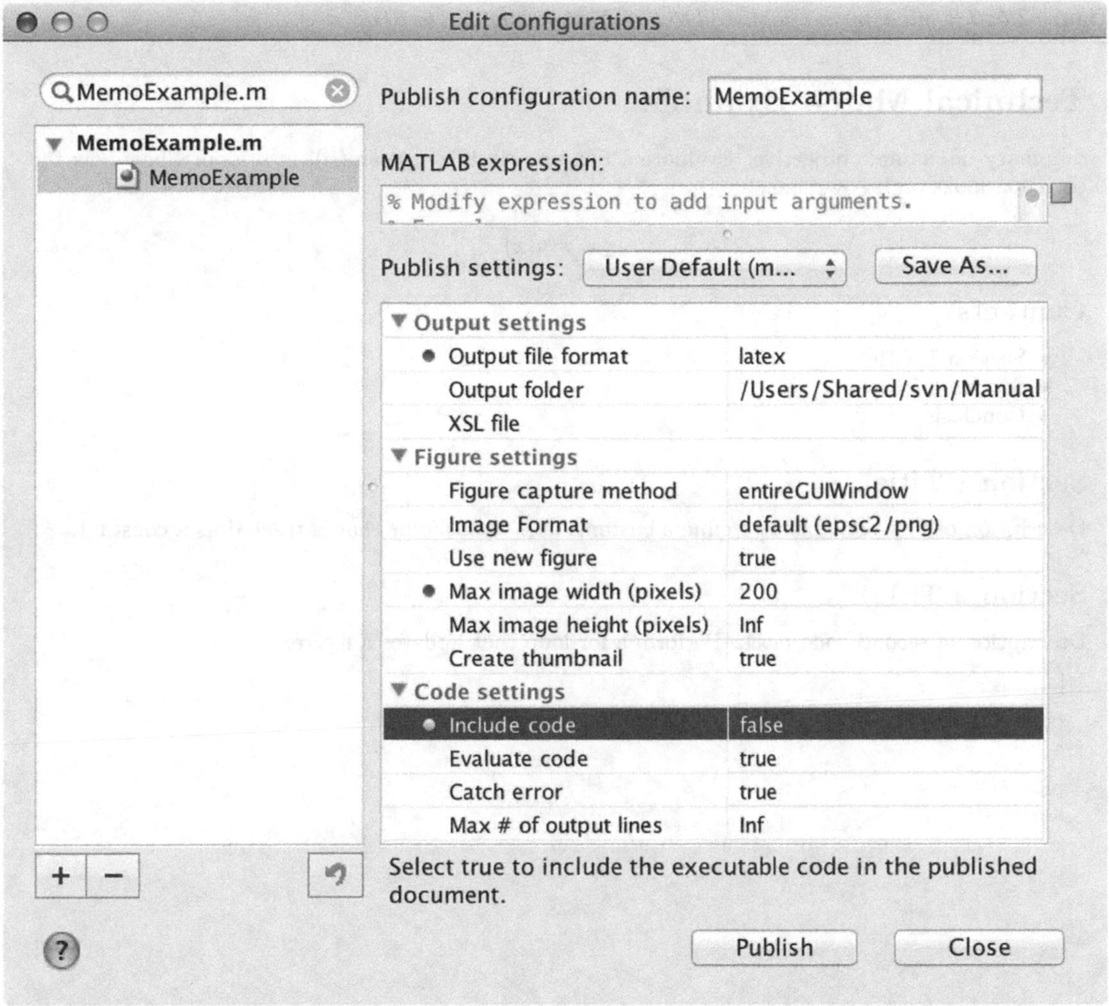

Figure 2.5: *Editing the publish settings for a file.*

expression for a function to include input arguments, rather than just running it as a built-in demo.

Figure 2.5 shows the settings window with PDF selected as the output type. Note the `Save As...` button which allows you to save settings. We set the maximum width of the figure to 200 pixels to enable the memo to fit on one page, for the purposes of this book.

Figure 2.6 shows a LaTeX memo generated and compiled for the preceding listing published without the code, with the figure generated in a loop. Note the table of contents, equation, and insertion of the graphic. We had to remove some extra `\vspace` commands that MATLAB added to the LaTeX to fit the memo on one page.

Figure 2.6: *Technical memo published to LaTeX and compiled to PDF.*

Technical Memo Example

Summary of example objective. Evaluate a function, in this case $sin(x)$, in a loop. Show how the equation looks on its own line:

$$y = sin(x)$$

Contents

- Section 1 Title
- Section 2 Title
- Conclusions

Section 1 Title

Description of first code block. Define a customizable scale factor that is treated as a constant.

Section 2 Title

Description of second code block. Perform a for loop that updates a figure.

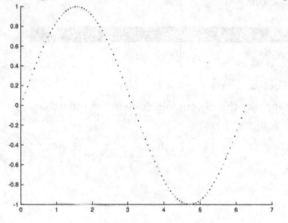

Conclusions

You can add additional text throughout your script. You can insert lists, HTML, links, images, etc.

2.10 Integrating Toolbox Documentation into the MATLAB Help System

Problem

You would like to write a user's guide and provide it with your toolbox.

Solution

If you write HTML help files, you can in fact include them with your toolbox when you distribute it, and the help will show up in MATLAB's help browser.

How It Works

You are not limited to the command-line help when providing documentation for your code or toolbox. MATLAB now provides an API for writing HTML documentation and displaying it to users in the help browser. You can write an entire HTML manual and include published versions of your demos.

In order to integrate your HTML help files into the MATLAB help system, you need to generate a few XML files. One provides a top-level table of contents for your toolboxes. Another provides a list of the demos or examples. The third identifies your product. The help topics to read are "Display Custom Documentation" and "Display Custom Examples." The help for third-party products is displayed in a separate section of the MATLAB help browser entitled "Supplemental Software." The files you need to generate are

info.xml – Identify your documentation

helptoc.xml – Table of contents

demos.xml – Table of examples

The MATLAB documentation describes the XML tags you need and provides template documents. Comments can be included within the files using standard HTML comments with `<!--` and `-->`.

The main purpose of the info.xml file is to provide a name for your toolbox, identify it as a toolbox or blockset, and provide a path to the remaining HTML documentation. The following is an example for our Recipes code.

EXAMPLE INFO.XML

```
1  <productinfo xmlns:xsi="http://www.w3.org/2001/XMLSchema-instance"
2      xsi:noNamespaceSchemaLocation="optional">
3      <?xml-stylesheet type="text/xsl" href="optional"?>
4
5      <matlabrelease>R2019b</matlabrelease>
6      <name>MATLAB Recipes</name>
7      <type>toolbox</type>
```

```
8      <icon></icon>
9      <help_location>Documentation</help_location>
10     <help_contents_icon>$toolbox/matlab/icons/bookicon.gif</
       help_contents_icon>
11
12   </productinfo>
```

The table of contents file, helptoc.xml, must provide a listing of all the HTML files in your help. This is accomplished with a `<tocitem>` tag that can be nested. You are generally providing a starting or main page for your toolbox, a getting started page, users guide pages, release notes, and further pages listing the functions provided. `<tocitem>` can have references to HTML anchors; they do not all need to refer to separate HTML files.

A small set of icons is included that can be displayed in the help contents. Consider the following helptoc.xml:

EXAMPLE HELPTOC.XML

```
1    <?xml version='1.0' encoding="utf-8"?>
2    <toc version="2.0">
3    <!-- First tocitem specifies top level page in Help browser Contents
         -->
4        <tocitem target="index.html">Recipes Toolbox
5           <!-- A Getting Started page is generally first -->
6           <tocitem target="getting_started.html" image="HelpIcon.
               GETTING_STARTED">
7              Getting Started
8              <tocitem target="requirements.html">System Requirements</
                  tocitem>
9              <tocitem target="features.html">Features
10                <!-- TOC levels may include anchor IDs -->
11                <tocitem target="features.html#10187">Feature 1</
                     tocitem>
12                <tocitem target="features.html#10193">Feature 2</
                     tocitem>
13             </tocitem>
14          </tocitem>
15          <!-- There is a special icon for the User Guide -->
16          <tocitem target="guide_intro.html"
17             image="HelpIcon.USER_GUIDE">Recipes User Guide
18             <tocitem target="setup.html">Setting Up</tocitem>
19             <tocitem target="data_processing.html">Processing Data</
                  tocitem>
20             <tocitem target="verification.html">Verifying Outputs
21                <tocitem target="test_failures.html">Handling Test
                     Failures</tocitem>
22             </tocitem>
23          </tocitem>
24          <!-- The function reference is next with the FUNCTION icon -->
```

```
25         <!-- First item is page describing function categories, if any
             -->
26         <tocitem target="function_categories.html"
27               image="HelpIcon.FUNCTION">Function Reference
28           <tocitem target="function_categories.html#1">Double
               Integrator
29             <!-- Inside category, list the functions -->
30             <tocitem target="function_1.html">function_1</tocitem>
31             <tocitem target="function_2.html">function_2</tocitem>
32             <!-- ... -->
33           </tocitem>
34           <tocitem target="function_categories.html#2">Aircraft
35             <tocitem target="function_3.html">function_3</tocitem>
36             <tocitem target="function_4.html">function_4</tocitem>
37           </tocitem>
38           <tocitem target="function_categories.html#3">Spacecraft
39             <!-- ... -->
40           </tocitem>
41         </tocitem>
42         <!-- Web links with the webicon.gif -->
43         <tocitem target="http://www.psatellite.com"
44               image="$toolbox/matlab/icons/webicon.gif">
45         Web Site (psatellite.com)
46         </tocitem>
47       </tocitem>
48   </toc>
```

This produces the contents listing in the help browser shown in Figure 2.7. The major icons helping to delineate the help sections are used. Anchor IDs are used for both **features.html** and **function_categories.html**. There is even a reference to an external website. Note that this means you need to have written the following HTML files:

- index.html
- getting_started.html
- requirements.html
- features.html
- guide_intro.html
- setup.html
- data_processing.html
- verification.html
- test_failures.html

Figure 2.7: *Custom toolbox table of contents.*

- function_categories.html
- function_1.html
- function_2.html
- ...

Clearly, generating a function list for a large toolbox by hand could be cumbersome. At PSS, we have functions to generate this XML automatically from a directory, using `dir`. You can use the functional form of `publish` to publish your functions and scripts to HTML automatically as well.

The demos file is similar to the `toc` file in that it provides a nested list of demos or examples. There are two main tags, `<demosection>` and `<demoitem>`. Items can be m-files or videos. Published demos will display a thumbnail for one of the figures from the demo, if any exist; the thumbnail image will have the same name as the HTML file but a different extension. The demos are completely independent from the HTML table of contents, and you can implement an examples listing without creating any other HTML help pages.

Here is a short example from our Cubesat Toolbox that includes a published demo called `MagneticControlDemo`.

EXAMPLE DEMOS.XML

```
1  <?xml version="1.0" encoding="utf-8"?>
2  <demos>
3      <name>CubeSat</name>
4      <type>toolbox</type>
5      <icon>$toolbox/matlab/icons/demoicon.gif</icon>
6      <description>Contains all the demo files for the CubeSat</
           description>
7      <website>
8          <a href="http://www.psatellite.com">For more info see psatellite.
               com</a>
9      </website>
```

```
10    <demosection>
11        <label>AttitudeControl</label>
12        <demoitem>
13            <label>MagneticControlDemo: Magnetic control demand analysis</
                  label>
14            <callback>MagneticControlDemo</callback>
15            <file>../CubeSat/Demos/AttitudeControl/html/
                  MagneticControlDemo.html</file>
16        </demoitem>
17    </demosection>
18  </demos>
```

Once you have created a set of HTML files, you can create a database that will allow MATLAB to search them efficiently. To do this, you use `builddocsearchdb` with a path to the folder containing your help files, that is, the same path you enter in your info.xml file. This function will create a subfolder called helpsearch containing the database. With this subfolder added to your help installation, users will get results from your documentation when they search in the help browser. Figure 2.8 shows a complete `Documentation` folder including the `helpsearch` database.

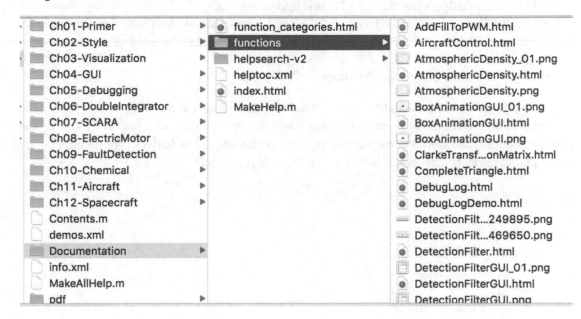

Figure 2.8: *Documentation including the search database for this toolbox.*

2.11   Structuring a Toolbox

Problem

You have a jumble of functions and scripts that you would like to organize into a toolbox that you can distribute to others.

Solution

A previous recipe showed how you can create or generate `Contents.m` files for individual folders in your toolbox. You can also create a top-level `Contents.m` file. We will describe our usual toolbox structure including the placement of these files.

How It Works

We have a fixed structure for our commercial toolboxes that is used by our build tools and testing routines.

- Group related functions together in folders.
- Place scripts in separate folders.
- Place script folders together in a Demos folder.
- Use the same name for the function folder and corresponding demos folder, for example, Mechanics/ being a folder with functions and Demos/Mechanics/ holding the corresponding demos.
- Organize folder groups into Modules or Toolboxes.

Once you create the help files as described in the previous recipes, they will appear in the directory structure as shown in the following – not in literal alphabetical order. Note that the published demos are stored in the html directories within the demo folders. We do not display them all, but every folder should have its own `Contents.m` file.

```
Module
|   Contents.m
|   Folder1
|   |      Contents.m
|   |      Function1.m
|   Folder2
|   |      Function2.m
|   Demos
|   |    Folder1
|   |    |    Function1Demo.m
|   |    |    html
|   |    Folder2
|   |    |    Function2Demo.m
|   |    |    html
|   |    CombinedDemos
```

```
|       |       |       SuperDemo.m
|       |       |       html
|   Documentation
|       |       demos.xml
|       |       info.xml
|       |       ToolboxHelp
|       |       |       helptoc.xml
|       |       |       GettingStarted.html
|       |       |       ...
```

You will note that there is a top-level `Contents.m` file within the Module, as the same level as the folders. MATLAB does not have any automated utility to make this for you. You can create one with a version line, the name of your toolbox, and any other information you would like to be displayed when the user types "help Module"; we generate a list of folders within the module using `dir`. Here is an example, noting that all lines in a `Contents.m` file are comments:

A Sample Contents.m

```
1  %  PSS Toolbox Folder NewModule
2  %  Version 2015.1       05-Mar-2015
3  %
4  %  Directories:
5  %  Folder1
6  %  Folder1
7  %  Demos
8  %  Demos/Folder1
9  %  Demos/Folder1
```

Your toolbox module will now appear when the user types `ver` at the command, for example:

```
>> ver
------------------------------------------------------------------
MATLAB Version: 8.4.0.150421 (R2014b)
MATLAB License Number: 6xxxxx
Operating System: Mac OS X  Version: 10.9.5 Build: 13F1066
Java Version: Java 1.7.0_55-b13 with Oracle Corporation Java HotSpot(TM)
    64-Bit Server VM mixed mode
------------------------------------------------------------------
MATLAB                                            Version 8.4
    (R2014b)
PSS Toolbox Folder NewModule                      Version 2015.1
```

Summary

In this chapter, we reviewed style guidelines for writing MATLAB code and highlighted some differences between styles for MATLAB and other languages. When establishing guidelines for your own toolboxes, consider the features you may want to use, such as automatic generation of contents files, publishing your results to HTML or Word, and even incorporating HTML help in the web browser. Also take the time to create proper headers and initialization when you generate code to avoid unpleasant surprises down the road!. Table 2.1 lists the code developed in the chapter.

Table 2.1: *Chapter Code Listing*

File	Description
Dot	Dot product header example
MemoExample	Example of a technical memo for publishing
OverloadedFunction	An internally overloaded function
ScriptDemo	Demo template for a script layout

CHAPTER 3

■ ■ ■

Visualization

MATLAB provides extensive capabilities for visualizing your data. You can produce 2D plots, 3D plots, and animations; view images; and create histograms, contour and surface plots, as well as other graphical representations of your data. You are probably familiar with making simple 2D plots with lines and markers, and pie and bar charts, but you may not be aware of the additional possibilities made available by the MATLAB low-level routines that underpin the frequently used functions like `plot`. There are also interactive capabilities for editing plots and figures and adding annotations before printing or exporting them.

MATLAB excels in scientific visualization and in engineering visualization of 3D objects. Three-dimensional visualization is used to visualize data that is a function of two parameters, for example, the height on the surface of the Earth, or to visualize objects. The former is used in all areas of science and engineering. The latter is particularly useful in the design and simulation of any kind of machine including robots, aircraft, automobiles, and spacecraft.

Three-dimensional visualization of objects can be further divided into engineering visualization and photo-realistic visualization. The latter helps you understand what an object looks like and how it is engineered. When the inside of an object is considered, we move into the realm of solid modeling which is used for creating models suitable for the manufacturing of the object. Photo-realistic rendering focuses on the interaction of light with the object and the eye. MATLAB does provide some capabilities for lighting and camera interaction but does not provide true photo-realistic rendering.

The main plotting routines are organized into several categories in the command-line help:

graphics – Low-level routines for figures, axes, lines, text, and other graphics objects.

graph2d – Two-dimensional graphs like linear plots, log scale plots, and polar plots.

graph3d – Three-dimensional graphs like lines, meshes, and surfaces; control of color, lighting, and the camera.

specgraph – Specialized graphs, the largest category. Special 2D graphs like bar and pie charts and histograms, contour plots, special 3D plots, volume and vector visualization, image display, movies, and animation.

© Michael Paluszek and Stephanie Thomas 2020
M. Paluszek and S. Thomas, *MATLAB Recipes*,
https://doi.org/10.1007/978-1-4842-6124-8_3

The online help has an entire top-level section devoted to graphics, including plots, formatting and annotation, images, printing and saving, graphics objects and performance, and major changes to plotting internals that occurred in R2014b.

A good command of these functions allows you to create very sophisticated graphics as well as to adapt them to different publication media, whether you need to adjust the dimensions, color, or font attributes of your plot. In this chapter, we will present recipes that cover what you need to know to use MATLAB graphics effectively. We don't have space to discuss every available plotting routine, and that is well covered in the available help, but we will cover the basic functionality and provide recipes for common usage.

3.1   Plotting Data Interactively from the MATLAB Desktop

Problem

You would like to plot data in your workspace but aren't sure of the best method for visualizing it.

Solution

You can use the PLOTS tab in the MATLAB desktop to plot data directly by selecting variables in the Workspace display as shown in Figure 3.1. You select from a variety of plot options, and MATLAB automatically only shows you those which are applicable to the selected data set.

How It Works

Let's create some sample data to demonstrate this interactive capability, which is a fairly new feature in MATLAB. We'll start with some trigonometric functions to create sample data that oscillates.

```
theta = linspace(0,4*pi);
y = sin(theta).*cos(2*theta) + 0.05*theta;
```

We now have two vector variables available in the workspace. Select the PLOTS tab in the desktop as shown in Figure 3.1, then select the y variable in the Workspace display. The variable will appear on the far left of the PLOTS tab area, and various plot icons in the ribbon

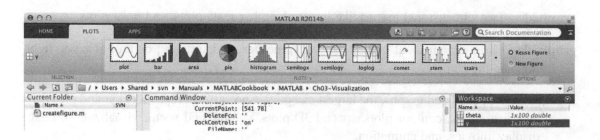

Figure 3.1: *PLOTS tab with plot icon ribbon.*

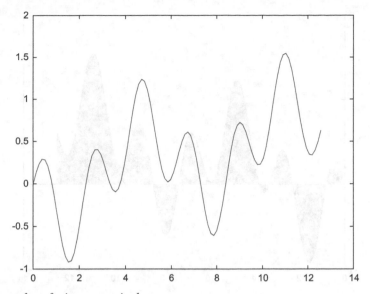

Figure 3.2: *Linear plot of trigonometric data.*

will become active: plot, bar, area, pie, and so on. Note the radio buttons on the far left for either reusing the current figure for the plot or creating a new figure.

Close all open figures with a `close all` and click the plot icon to create a new figure with a simple 2D plot of the data. Note that clicking the icon results in the plot command being printed to the command line:

```
>> plot(y)
```

The data is printed with linear indices along the x axis, as shown in Figure 3.2.

You simply click another plot icon to replot the data using a different function, and again the function call will be printed to the command line. The plot icons that are displayed are not all the plots available, but simply the default favorites from among all the many options; to see more icons, click the pop-up arrow at the right of the icon ribbon. The available plot types are organized by category, and there is a Catalog button that you can press to bring up a dedicated plot catalog window with the documentation for each function.

To plot our data `y` against our input `theta`, you need to select both variables in the workspace view. They will both be displayed in the plot ribbon with a button shown to reverse their order. Now click an area plot to get a plot with the angle on the x axis as shown in Figure 3.3.

```
>> area(theta,y)
```

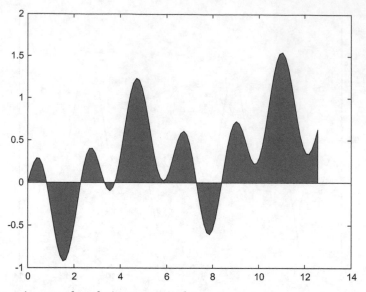

Figure 3.3: *Parametric area plot of trigonometric data.*

Note how this time the x-axis range is from 0 to 4π as expected.

You can annotate the plot interactively with arrows and text, add subplots, change line properties, and more using the Plot Edit toolbar and Figure Palette window shown in Figure 3.4. These are available from the View menu of the figure window and by clicking the "Show Plot Tools" button in the standard Figure Toolbar. For example, using the plot tools, we can select the axes, double-click it to open the property editor, type in an X Label, and turn on grid lines. We can add another subplot, plot the values of theta against linear indices, and then change the plot type to a stem plot, all from this window. See Figure 3.4.

The same changes can be made programmatically as will be shown in the following recipes. In fact, you can generate code from the Figure Palette, and MATLAB will create a function with all the commands necessary to replicate your figure from your data. The Generate Code command is under the File menu of the window. This allows you to interactively create a visualization that works with some example data and then programmatically adapt it to your toolbox. MATLAB calls the new autogenerated function `createfigure`. You can see the use of the following functions: `figure`, `axes`, `box`, `hold`, `ylabel`, `xlabel`, `title`, `area`, `stem`, and `annotation`.

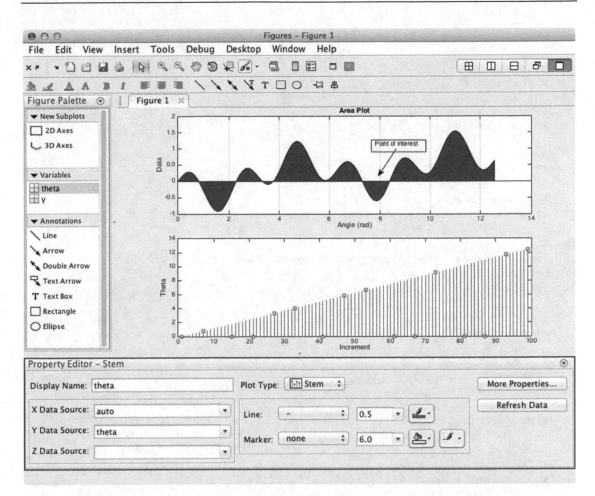

Figure 3.4: *Plot of trigonometric data in the Figure Palette.*

createfigure.m

```
1  function createfigure(X1, yvector1)
2  %CREATEFIGURE(X1, YVECTOR1) Autogenerated figure code.
3  %  X1:  area x
4  %  YVECTOR1:  area yvector
5
6  %  Auto-generated by MATLAB on 03-Jun-2015 14:32:43
7
8  % Create figure
9  figure1 = figure;
10
11 % Create axes
12 axes1 = axes('Parent',figure1,'XGrid','on','OuterPosition',[0 0.5 1
       0.5]);
13 box(axes1,'on');
```

105

```
14  hold(axes1,'on');
15
16  % Create ylabel
17  ylabel('Data');
18
19  % Create xlabel
20  xlabel('Angle (rad)');
21
22  % Create title
23  title('Area Plot');
24
25  % Create area
26  area(X1,yvector1,'DisplayName','Area','Parent',axes1);
27
28  % Create axes
29  axes2 = axes('Parent',figure1,'OuterPosition',[0 0 1 0.5]);
30  box(axes2,'on');
31  hold(axes2,'on');
32
33  % Create ylabel
34  ylabel('Theta');
35
36  % Create xlabel
37  xlabel('Increment');
38
39  % Create stem
40  stem(X1,'DisplayName','theta','Parent',axes2,'Marker','none',...
41      'Color',[0 0.447 0.741]);
42
43  % Create textarrow
44  annotation(figure1,'textarrow',[0.609822646657571
          0.568894952251023],...
45      [0.827828828828829 0.717117117117118]);
46
47  % Create textbox
48  annotation(figure1,'textbox',...
49      [0.553888130968622 0.814895792699917 0.120787482806052
          0.0489690721649485],...
50      'String',{'Point of interest'});
```

Note that this code did not in fact use the subplot function, but rather the option to specify the exact axes location in the figure with the 'OuterPosition' property. Note also how the units of the axes position and of the annotations are between 0 and 1, that is, normalized. This is in fact an option for axes, as can be seen by the following call using gca to get the handle to the current axes:

```
>> set(gca,'units')
    'inches'
    'centimeters'
    'characters'
```

```
    'normalized'
    'points'
    'pixels'
```

Using other units may be helpful for certain applications, but normalized units are always the default.

There are additional interactive buttons in the Figure Toolbar we should mention:

- Zoom in
- Zoom out
- Hand tool – Move an object in the plane of the figure
- Rotate tool – Rotate the view
- Data cursor
- Brush/select data
- Colorbar
- Legend

The hand and rotate tools are very helpful with 3D data. The data cursor displays the values of a plot point right in the figure. The brush highlights a segment of data using a contrast color of your choosing using the colors pop-up. The colorbar and legend buttons serve as on/off switches.

3.2   Incrementally Annotate a Plot

Problem

You need to annotate a curve in a plot at a subset of points on the curve.

Solution

Use the `text` function to annotate the plot.

How It Works

We will call `text` within a for loop in `AnnotatePlot`. Use `sprintf` to create the text for the annotations, which gives you control over the formatting of any numbers. In this case, we will use %d for integer display. `linspace` creates an evenly spaced index array into the data to give us the selected points to annotate, in this case, five points. `linspace` is used to produce evenly spaced points.

AnnotatePlot.m

```
 8   %% Parameters
 9   nPoints = 5; % Number of plot points to have annotations
10
11   %% Create the line
12   v        = [1;2;3];
```

```
13  t         = linspace(0,1000);
14  r         = [v(1)*t;v(2)*t;v(3)*t];
15
16  %% Create the figure and plot
17  s = 'Annotated Plot';
18  h = figure('name',s);
19  plot3(r(1,:),r(2,:),r(3,:));
20  xlabel('X');
21  ylabel('Y');
22  zlabel('Z');
23  title(s)
24  grid
25
26  %% Add the annotations
27  n     = length(t);
28  j     = ceil(linspace(1,n,nPoints));
29
30  for k = j
31    text(r(1,k), r(2,k), r(3,k), sprintf('- Time %d',floor(t(k))));
32  end
```

Note that we passed the index array j directly to the loop index k. Figure 3.5 shows the annotated plot. We create a three-dimensional straight line to annotate.

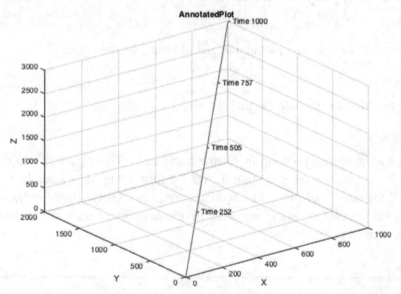

Figure 3.5: *Annotated three-dimensional plot.*

108

3.3   Create a Custom Plot Page with Subplot

Problem

You need multiple plots of your data for a particular application, and as you rerun your script, they are cluttering your screen and hogging memory. We often create many dozens of plots as we work on our commercial toolboxes.

Solution

Create a single plot with several subplots on it so you only need one figure to see the results of one run of your application.

How It Works

The `subplot` function allows you to create a symmetric array of plots in a figure in two dimensions. You generate an m-by-n array of small axes which are spaced in the figure automatically. A good example is a 3D trajectory with views from different angles. We can create a plot with a 2 x 2 array of axes, with the 3D plot in the lower left-hand corner and views from each direction around it. The function is `QuadPlot`. It has a built-in demo creating the figure in Figure 3.6.

Note that you must use the size of your axes array, in this case (2,2), in each call to `subplot`.

QuadPlot.m

```
1   %% QUADPLOT Create a quad plot page using subplot.
2   % This creates a 3D view and three 2D views of a trajectory in one
        figure.
3   %% Form
4   %   QuadPlot( x )
5   %% Input
6   %   x   (3,:)      Trajectory data
7   %
8   %% Output
9   % None. But you may want to return the graphics handles for further
        programmatic
10  % customization.
11  %
12
13  function QuadPlot(x)
14
15  if nargin == 0
16    disp('Demo of QuadPlot');
17    th = logspace(0,log10(4*pi),101);
18    in = logspace(-1,0,101);
19    x  = [sin(th).*cos(in);cos(th).*cos(in);sin(in)];
20    QuadPlot(x);
21    return;
```

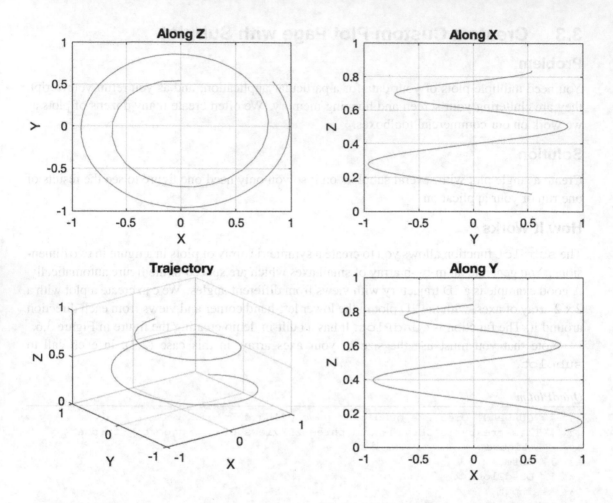

Figure 3.6: *QuadPlot using subplot for axes placement.*

```
22   end
23
24   h = figure('Name','QuadPage');
25   set(h,'InvertHardcopy','off')
26
27   % Use subplot to create plots
28   subplot(2,2,3)
29   plot3(x(1,:),x(2,:),x(3,:));
30   xlabel('X')
31   ylabel('Y')
32   zlabel('Z')
33   grid on
34   title('Trajectory')
35   rotate3d on
36
37   subplot(2,2,1)
```

```
38  plot(x(1,:),x(2,:));
39  xlabel('X')
40  ylabel('Y')
41  grid on
42  title('Along Z')
43
44  subplot(2,2,2)
45  plot(x(2,:),x(3,:));
46  xlabel('Y')
47  ylabel('Z')
48  grid on
49  title('Along X')
50
51  subplot(2,2,4)
52  plot(x(1,:),x(3,:));
53  xlabel('X')
54  ylabel('Z')
55  grid on
56  title('Along Y')
```

In the latest versions of MATLAB, you can easily access figure and axes properties using field names. For instance, let's get the figure generated by the demo using gcf, then look at the children, which should include our four subplots.

```
>> h = gcf

h =

  Figure (5: PlotPage) with properties:

      Number: 5
        Name: 'PlotPage'
       Color: [0.94 0.94 0.94]
    Position: [440 378 560 420]
       Units: 'pixels'

  Show all properties

>> h.Children

ans =

  5x1 graphics array:

  ContextMenu
  Axes            (Along Y)
  Axes            (Along X)
  Axes            (Along Z)
  Axes            (Trajectory)
```

Note that the titles of our axes are helpfully displayed. If you wanted to add additional objects or change the properties of the axes, you could access the handles this way. Or, you might want to provide the handles as an output for your function. You can also make a subplot in a figure the current axes just by calling subplot again with the array size and ID.

```
1  subplot(2,2,1)
```

3.4   Create a Heat Map

Problem

You would like to create a heat map from data. A heat map shows the variation of magnitude using color in a two-dimensional image.

Solution

You can create a heat map using the heatmap function.

How It Works

We'll create a random set of data and two cell arrays for the x and y names.

HeatMapDemo.m

```
1  %% Heat map
2  % Heat map plot from random data
3
4  cD       = rand(4,3);
5  xV       = {'1' '2' '3'};
6  yV       = {'a' 'b' 'c' 'd'};
7
8  NewFigure('Heat Map')
9
10 heatmap(xV,yV,cD)
```

heatmap generates the map from the data shown in Figure 3.7.

```
>> HeatMapDemo

ans =

  Figure (3: Heat Map) with properties:

      Number: 3
        Name: 'Heat Map'
       Color: [0.9400 0.9400 0.9400]
    Position: [616 598 560 420]
       Units: 'pixels'

  Show all properties
```

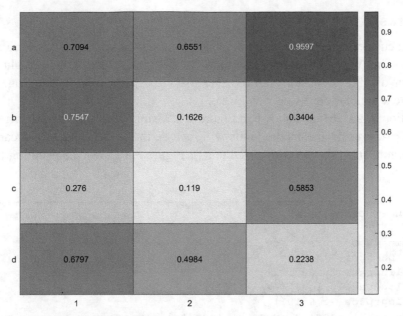

Figure 3.7: *A heat map from random data.*

```
ans =

  HeatmapChart with properties:

        XData: {3x1 cell}
        YData: {4x1 cell}
    ColorData: [4x3 double]

  Show all properties
```

3.5   Create a Plot Page with Custom-Sized Axes

Problem

You would like to group some plots together in one figure but not as evenly spaced subplots.

Solution

You can create custom-sized axes using the `'OuterPosition'` property of the axes, placing them anywhere in the figure you wish.

113

How It Works

We'll create a custom figure with two plots, one spanning the width of the figure and a second smaller axes. This will leave room for a block of descriptive text, which might describe the figure itself or display the results. In order to make the plots more interesting, we will add markers and text annotations using num2str.

The function is PlotPage shown in Figure 3.8. Using 'OuterPosition' for the axes instead of 'Position' means the limits will include the axes labels, so we can use the full range of the figure from 0 to 1 (normalized units). Figure 3.8 shows the resulting figure.

PlotPage.m

```
18  function PlotPage(t, x)
19
20  if nargin == 0
21    disp('Demo of PlotPage');
22    t  = linspace(0,100,101);
23    th = logspace(0,log10(4*pi),101);
24    in = logspace(-1,0,101);
25    x = [sin(th).*cos(in);cos(th).*cos(in);sin(in)];
26    PlotPage(t,x);
27    return
28  end
29
30  h = figure('Name','PlotPage');
31  set(h,'InvertHardcopy','off')
32
33  % Specify the axes position as [left, bottom, width, height]
34  axes('outerposition',[0.5 0 0.5 0.5]);
35  plot(t,x);
36  xlabel('Time')
37  grid on
38
39  % Specify an additional axes and make a 3D plot
40  axes('outerposition',[0 0.5 1 0.5]);
41  plot3(x(1,:),x(2,:),x(3,:));
42  xlabel('X')
43  ylabel('Y')
44  zlabel('Z')
45  grid on
46
47  % add markers evenly spaced with time
48  hold on
49  for k=1:10:length(t)
50    plot3(x(1,k),x(2,k),x(3,k),'x');
51    % add a text label
52    label = ['  ' num2str(t(k)) ' s'];
53    text(x(1,k),x(2,k),x(3,k),label);
54  end
55  hold off
56
```

114

```
57  uh = uicontrol('Style','text','String','Description of the plots',...
58          'units','normalized','position',[0.05 0.1 0.35 0.3]);
59  set(uh,'string',['You may wish to provide a detailed description '...
60              'of the visualization of your data or the results
                right on the figure '...
61              'itself in a uicontrol text box such as this.']);
62  set(uh,'fontsize',14);
63  set(uh,'foregroundcolor',[1 0 0]);
```

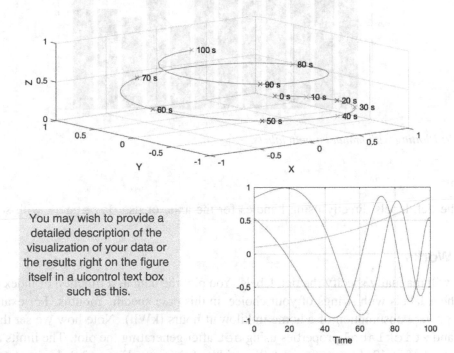

Figure 3.8: `PlotPage` *with custom-sized plots.*

3.6 Plotting with Dates

Problem

You want to plot data as a function of time using dates on the *x* axis.

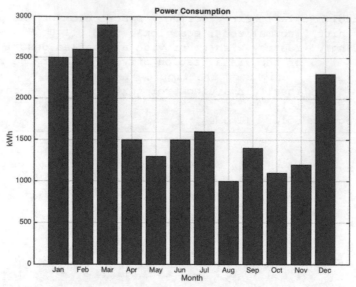

Figure 3.9: *Plotting with manual month labels.*

Solution

Access the tick labels directly using handles for the axis, or use `datetick` with serial date numbers.

How It Works

First, we will manually specify the tick labels. You plot the data as a function of index and then replace the x labels with strings of your choice, in this case specific months. For example, we will plot power consumption of a home in kilowatt hours (kWh). Note how we set the `xlim`, `xtick`, and `xticklabel` properties using `set` after generating the plot. The limits are set to [0 13] instead of [1 12] to accommodate the width of the bars. Figure 3.9 shows plotting with month labels.

PlottingWithDates.m

```
1   %% Plot using months as the x label
2   % First we will set the labels manually. Then we will use MATLAB's
        serial date
3   % numbers to set the labels automatically.
8
9   %% Specify specific months as labels
10  kWh   = [   2500 2600 2900 1500 1300 1500 1600 1000 1400 1100 1200
        2300];
11  month = {'Jan' 'Feb' 'Mar' 'Apr' 'May' 'Jun' 'Jul' 'Aug' 'Sep' 'Oct' '
        Nov' 'Dec'};
12
13  figure('Name','Plotting With Manual Date Labels');
14  bar(1:12,kWh)
15  xlabel('Month');
```

116

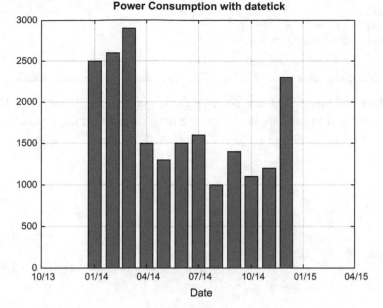

Figure 3.10: *Plotting using* `datetick` *with serial dates.*

```
16  ylabel('kWh')
17  title('Power Consumption');
18  grid on
19
20  set(gca,'xlim',[0 13],'xtick',1:12,'xticklabel',month);
```

If you are plotting data against complete dates, you can also use MATLAB's serial date numbers, which can be automatically displayed as tickmarks using `datetick`. You can convert between calendar dates and serial numbers using `datestr`, `datenum`, and `datevec`. A date vector is the six-component date as [year month day hour minute second]. So, for instance, let's assign our data in the preceding example to actual dates in the year 2014. The default date tickmarks will show months just like in our manual example, but for demonstration purposes, we specify a format including the year: `'mmmyy'`. Figure 3.10 shows plotting with serial dates.

```
22  %% Specify full dates and use serial dates to automatically produce
        labels
23  % Specifying only the month will use the current year by default. We
        will set
24  % the year to 2014 by using datevec.
25  N = datenum(month,'mmm');
26  V = datevec(N);
27  V(:,1) = 2014;
28  N = datenum(V);
29
30  figure('Name','Plotting With Serial Dates');
```

117

```
31  bar(N,kWh)
32  xlabel('Date');
33  title('Power Consumption with datetick');
34  datetick('x','mm/yy')
35  grid on
```

Note that the ticks themselves are no longer one per month; if you want to specify them manually, you now need to use date numbers. We have printed out the properties using get to show the XTicks used.

```
>> get(gca)
...
            XLim: [735508 735965]
        XLimMode: 'manual'
      XMinorGrid: 'off'
      XMinorTick: 'off'
          XScale: 'linear'
           XTick: [735508 735600 735690 735781 735873 735965]
      XTickLabel: [6x5 char]
```

MATLAB's serial date numbers do not correspond to other serial date formats like Julian date. MATLAB simply counts days from Jan-1-0000, so the year 2000 starts at a serial number of 2000*365 = 730,000. The following quick example demonstrates this as well as using now to get the current date:

```
>> v = datevec(now)
v =
          2015             7            31            11            37
              0.6198
>> n = datenum(v)
n =
    7.3618e+05
>> s = datestr(n,'local')
s =
31-Jul-2015 11:37:00
```

3.7   Generating a Color Distribution

Problem

You want to assign colors to markers or lines in your plot.

Solution

Specify the HSV components algorithmically from around the color wheel and convert to RGB.

How It Works

`ColorDistribution` chooses n colors from around the color wheel. The colors are selected using the hue component of HSV, with a full range from 0 to 1. Parameters allow the user to separately specify the saturation and value for all the colors generated. You could alternatively use these components to select a variety of colors of one hue.

Reducing the saturation (`sat`) lightens the colors while remaining on the same "spoke" of the color wheel. A saturation of 0 produces all grays. The value (`val`) keeps the ratio between RGB components remain the same, but lowering the magnitude makes colors darker, for example, [1 0.85 0] and [0.684 0.581 0]. See Figure 3.11.

ColorDistribution.m

```
1   %% Demonstrate a color distribution for an array of lines.
2   % Colors are calculated around the color wheel using hsv2rgb.
3
4   val      = 1;
5   sat      = 1;
6   n        = 100;
7   dTheta   = 360/n;
8   thetaV   = linspace(0,360-dTheta,n);
9
10  h        = linspace(0,1-1/n,n);
11  s        = sat*ones(1,n);
12  v        = val*ones(1,n);
13  colors   = hsv2rgb([h;s;v]');
14  y        = sin(thetaV*pi/180);
15  hF       = figure;
16  hold on;
17  set(hF,'name','Color Wheel')
18  l = gobjects(n);
19  for k = 1:n
20          l(k) = plot(thetaV,k*y);
21  end
22  set(gca,'xlim',[0 360]);
23  grid on
24  pause
25
26  for k = 1:n
27    set(l(k),'color',colors(k,:)*val);
28  end
```

Figure 3.11 plots a color distribution.

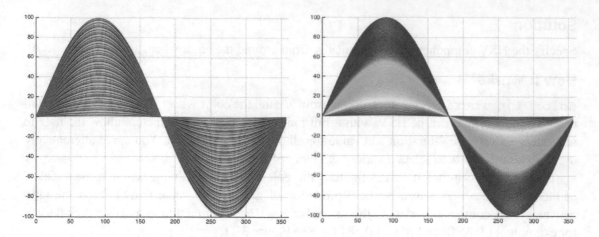

Figure 3.11: *Original lines and lines with a color distribution with values and saturation of 1.*

3.8 Visualizing Data over 2D or 3D Grids

Problem

You need to perform a calculation over a grid of data and view the results.

Solution

The function `meshgrid` produces grids over x and y that can be used for calculations and subsequently input to `surf`. This is also useful for `contour` and `quiver` plots.

How It Works

Our solution is in `GridVisualization.m`. First, you define the vectors in x and y that define your grid. You can perform your calculations in a for loop or in a vectorized function. The vectors do not have to be physical dimensions; indeed, in general, they are quite different quantities involved in a parametric study. The classic example is an exponential function of two variables, which is viewed as a surface in Figure 3.12.

GridVisualization.m

```
8   %% 2D example of meshgrid
9   figure('Name','2D Visualization');
10  xv = -1.5:0.1:1.5;
11  yv = -2:0.2:2;
12  [X,Y] = meshgrid(xv, yv);
13  Z = Y .* exp(-X.^2 - Y.^2);
14  surf(X,Y,Z,'edgecolor','none')
15  title('2D Grid Example')
16  zlabel('z = y exp( -x^2-y^2 )')
17  colormap hsv
18
```

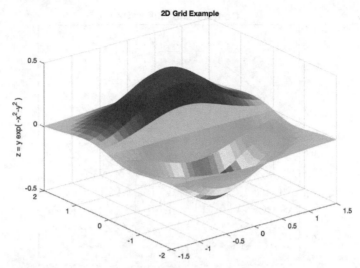

Figure 3.12: *3D surface generated over a 2D grid.*

```
19  size(X)
20  size(Y)
```

The generated matrices are square and consist of the input vector replicated in the correct dimension. You could achieve the same result by hand using `repmat`, but `meshgrid` eliminates the need to remember the details.

```
>> size(X)
ans =
    41    41
>> size(Y)
ans =
    41    41
>> X(1:5,1:5)
ans =
          -2         -1.9         -1.8         -1.7         -1.6
          -2         -1.9         -1.8         -1.7         -1.6
          -2         -1.9         -1.8         -1.7         -1.6
          -2         -1.9         -1.8         -1.7         -1.6
          -2         -1.9         -1.8         -1.7         -1.6
>> Y(1:5,1:5)
ans =
          -2           -2           -2           -2           -2
        -1.9         -1.9         -1.9         -1.9         -1.9
        -1.8         -1.8         -1.8         -1.8         -1.8
        -1.7         -1.7         -1.7         -1.7         -1.7
        -1.6         -1.6         -1.6         -1.6         -1.6
```

121

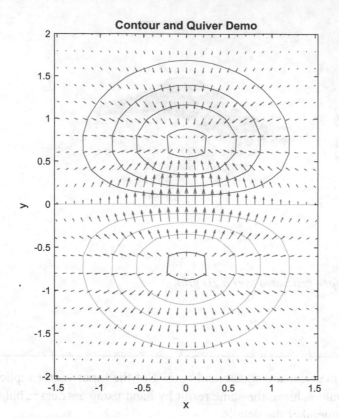

Figure 3.13: *3D surface visualized as contours.*

For fun, we can plot contours of the data as well. We can use the `gradient` function to calculate the slope and plot this using `quiver`. This uses `meshgrid` that returns a 2D mesh from x and y vectors. Figure 3.13 shows a contour plot.

```
22  figure('Name','Contour and Quiver')
23  [px,py] = gradient(Z,0.1,0.2);
24  contour(X,Y,Z), hold on
25  quiver(X,Y,px,py)
26  title('Contour and Quiver Demo')
27  xlabel('x')
28  ylabel('y')
29  colormap hsv
30  axis equal
```

You can also generate a 3D grid and compute data over the volume, for a fourth dimension. In order to view this extra data over the volume, you can use `slice`. This uses interpolation to draw slices at any location along the axes you specify. If you want to see the exact planes in your data, you can use `pcolor`, `surf`, or `contour` in individual figures. `quiver3` can be used to plot arrows in 3D space. We are going to generate five slices at three different x values and at two different z values. The result is shown in Figure 3.14.

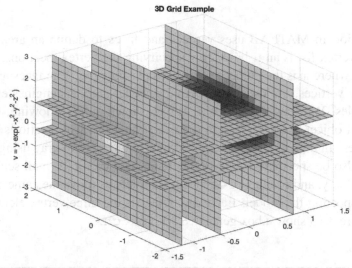

Figure 3.14: *3D volume with slices.*

```
34   %% 3D example of meshgrid
35   % meshgrid can be used to produce 3D matrices, and slice can display
         selected
36   % planes using interpolation.
37   figure('Name','3D Visualization');
38   zv = -3:0.3:3;
39   [x,y,z] = meshgrid(xv, yv, zv);
40   v = x .* exp(-x.^2 - y.^2 - z.^2);
41   slice(x,y,z,v,[-1.2 -0.5 0.8],[],[-0.25 1])
42   title('3D Grid Example')
43   zlabel('v = y exp( -x^2-y^2-z^2 )')
44   colormap hsv
```

3.9   Generate 3D Objects Using Patch

Problem

You would like to draw a 3D box.

Solution

You can create a 3D box using the `patch` function.

123

How It Works

The `patch` function in MATLAB uses vertices and faces to define an area in two or three dimensions. The vertex list is an n-by-3 array specifying the vertex locations. The faces array is an n by m array where m is the number of vertices per polygon. The faces array contains the row indices for the vertices. We usually set m to 3 since all graphics engines eventually reduce polygons to triangles. We draw a box in `BoxPatch` shown in the following. Generally, when drawing a physical object, we set `axis` to `equal` so that the aspect ratio is correct. `patch` has many properties. In this case, we just set the color of the faces to gray using RGB. The edge color, which can also be specified, is black by default. The `view(3)` call sets the camera to a position with equal x, y, and z values. `rotate3d` on lets us move the camera around. This is very handy for inspecting the model. Each line in face is the three vertex elements that form a triangle face. Figure 3.15 show a box generated with `patch`.

BoxPatch.m

```
9   %% Box design
10  x    = 3;
11  y    = 2;
12  z    = 1;
13
14  % Faces
15  f    = [2 3 6;3 7 6;3 4 8;3 8 7;4 5 8;4 1 5;2 6 5;2 5 1;1 3 2;1 4 3;5 6
        7;5 7 8];
16
17  % Vertices
18  v = [-x   x   x  -x  -x   x   x  -x;...
19        -y -y   y   y  -y -y   y   y;...
20        -z -z  -z  -z   z   z   z   z]'/2;
21
```

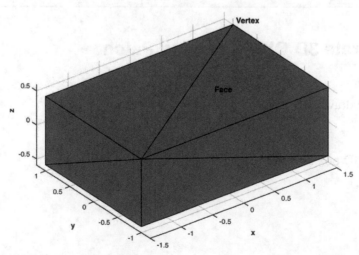

Figure 3.15: *Box generated using* `patch`.

```
22  %% Draw the object
23  h = figure('name','Box');
24  patch('vertices',v,'faces',f,'facecolor',[0.5 0.5 0.5]);
25  axis equal
26  grid on
27  axis([-3 3 -3 3 -3 3])
28  xlabel('x')
29  ylabel('y')
30  zlabel('z')
31  view(3)
32  rotate3d on
```

3.10   Working with Light Objects

Problem

You would like to illuminate the 3D box drawn in the previous recipe.

Solution

You can create ambient or directed light objects using the `light` function. Light objects affect both patch and surface objects, which are created by `surf`, `mesh`, `pcolor`, `fill`, `fill3`, and `patch`.

How It Works

The main properties for working with light objects are color, style, position, and visible. The style may be infinite, with the light shining in parallel rays from a specified direction, or local, with a point source shining in all directions. The position property has a different meaning for each of these styles. `PatchWithLighting` adds a local light to the box script. We modify the box surface properties using `material` to get different effects.

PatchWithLighting.m

```
1   %% Add lighting to the cube patch
2   % We use findobj to locate the patch drawn in Patch, then change its
        properties
3   % to be suitable for lighting. We add a local light.
8
9   %% Create the box patch object
10  BoxPatch;
11
12  %% Find and update the patch object
13  p = findobj(gcf,'type','patch');
14  c = [0.7 0.7 0.1];
```

```
15  set(p,'facecolor',c,'edgecolor',c,...
16       'edgelighting','gouraud','facelighting','gouraud');
17  material('metal');
18
19  %% Lighting
20  l = light('style','local','position',[10 10 10]);
```

Figure 3.16 shows dull and metal material with the same lighting. The lighting produced by MATLAB is limited by being an OpenGL lighting. Modern 3D graphics use textures and shaders for photo-realistic scene lighting. You also cannot generate shadows in MATLAB. The one on the right has a somewhat sharper color gradient at the corner.

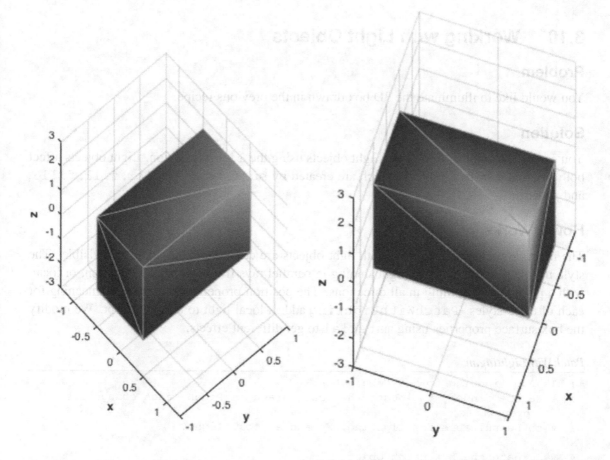

Figure 3.16: *Box illuminated with a local light object. The left box has "dull" material. The one on the right has "metal."*

The dull, shiny, and metal settings for `material` set the patch properties to produce these effects. We can easily print the effects to the command line using `get`.

```
>> material dull
>> get(p)
                DiffuseStrength: 0.8
...
      SpecularColorReflectance: 1
              SpecularExponent: 10
              SpecularStrength: 0

>> material metal
>> get(p)
                DiffuseStrength: 0.3
                     ...
      SpecularColorReflectance: 0.5
              SpecularExponent: 25
              SpecularStrength: 1

>> material shiny
>> get(p)
                DiffuseStrength: 0.6
                     ...
      SpecularColorReflectance: 1
              SpecularExponent: 20
              SpecularStrength: 0.9
```

Note that the `AmbientStrength` is 0.3 for all the material settings listed earlier. If you want to see the effect of only your light objects without ambient light, you have to manually set this to zero. In Figure 3.17, we have set the ambient strength to zero and applied the shiny material.

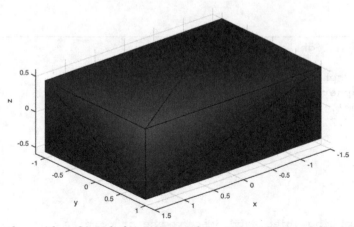

Figure 3.17: *Shiny box with ambient lighting removed (`AmbientStrength` set to 0) and a different camera viewpoint.*

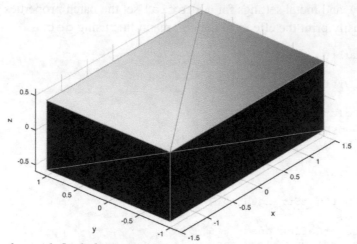

Figure 3.18: *Shiny box with flat lighting.*

MATLAB has a lighting function to control the lighting model with four settings: none, Gouraud, Phong, and flat. Gouraud interpolates the lighting across the faces and gives the most realistic effect. Note that setting the lighting to Gouraud for our box sets the FaceLighting property to gouraud but the EdgeLighting to none, which will give a different effect than in our script earlier where the edge lighting was also set to Gouraud via its property. Flat lighting gives each entire face a uniform lighting, as in Figure 3.18, where we set the view to (-50,30) and the lighting to flat.

The MATLAB recommendations are to use flat lighting for faceted objects and Gouraud lighting for curved objects. The easiest way to compare these is to create a sphere, which is simple using the sphere function and generating a surface. This is done in the following SphereLighting. The infinite light object shines from the x axis. See Figure 3.19 for the resulting plots.

SphereLighting.m

```
1   %% Create and light a sphere
2
3   %% Make the sphere surface in a new figure
4   [X,Y,Z] = sphere(16);
5   figure('Name','Sphere Demo')
6   s = surf(X,Y,Z);
7   xlabel('x')
8   ylabel('y')
9   zlabel('z')
10  axis equal
11  view(70,15)
12
13  %% Add a lighting object and display the properties
14  light('position',[1 0 0])
15  disp(s)
```

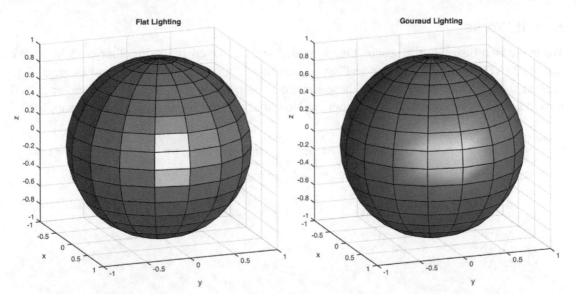

Figure 3.19: *Sphere illuminated with an infinite light object. The left sphere has flat lighting. The one on the right has Gouraud.*

```
16  title('Flat Lighting')
17  pause
18
19  %% Change to Gouraud lighting and display again
20  lighting gouraud
21  title('Gouraud Lighting')
22  disp(s)
```

In addition to a `sphere` function, MATLAB also provides `cylinder` and `ellipsoid`.

3.11 Programmatically Setting the Camera Properties

Problem

You would like to have a camera in your scene that can be pointed.

Solution

Use the MATLAB `cam` functions. These provide the same functionality as the buttons in the camera toolbar, but with repeatability and the ability to pass in variables for the parameters. We demonstrate this in the script `PatchWithCamera.m`.

129

How It Works

We make two boxes in the scene. One is scaled and displayed from the other by 5 in x. We use the MATLAB functions `camdolly`, `camorbit`, `campan`, `camzoom`, and `camroll` to control the camera. We put all of these functions in the `PatchWithCamera.m` script and provide examples of two sets of parameters. Note that without lighting, the edges disappear.

PatchWithCamera.m

```
1  %% Generate two cubes using patch and point a camera at the scene
2  % The camera parameters will be set programmatically using the cam
       functions.

7
8  %% Camera parameters
9  % Orbit
10 thetaOrbit    = 0;
11 phiOrbit   = 0;
12
13 % Dolly
14 xDolly     = 0;
15 yDolly     = 0;
16 zDolly     = 0;
17
18 % Zoom
19 zoom       = 1;
20
21 % Roll
22 roll       = 50;
23
24 % Pan
25 thetaPan   = 1;
26 phiPan     = 0;
27
28 %% Box design
29 x    = 1;
30 y    = 2;
31 z    = 3;
32
33 % Faces
34 f    = [2 3 6;3 7 6;3 4 8;3 8 7;4 5 8;4 1 5;2 6 5;2 5 1;1 3 2;1 4 3;5 6
       7;5 7 8];
35
36 % Vertices
37 v = [-x  x  x -x -x  x  x -x;...
38      -y -y  y  y -y -y  y  y;...
39      -z -z -z -z  z  z  z  z]'/2;
40
41 %% Draw the object
```

```
42   h = figure('name','Box');
43
44   c = [0.7 0.7 0.1];
45   patch('vertices',v,'faces',f,'facecolor',c,'edgecolor',c,...
46         'edgelighting','gouraud','facelighting','gouraud');
47
48   c = [0.2 0 0.9];
49   v       = 0.5*v;
50   v(:,1) = v(:,1) + 5;
51   patch('vertices',v,'faces',f,'facecolor',c,'edgecolor',c,...
52         'edgelighting','gouraud','facelighting','gouraud');
53
54   material('metal');
55   lighting gouraud
56   axis equal
57   grid on
58   xlabel('x')
59   ylabel('y')
60   zlabel('z')
61   view(3)
62   rotate3d on
63
64   %% Camera commands
65   campan(thetaPan,phiPan)
66   camzoom(zoom)
67   camdolly(xDolly,yDolly,zDolly);
68   camorbit(thetaOrbit,phiOrbit);
69   camroll(roll);
70
71   s = sprintf('Pan %3.1f %3.1f\nZoom %3.1f\nDolly %3.1f %3.1f %3.1f\
         nOrbit %3.1f %3.1f\nRoll %3.1f',...
72   thetaPan,phiPan,zoom,xDolly,yDolly,zDolly,thetaOrbit,phiOrbit,roll);
73
74   text(2,0,0,s);
```

Additional functions for interacting with the scene camera include `campos` and `camtarget`, which can be used to set the camera position and target. This can be used to image one object from the vantage point of another. `camva` sets the camera view angle, so you can model a real camera's field of view. `camup` specifies the camera "up" vector or the direction of the top of the frame.

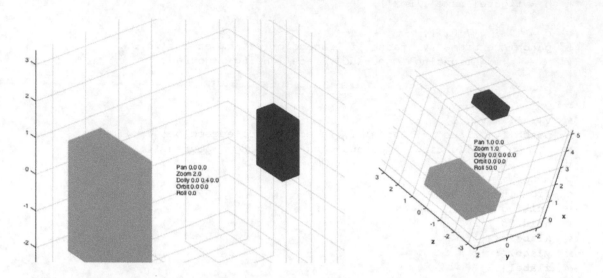

Figure 3.20: *Boxes with different camera parameters.*

3.12   Display an Image

Problem

You would like to draw an image.

Solution

You can read in an image directly from an image file and draw it in a figure window. MATLAB supports a variety of formats including GIF, JPG, TIFF, PNG, and BMP. Our solution is in the script ReadImage.m.

How It Works

We read in a black and while image using imread and display it using imagesc. imagesc scales the color data into the colormap. It is necessary to apply the grayscale colormap; otherwise, you'll get the colors in the default colormap. In parula, this is blue and yellow.

ReadImage.m

```
1   %% Draw a JPEG image in a figure multiple ways
2   % We will load and display an image of a mug.
3   %% See also
4   % imread, pcolor, imagesc, imshow, colormap
9
10  %% Read in the JPEG image
11  i = imread('Mug.jpg');
12
13  %% Draw the picture with imagesc
```

132

```
14   % This preserves an axes. Each pixel center of the image lies at
          integer
15   % coordinates ranging between 1 and M or N. Compare the result of
          imagesc to
16   % that of pcolor. axis image sets the aspect ratio so that tick marks
          on both
17   % axes are equal, and makes the plot box fit tightly around the data.
18   h = figure('name','Mug');
19   subplot(1,2,1)
20   pcolor(i)
21   shading('interp')
22   colorbar
23   axis image
24   title('pcolor with colorbar')
25   a = subplot(1,2,2);
26   % scale the image into the colormap
27   imagesc( i );
28   colormap(a,'gray')
29   axis image
30   grid on
31   title('imagesc with gray colormap')
```

Figure 3.21 shows the mug first using pcolor, which creates a pseudocolor plot of a matrix, which is really a surf with the view looking down from above. To highlight this fact, we added a colorbar. Then on the right, the image is drawn using imagesc with a gray

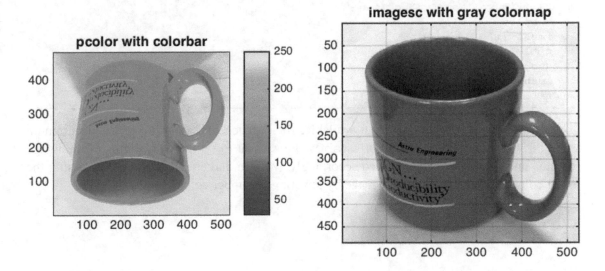

Figure 3.21: *Mug displayed using* pcolor *and* imagesc.

Figure 3.22: *Mug displayed using* imshow, *with color limits applied on the right.*

colormap. Observe that imagesc has changed the direction of the axes so that the image appears right-side up. Both plots have axes with tickmarks.

MATLAB has another image display function called imshow, which is considered the fundamental image display function. This optimizes the figure, axes, and image object properties for displaying an image. If you have the Image Processing toolbox, imtool extends imshow with additional features. Notice how the image is displayed without the axes box. This function scales and selects the gray colormap automatically. Figure 3.22 shows the use of imshow

```
33  %% Draw with imshow
34  % The axes will be turned off. The image will be scaled to fit the
        figure if it
35  % is too large.
36  f = figure('Name','Mug Image'); '
37  subplot(1,2,1)
38  imshow(i)
39  title('imshow')
40  subplot(1,2,2)
41  imshow(i,[30 200])
42  title('imshow with limits [30 200]')
```

Not all images use the full depth available; for instance, this mug image has a minimum value of 30 and a maximum of 250. imshow allows you to set the color limits of the image directly, and the pixels will be scaled accordingly. We can darken the image by increasing the lower color limit and brighten the image by lowering the upper color limit.

3.13   Graph and Digraph

Problem

We have a stochastic process for which we want a graphical representation.

Solution

Use graph and digraph in the script `RandomWalk.m`.

How It Works

Generate a transition matrix showing the probability of transition from one state to a second state.

The code in `RandomWalk.m` creates a digraph, graph, and a random walk. The first part creates a transition matrix.

RandomWalk.m

```
1   %% Demonstrate a digraph and graph
2
3   % Generate a transition matrix
4   % x ranges from -5 to 5
5   p = zeros(11,11);
6   for k = 2:10
7     p(k,k-1) = 0.5;
8     p(k,k+1) = 0.5;
9   end
10
11  p(1,2)    = 1;
12  p(11,10)  = 1;
13
14  fprintf('%4.1f%4.1f%4.1f%4.1f%4.1f%4.1f%4.1f%4.1f%4.1f%4.1f%4.1f\n',p);
```

When we run `RandomWalk` at the command line we get the below output:

```
>> RandomWalk
 0.0 0.5 0.0 0.0 0.0 0.0 0.0 0.0 0.0 0.0 0.0
 1.0 0.0 0.5 0.0 0.0 0.0 0.0 0.0 0.0 0.0 0.0
 0.0 0.5 0.0 0.5 0.0 0.0 0.0 0.0 0.0 0.0 0.0
 0.0 0.0 0.5 0.0 0.5 0.0 0.0 0.0 0.0 0.0 0.0
 0.0 0.0 0.0 0.5 0.0 0.5 0.0 0.0 0.0 0.0 0.0
 0.0 0.0 0.0 0.0 0.5 0.0 0.5 0.0 0.0 0.0 0.0
 0.0 0.0 0.0 0.0 0.0 0.5 0.0 0.5 0.0 0.0 0.0
 0.0 0.0 0.0 0.0 0.0 0.0 0.5 0.0 0.5 0.0 0.0
 0.0 0.0 0.0 0.0 0.0 0.0 0.0 0.5 0.0 0.5 0.0
 0.0 0.0 0.0 0.0 0.0 0.0 0.0 0.0 0.5 0.0 0.5 0.0
 0.0 0.0 0.0 0.0 0.0 0.0 0.0 0.0 0.0 0.5 0.0 1.0
 0.0 0.0 0.0 0.0 0.0 0.0 0.0 0.0 0.0 0.0 0.5 0.0
```

The next part of RandomWalk creates a digraph shown in Figure 3.23 and a graph shown in Figure 3.24.

135

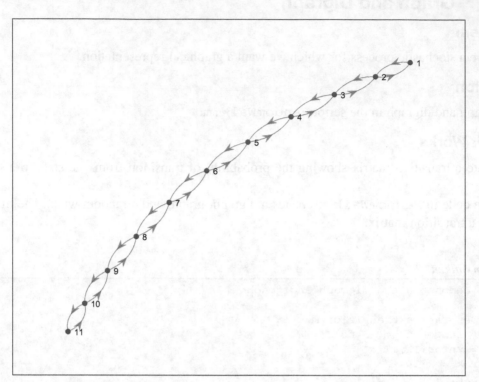

Figure 3.23: *Digraph for the random walk.*

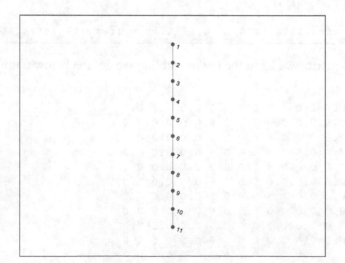

Figure 3.24: *Graph for the random walk.*

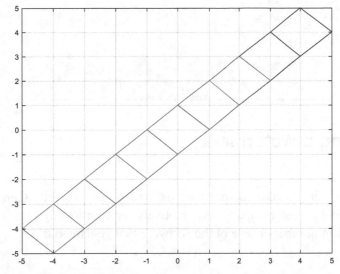

Figure 3.25: *The random walk. The lines show the connections between the nodes in the random walk. All possible paths are shown.*

```
16  g = digraph(p);
17
18  NewFigure('Digraph');
19  plot(g)
20  grid on
```

```
22  g = graph(p,'upper');
23
24  NewFigure('Graph');
25  plot(g)
26  grid on
```

The random walk based on the transition matrix is shown in Figure 3.25.

```
28  n = 100;
29  m = 50;
30
31  NewFigure('Random Walk');
32  for k = 1:m
33    x = zeros(1,n);
34    for j = 2:n
35      if(x(j-1) == -5)
36        x(j) = -4;
37      elseif( x(j-1) == 5 )
38        x(j) = 4;
39      else
40        x(j) = x(j-1) + sign(randn);
```

```
41      end
42    end
43    plot(x(1:n-1),x(2:n))
44    hold on
45  end
46  grid on
```

3.14   Adding a Watermark

Problem

You have a lot of great graphics in your toolbox, and you would like them to be marked as having been created by your company. Alternatively, or additionally, you may want to mark images with a version number or date of the software that generated them.

Solution

You can use low-level graphics functions to add a textual or image watermark to figures that you generate in your toolbox. The tricky part is adding the items to the figure at the correct time so they are not overridden.

How It Works

The best way to add watermarks is to make a special axis for each text or image item you want to add. You turn the axis box off so all that you see is the text or image. In the first example, we add an icon and text to the lower left-hand corner of the plot. We add a color for the edge around the text so that it is nicely delineated. This is shown in Figure 3.26 using the `Watermark.m` function. In the example, we set the hard copy inversion to off, so that when we print the figure, we will get a gray background – this makes it easier to see in the book.

```
>> h = figure;
>> set(h,'InvertHardCopy','off')
>> axes
>> Watermark(h)
```

Watermark.m

```
1  %% WATERMARK Add a watermark to a figure.
2  % This function creates two axes, one for the image and one for the
       text.
3  % Calling it BEFORE plotting can cause unexpected results. It will
       reset
4  % the current axes after adding the watermark. The default position is
5  % the lower left corner, (2,2).
6  %% Form
7  %   Watermark( fig, pos )
8  %% Inputs
9  %   fig          (1,1) Figure hangle
```

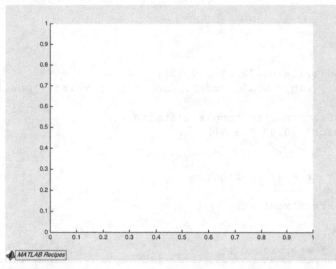

Figure 3.26: *Company watermark.*

```
10  %    pos            (1,2) Coordinates, (left, bottom)
11  %% Outputs
12  % None.
13
14  function Watermark( fig, pos  )
15
16  if (nargin<1 || isempty(fig))
17    fig = figure('Name','Watermark Demo');
18    set(fig,'color',[0.85 0.9 0.85]);
19  end
20
21  if (nargin<2 || isempty(pos))
22    pos = [2 2];
23  end
24
25  string = 'MATLAB Recipes';
26
27  % Save the current axes so we can restore it
28  aX = [];
29  if ~isempty(get(fig,'CurrentAxes'))
30    aX = gca;
31  end
32
33  % Draw the icon
34  %--------------
35  [d,map] = imread('matlabicon','gif');
36  posIcon = [pos(1:2) 16 16];
37  a = axes( 'Parent', fig, 'box', 'off', 'units', 'pixels', 'position',
        posIcon );
38  image( d );
39  colormap(a,map)
```

139

```
40  axis off
41
42  % Draw the text
43  %--------------
44  posText = [pos(1)+18 pos(2)+1 100 15];
45  axes( 'Parent', fig, 'box', 'off', 'units', 'pixels', 'position',
          posText );
46  t = text(0,0.5,string,'fontangle','italic');
47  set(t,'edgecolor',[0.87 0.5 0])
48  axis off
49
50  % Restore current axes in figure
51  if ~isempty(aX)
52    set(fig,'CurrentAxes',aX);
53  end
54
55  set(fig,'tag','Watermarked')
```

As an additional example, we added text along the left- and right-hand sides of a figure using text rotation in the function `DraftMark.m`. We gave the text a light color. This marks the figure as a draft. We create a blank figure and axis before adding the draft mark, as shown in Figure 3.27.

```
>> h = figure('Name','Draftmark Demo');
>> set(h,'color',[0.85 0.9 0.85]);
>> set(h,'InvertHardCopy','off')
>> axes;
>> Draftmark(h);
```

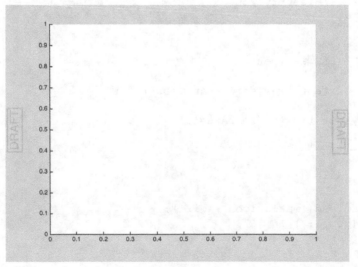

Figure 3.27: *Draft watermark.*

Draftmark.m

```matlab
1  %% DRAFTMARK Add a draft marking to a figure.
2  % This function creates two axes, one each block of text.
3  % Calling it BEFORE plotting can cause unexpected results. It will
      reset
4  % the current axes after adding the watermark. The default position is
5  % the lower left corner, (2,2).
6  %% Form
7  %    Draftmark( fig, pos  )
8  %% Inputs
9  %    fig          (1,1) Figure hangle
10 %    pos          (1,2) Coordinates, (left, bottom)
11 %% Outputs
12 % None.
13
14 function Draftmark( fig, pos  )
15
16 if (nargin<1 || isempty(fig))
17   fig = figure('Name','Draft Demo');
18   set(fig,'color',[0.85 0.9 0.85]);
19 end
20
21 if (nargin<2 || isempty(pos))
22   pos = [2 2];
23 end
24
25 string = 'DRAFT';
26
27 % Save the current axes so we can restore it
28 aX = [];
29 if ~isempty(get(fig,'CurrentAxes'))
30   aX = gca;
31 end
32
33 % Draw the text
34 %--------------
35 pf = get(fig,'position');
36 posText = [pos(1)+5 pos(2)+0.5*pf(4)-40 20 80];
37 axes( 'Parent', fig, 'box', 'on', 'units', 'pixels', 'outerposition',
      posText );
38 t1 = text(0,0,string,'fontsize',20,'color',[0.8 0.8 0.8]);
39 set(t1,'rotation',90,'edgecolor',[0.8 0.8 0.8],'linewidth',2)
40 axis off
41
42 posText = [pos(1)+pf(3)-25 pos(2)+0.5*pf(4)-40 20 80];
43 axes( 'Parent', fig, 'box', 'on', 'units', 'pixels', 'outerposition',
      posText );
44 t2 = text(0,1,string,'fontsize',20,'color',[0.8 0.8 0.8]);
45 set(t2,'rotation',270,'edgecolor',[0.8 0.8 0.8],'linewidth',2)
46 axis off
48
```

141

```
49  % Restore current axes in figure
50  if ~isempty(aX)
51    set(fig,'CurrentAxes',aX);
52  end
```

If you want to get very fancy, you could draw objects across the front of the figure and give them transparency, but it has to be fill or patch objects; text cannot be given transparency.

Summary

In this chapter, we reviewed key features of MATLAB visualization, from basic plotting to 3D visualization including objects and lighting. We demonstrated accessing figure and axes handles and setting properties programmatically, as well as using the interactive tools for figures. Creating helpful visualization routines is a key part of any toolbox. MATLAB provides excellent data management routines, including for large grids of data, and many options for colorization. Table 3.1 lists the code developed in the chapter.

Table 3.1: *Chapter Code Listing*

File	Description
AnnotatePlot	Add text annotations evenly spaced along a curve
BoxPatch	Generate a cube using patch
ColorDistribution	Demonstrate a color distribution for an array of lines
DraftMark	Add a draft marking to a figure
GridVisualization	Visualize data over 2D and 3D grids
PatchWithCamera	Generate two cubes using patch and point a camera at the scene
PatchWithLighting	Add lighting to the cube patch
PlotPage	Create a plot page with several custom plots in one figure
PlottingWithDates	Plot using months as the x label
QuadPlot	Create a quad plot page using subplot
ReadImage	Draw a JPEG image in a figure multiple ways
SphereLighting	Create and light a sphere
Watermark	Add a watermark to a figure

CHAPTER 4

■ ■ ■

Interactive Graphics

The previous chapter addressed generating static graphics. In this chapter, we provide some recipes for generating dynamic graphics. This includes animations of line and patch objects, utilizing `uicontrols` in figures, designing GUIs using App Designer, and deploying your GUI as a MATLAB app. Along the way, we present some tips for maximizing the performance of your dynamic graphics functions.

4.1 Creating a Simple Animation

Problem

You want a visualization which changes over time without generating hundreds of different figures.

Solution

You can create an animation by updating patch objects in a figure successively in a loop, as shown in Figure 4.1 for a simple box. We will do this in the script `PatchAnimation.m`.

How It Works

First, we will create a graphic involving our 3D box from the previous chapter. Then we will update it in a loop. This is most efficient if you can assign new data to the existing graphics object. This could be changing the color, style, or physical location of the object. Alternatively, you can delete and recreate the object in the current axes. In both cases, you need to store the graphics handle for the updates. Deleting and recreating the object is often much slower than just changing its properties. For example, to rotate and translate an object, you need only change the vertices.

In this case, we update the vertices by multiplying by a rotation matrix b. We then pass the vertices to the patch via the handle `set(p,'vertices',vK)`. Note the use of the transposes as the vertices are stored in an n-by-3 array. A light object makes the resulting animation more interesting. We set `'linestyle'` to `'none'` for the patch object to eliminate the lines between triangles.

© Michael Paluszek and Stephanie Thomas 2020
M. Paluszek and S. Thomas, *MATLAB Recipes*,
https://doi.org/10.1007/978-1-4842-6124-8_4

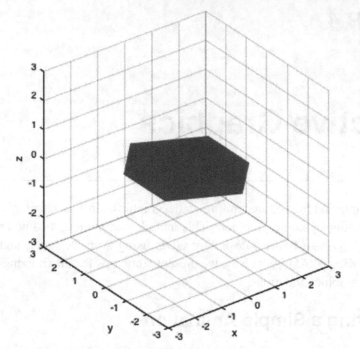

Figure 4.1: *One frame of an animation of a rotating box.*

PatchAnimation.m

```
1  %% Animate a cube using patch
2  % Create a figure and draw a cube in it. The vertices and faces are
      specified
3  % directly. We only update vertices to get a smooth animation.
8
9  %% Box design
10 x   = 3;
11 y   = 2;
12 z   = 1;
13
14 % Faces
15 f   = [2 3 6;3 7 6;3 4 8;3 8 7;4 5 8;4 1 5;2 6 5;2 5 1;1 3 2;1 4 3;5 6
      7;5 7 8];
16
17 % Vertices
18 v = [-x  x  x -x -x  x  x -x;...
19      -y -y  y  y -y -y  y  y;...
20      -z -z -z -z  z  z  z  z]'/2;
21
22 %% Draw the object
23 h = figure('name','Box');
24 p = patch('vertices',v,'faces',f,'facecolor',[0.5 0.5 0.5],...
25            'linestyle','none','facelighting','gouraud');
26 ax = gca;
```

```
27  set(ax,'DataAspectRatio',[1 1 1],'DataAspectRatioMode','manual')
28  axis([-3 3 -3 3 -3 3])
29  grid on
30  xlabel('x')
31  ylabel('y')
32  zlabel('z')
33  view(3)
34  rotate3d on
35  light('position',[0 0 1])
36
37  %% Animate
38  % We use tic and toc to time the animation. Pause is used with a
        fraction of a
39  % second input to slow the animation down.
40  tic
41  n = 10000;
42  a = linspace(0,8*pi,n);
43
44  c = cos(a);
45  s = sin(a);
46
47  for k = 1:n
48    b   = [c(k) 0 s(k);0 1 0;-s(k) 0 c(k)];
49    vK  = (b*v')';
50    set(p,'vertices',vK);
51    pause(0.001);
52  end
53  toc
```

The full animation of four rotations takes about 16 seconds on a Mac Pro laptop. We expect about 10 seconds of this to be from the pause command, that is, 10000 steps * 0.001 seconds. Using pause allows you to slow down an animation that would otherwise be too fast to be useful. Remember that pause commands can be temporarily disabled by using pause off; this is useful when testing graphics functions that use pause.

■ **TIP**   Use pause to slow down animations as needed, and remember to use pause off to disable the pausing and run full speed during testing.

Check the execution time on your computer. Run the script twice, with pause turned off the second time.

```
>> pause on
>> PatchAnimation;
Elapsed time is 15.641664 seconds.
>> pause off
>> PatchAnimation;
Elapsed time is 0.780012 seconds.
```

Note that `pause` flushes the system graphics queue, drawing the updated patch to the screen. When `pause` is off, the graphics don't update, and you will see just the initial frame in the window. Also note that the actual increase in time from the `pause` and the graphics updates was almost 15 seconds. To force a graphics update without using `pause`, use `drawnow`.

This script uses cells to allow the individual sections to be run independently. This means you can rerun the animation without recreating the figure, by reexecuting that cell. This can be done from the Run Section toolbar button or by a keyboard command, for example, Command-Enter on a Mac.

Suppose we want to add a text item to the animation that displays the angle of rotation. We can do this using `title` easily enough:

```
title(sprintf('Angle: %f deg',a(k)*180/pi));
```

However, you will be surprised at the performance impact – the animation now takes over 90 seconds! Displaying text is much less efficient than updating the graphics vertices. First, we can try using the handle to the axis title object directly and setting the string. However, this makes little difference, still taking about 90 seconds. Another solution would be to add an inner loop to update the title less often. Trial and error shows that an update every 50 steps, as shown in the code snippet below, has little impact on the runtime. The figure with the title is shown in Figure 4.2.

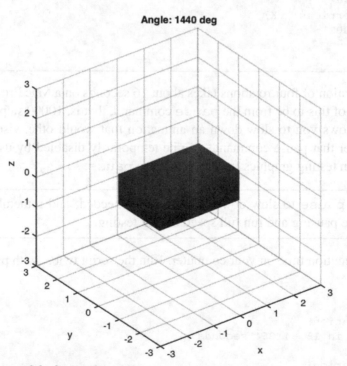

Figure 4.2: *Animation of the box with a changing title.*

```
if rem(k,50)==0
    set(ax.Title,'string',sprintf('Angle: %.5g deg',a(k)*180/pi));
end
```

Here is a summary of the execution times on the reference MacBook. We have given the times as printed from `toc`, and they are from a single run, not an average. Expect a variation in execution time of up to 10% across multiple runs. `set` is setting properties.

Table 4.1: *Execution Times for* `PatchAnimation`

No Title	
`pause off`	0.780154 sec
`pause on` (0.001 sec)	17.254507 sec
With Title	
`title`	90.761500 sec
set every step	90.213318 sec
set every 50 steps	20.218774 sec

4.2 Playing Back an Animation

Problem

We want to store and play back an animation.

Solution

Save each frame of the animation into an avi file using the `VideoWriter` class.

How It Works

In the following listing, we use `VideoWriter` to save the animation and read it into an avi file. This line of code opens an avi file:

```
1  vObj = VideoWriter('RotatingBox.avi');
```

The script is `PatchAnimationStorage.m`. We don't show the sections creating the box and figure as they are the same as the previous recipe. In this case, we only use 100 points for the four rotations; the execution time, including saving the movie, is about 4 seconds with pause off. Running the script in the Profiler shows that almost 90% of the execution time is spent in the `writeVideo` command.

PatchAnimationStorage.m

```
1  %% Animate a cube using patch and store as an AVI file
2  % The figure and box are created as in PatchAnimation. This time we use
      a
3  % VideoWriter to store the frames in a movie.
```

⋮

```
35  %% Animate
36  n = 100;
37  a = linspace(0,8*pi,n);
38  c = cos(a);
39  s = sin(a);
40
41  % Create a video file
42  vObj = VideoWriter('RotatingBox.avi');
43  open(vObj);
44
45  tic
46  for k = 1:n
47    pause(0.01);
48    b    = [c(k) 0 s(k);0 1 0;-s(k) 0 c(k)];
49    vK   = (b*v')';
50    set(p,'vertices',vK);
51    writeVideo(vObj,getframe(h));
52  end
53  toc
54
55  close(h)
56  close(vObj)
```

You can then play back your animation in any movie player shown in Figure 4.3.

4.3   Animate Line Objects

Problem

You would like to update a plot with line objects in a loop.

Solution

This is similar to Recipe 4.1, but we will update different properties of the graphics object. We'll use the quad plot from Chapter 3 and add animation of a marker along the trajectory. This will also demonstrate adding a menu to a figure using `uimenu`.

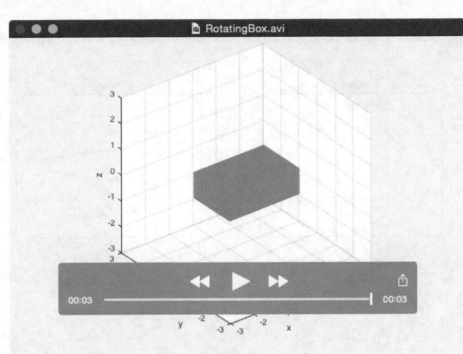

Figure 4.3: *Movie file playing back in a video player.*

How It Works

Start with the QuadPlot.m function. This creates four subplots to view a trajectory: one in 3D and three 2D views from different directions. We will do a few things to add a marker we can animate:

- Add a time input
- Add a marker to each subplot and save the handles
- Add a text uicontrol to display the current time as the animation progresses
- Store the trajectory data and handles in the figure UserData
- Turn off the regular figure menu, and add an Animate menu with a Start function

We will use a nargin check to determine if we are creating the figure using t and x data or entering a callback using the input 'update'. An alternative would be to place the callback in a separate function; the figure executing the callback can be identified using a Tag property or using gcbf, as done here. The new function is called QuadAnimator.m.

QuadAnimator.m

```
1  %% QUADANIMATOR Create a quad plot page with animation.
2  % This creates a 3D view and three 2D views of a trajectory in one
       figure. A
3  % menu is provided to animate the trajectory over time.
4  %% Form
5  %   QuadAnimator( t, x )
```

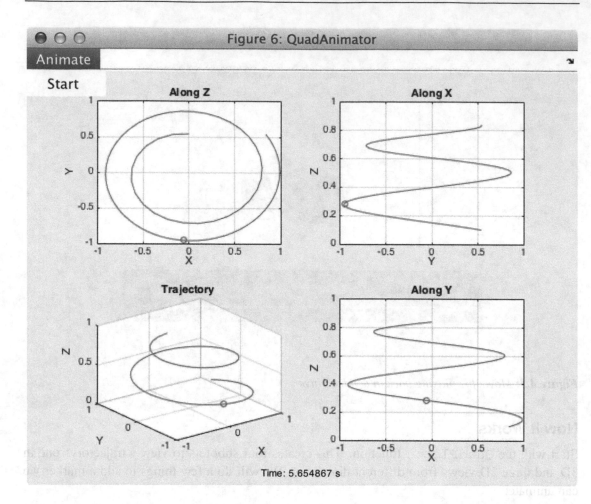

Figure 4.4: *Frame of an animation of the quad plot.*

```
6   %% Input
7   %   t    (1,:)     Time data
8   %   x    (3,:)     Trajectory data
9   %
10  %% Output
11  % None.
17  function QuadAnimator(t,x)
18
19  if nargin == 0
20    disp('Demo of QuadAnimator');
21    t = linspace(0,4*pi,101);
22    th = logspace(0,log10(4*pi),101);
23    in = logspace(-1,0,101);
24    x = [sin(th).*cos(in);cos(th).*cos(in);sin(in)];
25    QuadAnimator(t,x);
26    return;
```

```
27  end
28
29  if nargin==2
30    h = figure('Name','QuadAnimator');
31    set(h,'InvertHardcopy','off','menubar','none')
32    ma = uimenu(h,'Label','Animate');
33    ms = uimenu(ma,'Label','Start','Callback','QuadAnimator(''update'')')
        ;
34    m = Plot(x);
35    p = get(h,'position');
36    ut = uicontrol('Style','text','String','Time: 0.0 s',...
37                    'Position',[0 0 p(3) 20]);
38    d.t = t;
39    d.x = x;
40    d.m = m;
41    d.ut = ut;
42    set(h,'UserData',d);
43  else
44    h = gcbf;
45    d = get(h,'UserData');
46    Animate(d);
47  end
```

As can be seen, two subfunctions segregate the plotting and animating functionality. The animation sets the XData, YData, and ZData of the markers for the current time and updates the text control. We use a `drawnow` to flush the events queue. The animation runs at a nice speed without requiring a `pause` command, but this may be different on your computer, so be prepared to experiment!

The function `Plot` sets up the four subplots and returns a handle to the plots. This handle will be used by the `set` calls in `Animate`.

```
50  function m = Plot(x)
51  % Use subplot to create four plots of a trajectory
52
53  subplot(2,2,3)
54  plot3(x(1,:),x(2,:),x(3,:));
55  hold on
56  m(3) = plot3(x(1,1),x(2,1),x(3,1),'o');
57  hold off
58  xlabel('X')
59  ylabel('Y')
60  zlabel('Z')
61  grid on
62  title('Trajectory')
63
64  subplot(2,2,1)
65  plot(x(1,:),x(2,:));
66  hold on
67  m(1) = plot(x(1,1),x(2,1),'o');
68  hold off
```

```
69  xlabel('X')
70  ylabel('Y')
71  grid on
72  title('Along Z')
73
74  subplot(2,2,2)
75  plot(x(2,:),x(3,:));
76  hold on
77  m(2) = plot(x(2,1),x(3,1),'o');
78  hold off
79  xlabel('Y')
80  ylabel('Z')
81  grid on
82  title('Along X')
83
84  subplot(2,2,4)
85  plot(x(1,:),x(3,:));
86  hold on
87  m(4) = plot(x(1,1),x(3,1),'o');
88  hold off
89  xlabel('X')
90  ylabel('Z')
91  grid on
92  title('Along Y')
```

The last function animates the plots using set.

```
95  function Animate( d )
96  % Animate the markers on the subplots over time
97
98  for k = 1:length(d.t)
99    x = d.x(:,k);
100   set(d.m(3),'XData',x(1),'YData',x(2),'ZData',x(3));
101   set(d.m(1),'XData',x(1),'YData',x(2));
102   set(d.m(2),'XData',x(2),'YData',x(3));
103   set(d.m(4),'XData',x(1),'YData',x(3));
104   set(d.ut,'string',sprintf('Time: %f s',d.t(k)));
105   drawnow;
106 end
```

4.4 Implementation of a uicontrol Button

Problem

We want to use a button in a dialog box to stop a script as it is running.

152

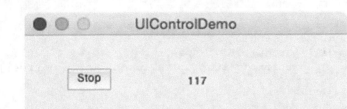

Figure 4.5: *UIControlDemo window.*

Solution

Use uicontrol and figure to create a pop-up window with control, as shown in Figure 4.5.

How It Works

We use two uicontrol calls in the script, UIControlDemo.m. The first puts a text box in the window. The second puts a button in the window. We don't have to specify a style for a button, as this is the default style, but you may choose to specify it for clarity. The button control has a callback. A callback can be any MATLAB code or a function handle. In this case, we just set the global stop to true to stop the loop. Note that we use true and false for the global boolean for clarity, although 1 and 0 will work.

speed and step are handles to the uicontrols. Units are set to pixels. This means that numbers, such as bottom, will be interpreted as pixels as opposed to mm or inches. Note that menubar and figure are also graphics objects.

UIControlDemo.m

```
1  %% Demonstrate the use of a uicontrol button with a callback
2  % Create a window with a button that interacts with a global variable
       in the
3  % script.
8
9  %% Build the GUI
10 % This is a global to communicate the button push from the GUI
11 global stop;
12 stop = false;
13
14 % Build the GUI
15 set(0,'units','pixels')
16 p      = get(0,'screensize');
17 bottom = p(4) - 190;
18 fig    = figure('name','UIControlDemo','position',[340 bottom 298
       90],...
19                 'NumberTitle','off','menubar','none',...
20                 'resize','off');
21
22 % The display text
23 speed = uicontrol( 'Parent', fig, 'position', [ 20 40 280 15],...
24                    'style', 'text','string','Waiting to start.');
```

153

```
25
26  % This has a callback setting stop to 1
27  step = uicontrol( 'parent', fig, 'position',[ 40 40 40 20],...
28                    'string','Stop', 'callback','stop = true;');
29
30  %% Run the GUI
31  for k = 1:1000
32    pause(0.01)
33    set( speed, 'String', k );
34    if( stop )
35      break; %#ok<UNRCH>
36    end
37    %drawnow   % alternative to pause
38  end
```

The position input is defined as [left bottom width height]

We obtained the computer's screen size to place the window near the top of the screen, by assigning the bottom position parameter to the screen size minus the figure size (90 pixels tall), plus a 100 pixel margin, that is, p(4)-190.

MATLAB's code analyzer will place an alert on the line with the break saying that the line is unreachable when stop is false. This is in fact the case, but we have our uicontrol to change that parameter. MATLAB can't ascertain that, so we add the %#ok<UNRCH> comment to suppress the warning. This comment can be automatically added by MATLAB, that is, Autofixed, if you right-click the line with the warning and select Suppress "This statement..." from the pop-up menu.

■ **TIP**   Suppress warnings on lines with code that is reachable only by changes in your boolean logic.

Let's check the execution time. Run the script with tic and toc twice, with pause turned off the second time.

```
>> tic; UIControlDemo; toc
Elapsed time is 11.601678 seconds.
>> pause off
>> tic; UIControlDemo; toc
Elapsed time is 0.147663 seconds.
```

If you want the animation to last more or less than the 11 seconds we got, you can adjust the pause time. We can see that the graphics loop alone takes only a small fraction of a second, despite updating 1000 times! This is because with pause off, the graphics are not forced to update every step of the loop. MATLAB will flush the graphics only when the script ends, unless you have one of the commands that flushes the system queue, such as pause, drawnow, or getframe. If you don't want or need pause, use drawnow to force a graphics update every step of the loop. The following table shows the execution times with pause on or off and using drawnow instead. These are times from a single run on the reference MacBook, not an average; expect a variation in runtimes of up to 10%.

Table 4.2: *Execution Times for* `PatchAnimation`

pause off	0.147663 sec
pause on (0.01 sec)	11.601678 sec
drawnow	2.642504 sec
pause AND drawnow	13.785564 sec

4.5 Display Status of a Running Simulation or Loop

Problem

We want to display the time remaining for a time-consuming task done in a loop.

Solution

Create a window with a text `uicontrol` to display the time remaining, as in Figure 4.6.

How It Works

`TimeDisplayGUI` implements the time window. It uses three actions with `varargin`. A persistent variable, `hGUI`, stores the steps and increments automatically for every **update** call. Some things to notice in this function are

- The MATLAB function `now` is used to get the current date for timing purposes.
- The number of steps completed is stored in `hGUI.stepsDone`.
- The GUI only updates the text string every half second of real time.
- It calculates an estimated amount of real time until script completion, assuming all steps take the same amount of time.
- The built-in demo uses `pause`.

TimeDisplayGUI.m

```
1  %% TIMEDISPLAYGUI  Displays an estimate of time to go in a loop.
2  % Call TimeDisplayGUI('update') each step; the step counter is
     incremented
3  % automatically using a persistent variable. Updates at 0.5 sec
     intervals.
```

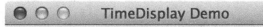

Figure 4.6: *Time display window.*

```
4    %
5    %    TimeDisplayGUI( 'initialize', nameOfGUI, totalSteps )
6    %    TimeDisplayGUI( 'update' )
7    %    TimeDisplayGUI( 'close' )
8    %
9    % You can only have one TimeDisplayGUI operating at once. The built-in
         demo uses
10   % pause to run for about 5 seconds.
11   %% Form:
12   %    TimeDisplayGUI( action, varargin )
13   %% Inputs
14   %    action         (1,:)    'initialize', 'update', or 'close'
15   %    nameOfGUI      (1,:)    Name to display
16   %    totalSteps     (1,1)    Total number of steps
17   %
18   %% Outputs
19   %    None
25   function TimeDisplayGUI( action, varargin )
26
27   persistent hGUI
28
29   if nargin == 0
30     % Demo
31     disp('Initializing demo window with 100 steps.')
32     TimeDisplayGUI( 'initialize', 'TimeDisplay Demo', 100 );
33     for k = 1:100
34       pause(0.05)
35       TimeDisplayGUI( 'update' );
36     end
37     return;
38   end
39
40   switch action
41     case 'initialize'
42       hGUI            = BuildGUI( varargin{1} );
43       hGUI.totalSteps = varargin{2};
44       hGUI.stepsDone  = 0;
45       hGUI.date0      = now;
46       hGUI.lastDate   = now;
47     case 'update'
48       if( isempty( hGUI ) )
49         return
50       end
51       hGUI.stepsDone = hGUI.stepsDone + 1;
52       hGUI = Update( hGUI );
53     case 'close'
54       if ~isempty(hGUI) && ishandle(hGUI.fig)
55         delete( hGUI.fig );
56       else
57         delete(gcf)
58       end
59       hGUI = [];
```

```
60   end
62
63   function hGUI = Update( hGUI )
64   % Update the display
65
66   thisDate = now;
67   dTReal    = thisDate-hGUI.lastDate; % days
68   if (dTReal > 0.5/86400)
69     % Increment every 1/2 second
70     stepPer   = hGUI.stepsDone/(thisDate - hGUI.date0);
71     stepsToGo = hGUI.totalSteps - hGUI.stepsDone;
72     tToGo     = stepsToGo/stepPer;
73     datev     = datevec(tToGo);
74     str       = FormatString( hGUI.stepsDone/hGUI.totalSteps, datev );
75
76     set( hGUI.percent, 'String', str );
77     drawnow;
78     hGUI.lastDate = thisDate;
79   end
81
82   function h = BuildGUI( name )
83   % Initialize the GUIs
84
85   set(0,'units','pixels')
86   p            = get(0,'screensize');
87   bottom       = p(4) - 190;
88   h.fig        = figure('name',name,'Position',[340 bottom 298 90],'
        NumberTitle','off',...
89                          'menubar','none','resize','off','closerequestfcn'
                            ,...
90                          'TimeDisplayGUI(''close'')');
91
92   v            = {'Parent',h.fig,'Units','pixels','fontunits','pixels'};
93
94   str = FormatString( 0, [0 0 0 0 0 0] );
95   h.percent    = uicontrol( v{:}, 'Position',[ 20 35 260 20], 'Style','
        text',...
96                          'fontsize',12,'string',str,'Tag','StaticText2'
                            );
97   drawnow;
99
100  function str = FormatString( fSteps, date )
101  % Format the time to go string
102
103  str = sprintf('%4.2f%% complete with %2.2i:%2.2i:%5.2f to go',...
104                    100*fSteps,date(4),date(5),date(6));
```

The following script, TimeDisplayDemo, shows how the function is used. Figure 4.6 shows the resulting window.

TimeDisplayDemo.m

```
10  n   = 10000;
11  dT  = 0.1;
12  a   = rand(10,10);
13
14  %% Initialize the time display
15  TimeDisplayGUI( 'initialize', 'SVD', n )
16
17  %% Loop
18  for j = 1:n
19
20    % Do something time consuming
21    for k = 1:100
22      svd(a);
23    end
24
25    % Display the status message
26    TimeDisplayGUI( 'update' );
27
28  end
29
30  %% Finish
31  TimeDisplayGUI( 'close' );
```

4.6   Create a Custom GUI with App Designer

Problem

You have a repeating workflow and you would like to build a GUI to avoid changing parameters in your script repeatedly. For example, let's take our rotating cube animation from Recipe 4.1 and put it in a GUI, so we can easily see the effect of different pause lengths.

Solution

We will use appdesigner to create our GUI, starting from a blank figure. We will need an edit box for the pause length plus buttons for operation.

How It Works

Click the "Design App" button in the APPS toolbar to open a new window, as shown in Figure 4.7.

The App Designer interface is shown in Figure 4.8. The interface gives you several templates that you can use as starting points for your app. We will use the blank app to start. Recent apps are on the left. In our case, there is the Detection Filter GUI, DFGUI, that we will write in a later chapter. There is also the MATLAB "Mortgage" demonstration app.

A blank GUI is shown in Figure 4.9. The Component Library is on the left. You drag and drop components onto the GUI from this list. Properties of a selected component are displayed on the right. When you drag a component to the GUI, you can view the associated properties

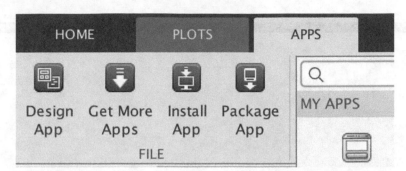

Figure 4.7: *Click "Design App."*

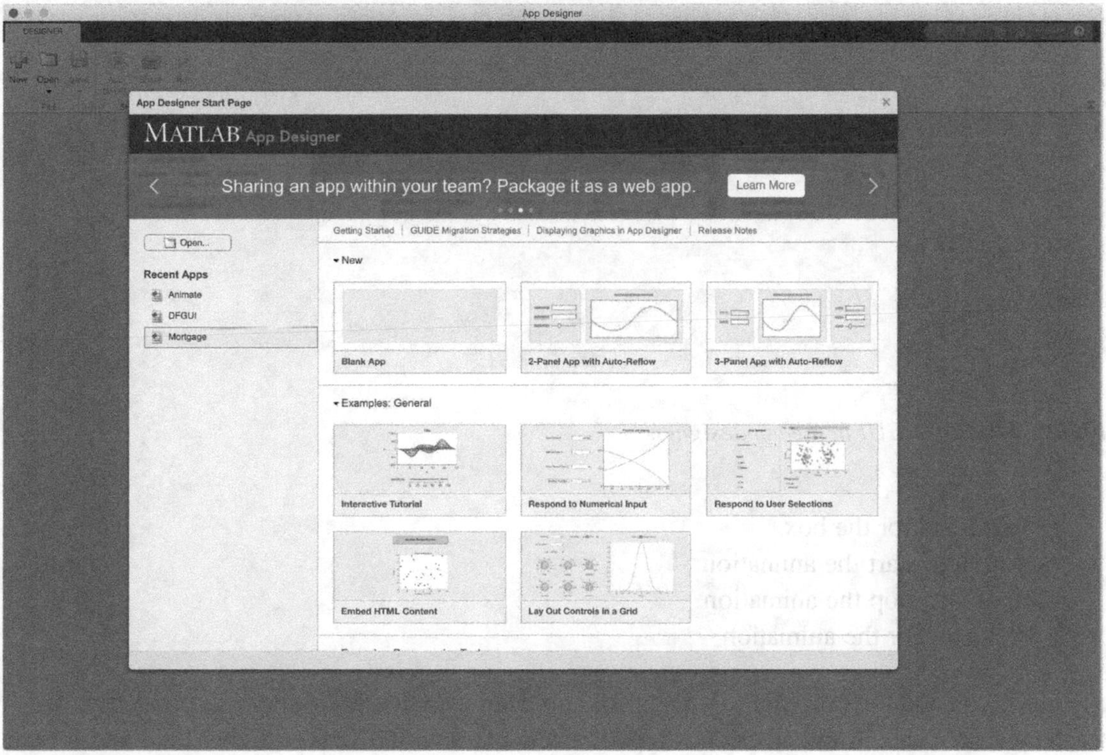

Figure 4.8: *App Designer interface.*

on the right. Note the tabs in the center window for Design View and Code View, which you use to switch between a view of the GUI layout and the code.

We will create the following items in the window by dragging and dropping icons from the palette on the left.

- Edit box, numeric, for entering the pause time
- Text label for the box
- Edit box, numeric, for entering the number of steps

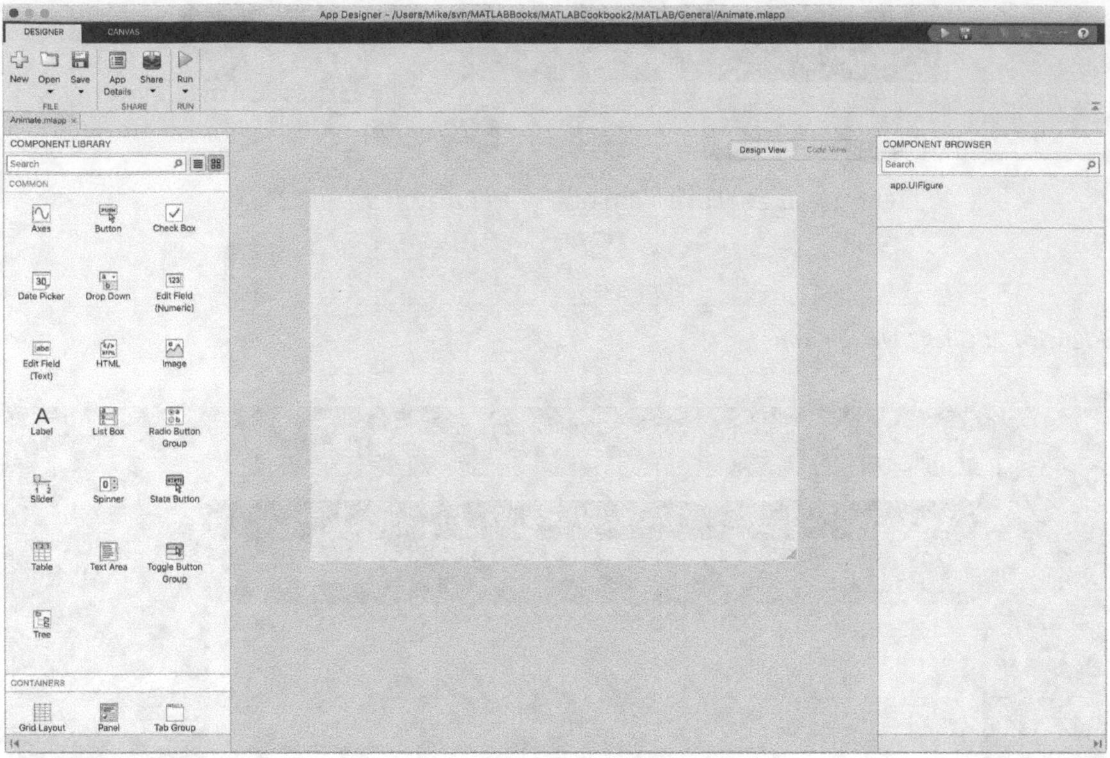

Figure 4.9: *Blank GUI in the appdesigner.*

- Text label for the box
- Button to start the animation
- Button to stop the animation
- Plot axes for the animation

Once you add the preceding items to the app window, click each in the component browser to change its name. In the inspector, type the new name in the Text or Label field and hit tab. We used the names: PauseDuration, Seconds, Start, Stop, and NumberofSteps. As you edit the GUI, if you change a component's name, the associated callback functions will be renamed automatically.

Your figure should now look something like Figure 4.10. You don't have to finish the GUI before starting the code. You can go between the code and design views whenever you want and run the app anytime. The MATLAB debugger will bring you to any line of code with a bug.

We will implement a signal to stop the animation in the Stop button callback via a *property*. First, switch to the Code View of the app. On the left, there is now a code browser where the Component Library used to be, Figure 4.11. Click the Properties tab and click the plus sign pop-up menu. There are two choices here: you may add a private or public property. Add a public property; access needs to be public for other callbacks to see the property.

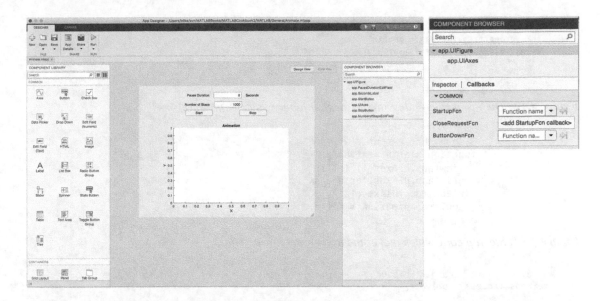

Figure 4.10: *Newly saved GUI. On the right, the Callbacks tab of the figure.*

Figure 4.11: *Properties are added in the Code View.*

This will add a properties block in the code of the app and place the cursor there. Select the default name, "Property," and rename it `stopAnimation`. Set the initial value to false as shown in Figure 4.12. Note that you do not preface the property name with `app`.

Return to the Design View to work on the plot axes. The axes component defaults to 2D. We want a 3D animation, but there is not a way to make it 3D directly in the properties list. To make the axes 3D, you need to add a startup function to add the 3D commands. First, make sure

```
properties (Access = public)
  stopAnimation = false; % Stop button state
end
```

Figure 4.12: *The properties code with our single property,* **stopAnimation.**

```
% Code that executes after component creation
function startupFcn(app)
  view(app.UIAxes,[1,1,1]);
  grid(app.UIAxes,'grid','on')
  zlabel(app.UIAxes,'Z');
  app.stopAnimation = false;
end
```

Figure 4.13: *Startup code which makes the axes three-dimensional.*

```
% Callbacks that handle component events
methods (Access = private)

  % Button pushed function: StartButton
  function StartButtonPushed(app, event)

  end
end
```

Figure 4.14: *Start button code section highlighted.*

that the figure, app.UIFigure, is selected in the component browser. Click the "Callbacks" tab as in the image on the right of Figure 4.10, click the pop-up next to StartupFcn, and add a startup function. Select "Code View" and add the lines of code shown in Figure 4.13. Note that every call to an axes or plot function in an app needs to have the axes handle, `app.UIAxes`, passed in at the start of the function call. Just adding the `view` function call with a three-element vector is enough to make the axes 3D. We also turn the grid on, label the Z axis, and initialize the stopAnimation property (discussed later) to false.

The animation portion of the code needs to be added in the Start button callback. Click the button in the component browser, click the Callbacks tab, and use the pop-up menu to add a callback. The function `StartButtonPushed` will be added, and the area for the start button code will be highlighted as shown in Figure 4.14.

162

```
% Button pushed function: StartButton
function StartButtonPushed(app, event)
  %% Box design
  x   = 3; y = 2; z = 1;
  f   = [2 3 6;3 7 6;3 4 8;3 8 7;4 5 8;4 1 5;2 6 5;2 5 1;1 3 2;1 4 3;5 6 7;5 7 8];
  v   = [-x x x -x -x x x -x; -y -y y y -y -y  y  y; -z -z -z -z z z z z]'/2;
  p   = patch(app.UIAxes,'vertices',v,'faces',f,'facecolor',[0.5 0.5 0.5],...
      'linestyle','none','facelighting','gouraud');
  set(app.UIAxes,'DataAspectRatio',[1 1 1],'DataAspectRatioMode','manual')
  axis(app.UIAxes,[-8 8 -8 8 -8 8])
  light(app.UIAxes,'position',[0 0 1])

  % Animate
  n = app.NumberofStepsEditField.Value;
  a = linspace(0,8*pi,n); c = cos(a); s = sin(a);

  app.stopAnimation = false;
  for k = 1:n
    b   = [c(k) 0 s(k);0 1 0;-s(k) 0 c(k)];
    vK  = (b*v')';
    set(p,'vertices',vK);
    pause(app.PauseDurationEditField.Value);
    if( app.stopAnimation )
      break
    end
  end
  delete(p);
end
```

Figure 4.15: *The animation code in the Start button callback.*

You enter all the animation code in this section as in Figure 4.15. The first part draws the box and uses patch to create the box. The coloring and lighting are also set in the patch call. We use Gouraud shading that gives somewhat natural lighting. We set the data aspect ratio manually. The first argument to patch is the axis handle, app.UIAxes. We then get the number of steps from the edit text box using its **Value** field. The loop updates the patch by changing just the vertices. b is the single-axis rotation matrix. The pause function is passed the value of the duration edit text box. The **stopAnimation** property is used to break the loop. After the loop, we delete the patch so that we can rerun the animation.

Add a Stop button callback to set the property we created earlier, **stopAnimation**, to true when pushed, as shown in Figure 4.16. The function is automatically named StopButtonPushed.

```
% Button pushed function: StopButton
function StopButtonPushed(app, event)
    app.stopAnimation = true;
  end
end
```

Figure 4.16: *The* `StopButtonPushed` *callback.*

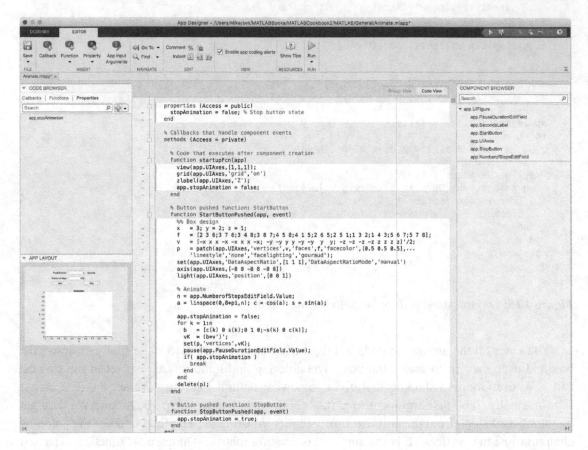

Figure 4.17: *App Designer with our completed code.*

The complete code is shown in Figure 4.17. The places where we added code have white backgrounds. When the animation starts, the app stays in the loop until the input number of steps is complete. Any button push needs to be handled in the loop, as is shown.

Push the green run button to run the app. The app is shown in Figure 4.18. You need to hit start to begin the animation. When it completes, it deletes the box.

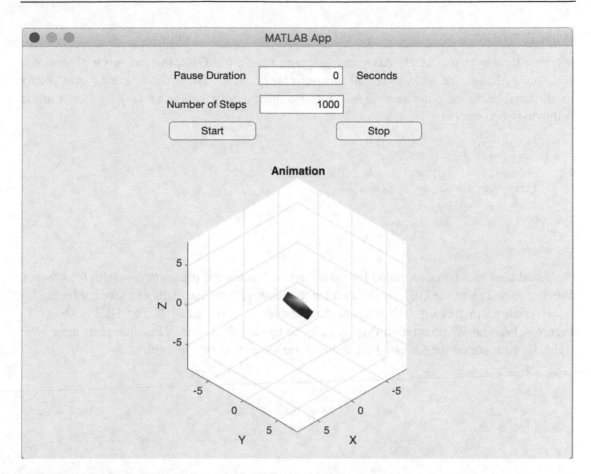

Figure 4.18: *The box animation app in operation.*

4.7 Build a Data Acquisition GUI

Problem

Build a data acquisition GUI to display the real-time data and output it into training sets without using App Designer.

Solution

Use nested functions to create a GUI.

How It Works

We aren't going to use MATLAB's appdesigner to build our GUI. Instead, we will write the code out by hand. We will use nested functions for the GUI. The inner functions have access to all variables in the outer functions. This also makes using callbacks easy as shown in the following code snippet:

```
function DancerGUI( file )
function DrawGUI(h)
 uicontrol( h.fig,'callback',@SetValue);
   function SetValue(hObject, ~, ~ )
   % do something
  end
end
end
```

A callback is a function called by a `uicontrol` when the user interacts with the control. When you first open the GUI, it will look for the Bluetooth device. This can take a while.

Everything in `DrawGUI` has access to variables in `DancerGUI`. The GUI is shown in Figure 4.19. The 3D orientation display is in the upper-left corner. Real-time plots are on the right. Buttons are on the lower left, and the movie window is on the right.

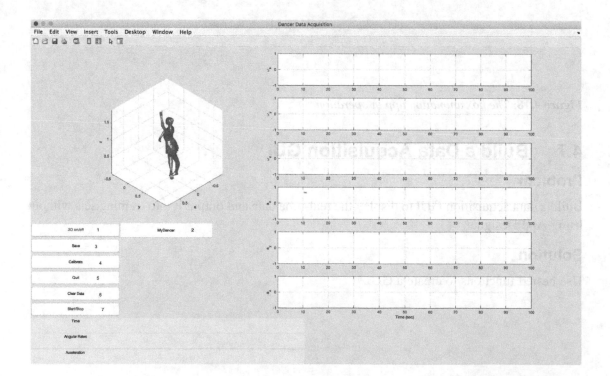

Figure 4.19: *Data acquisition GUI.*

166

The upper-left picture shows the dancer's orientation. The plots on the right show angular rates and accelerations from the IMU. From top to bottom of the buttons

1. Turn the 3D on/off. The default model is big, so unless you add your own model with fewer vertices, it should be set to off.

2. The text box to its right is the name of the file. The GUI will add a number to the right of the name for each run.

3. Save saves the current data to a file.

4. Calibrate sets the default orientation and sets the gyro rates and accelerations to whatever it is reading when you hit the button. The dancer should be still when you hit calibrate. It will automatically compute the gravitational acceleration and subtract it during the test.

5. Quit closes the GUI.

6. Clear data clears out all the internal data storage.

7. Start/Stop starts and stops the GUI.

The remaining three lines display the time, the angular rate vector, and the acceleration vector as numbers. This is the same data that is plotted.

The first part creates the figure and draws the GUI. It initializes all the fields for GUIPlots. It reads in a default picture for the movie window as a placeholder.

DancerGUI.m

```
16  function DancerGUI( file )
17
18  % Demo
19  if( nargin < 1 )
20    DancerGUI('Ballerina.obj');
21    return
22  end
23
24  % Storage of data need by the deep learning system
25  kStore       = 1;
26  accelStore   = zeros(3,1000);
27  gyroStore    = zeros(3,1000);
28  quatStore    = zeros(4,1000);
29  timeStore    = zeros(1,1000);
30  time         = 0;
31  on3D         = false;
32  quitNow      = false;
33
34  sZ = get(0,'ScreenSize') + [99 99 -200 -200];
35
36  h.fig = figure('name','Dancer Data Acquisition','position',sZ,'units','
          pixels',...
```

```
37   'NumberTitle','off','tag','DancerGUI','color',[0.9 0.9 0.9]);
38
39   % Plot display
40   gPlot.yLabel   = {'\omega_x' '\omega_y' '\omega_z' 'a_x' 'a_y' 'a_z'};
41   gPlot.tLabel   = 'Time (sec)';
42   gPlot.tLim     = [0 100];
43   gPlot.pos      = [0.45    0.88    0.46    0.1];
44   gPlot.color    = 'b';
45   gPlot.width    = 1;
46
47   % Calibration
48   q0             = [1;0;0;0];
49   a0             = [0;0;0];
50
51   dIMU.accel     = a0;
52   dIMU.quat      = q0;
53
54   % Initialize the GUI
55   DrawGUI;
```

The notation

```
1   '\omega_x'
```

is a LaTeX format. This will generate ω_x.

The next part tries to find Bluetooth. It first sees if Bluetooth is available at all. It then enumerates all Bluetooth devices. It looks through the list to find our IMU.

```
57   % Get bluetooth information
58   instrreset; % Just in case the IMU wasn't close properly
59   btInfo  = instrhwinfo('Bluetooth');
61
62   if( ~isempty(btInfo.RemoteIDs) )
63     % Display the information about the first device discovered
64     btInfo.RemoteNames(1)
65     btInfo.RemoteIDs(1)
66     for iB = length(btInfo.RemoteIDs)
67       if( strcmp(btInfo.RemoteNames(iB),'LPMSB2-4B31D6') )
68         break;
69       end
70     end
71     b       = Bluetooth(btInfo.RemoteIDs{iB}, 1);
72         fopen(b); % No output allowed for some reason
73     noIMU   = false;
74         a       = fread(b,91);
75         dIMU    = DataFromIMU( a );
76   else
77     warndlg('The IMU is not available.', 'Hardware Configuration')
78     noIMU   = true;
79   end
```

The following is the run loop. If no IMU is present, it synthesizes data. If the IMU is found, the GUI reads data from the IMU in 91 byte chunks. The `uiwait` is to wait until the user hits the start button. When used for testing, the IMU should be on the dancer. The dancer should remain still when the start button is pushed. It will then calibrate the IMU. Calibration fixes the quaternion reference and removes the gravitational acceleration. You can also hit the calibration button at any time.

```matlab
81   % Wait for user input
82   uiwait;
83   % The run loop
84   time   = 0;
85   tic
86   while(1)
87     if( noIMU )
88       omegaZ      = 2*pi;
89       dT          = toc;
90       time        = time + dT;
91       tic
92       a           = omegaZ*time;
93       q           = [cos(a);0;0;sin(a)];
94       accel       = [0;0;sin(a)];
95       omega       = [0;0;omegaZ];
96     else
97       % Query the bluetooth device
98       a        = fread(b,91);
99       pause(0.1); % needed so not to overload the bluetooth device
100
101      dT      = toc;
102      time    = time + dT;
103      tic
104
105      % Get a data structure
106      if( length(a) > 1 )
107        dIMU    = DataFromIMU( a );
108      end
109      accel   = dIMU.accel - a0;
110      omega   = dIMU.gyro;
111      q       = QuaternionMultiplication(q0,dIMU.quat);
112
113      timeStore(1,kStore)    = time;
114      accelStore(:,kStore)          = accel;
115      gyroStore(:,kStore)    = omega;
116      quatStore(:,kStore)    = q;
117      kStore = kStore + 1;
118    end
```

The following closes down the GUI. It uses a variable set in one of the callbacks.

```matlab
120    if( quitNow )
121      close( h.fig )
```

```
122      return
123    else
124      if( on3D )
125        QuaternionVisualization( 'update', q );
126      end
127      set(h.text(1),'string',sprintf('[%5.2f;%5.2f;%5.2f] m/s^2',accel));
128      set(h.text(2),'string',sprintf('[%5.2f;%5.2f;%5.2f] rad/s',omega));
129      set(h.text(3),'string',datestr(now));
130      gPlot = GUIPlots( 'update', [omega;accel], time, gPlot );
131    end
```

This code displays the IMU data `accel` and `omega`:

```
127      set(h.text(1),'string',sprintf('[%5.2f;%5.2f;%5.2f] m/s^2',accel));
128      set(h.text(2),'string',sprintf('[%5.2f;%5.2f;%5.2f] rad/s',omega));
```

The drawing code uses `uicontrol` to create all the buttons. `GUIPlots` and `Quatern ionVisualization` are also initialized. A `uicontrol` that requires an action has call-backs.

```
135    function DrawGUI
136
137    % Plots
138    gPlot = GUIPlots( 'initialize', [], [], gPlot );
139
140    % Quaternion display
141    subplot('position',[0.05 0.5 0.4 0.4],'DataAspectRatio',[1 1 1],'
           PlotBoxAspectRatio',[1 1 1] );
142    QuaternionVisualization( 'initialize', file, h.fig );
143
144    % Buttons
145    f   = {'Acceleration', 'Angular Rates' 'Time'};
146    n   = length(f);
147    p   = get(h.fig,'position');
148    dY  = p(4)/20;
149    yH  = p(4)/21;
150    y   = 0.5;
151    x   = 0.15;
152    wX  = p(3)/6;
153
154    % Create pushbuttons and defaults
155    for k = 1:n
156      h.pushbutton(k) = uicontrol( h.fig,'style','text','string',f{k},'
             position',[x    y wX yH]);
157      h.text(k)       = uicontrol( h.fig,'style','text','string','',  '
             position',[x+wX y 2*wX yH]);
158      y               = y + dY;
159    end
160
```

```
161    h.onButton          = uicontrol( h.fig,'style','togglebutton','string',
          'Start/Stop',...
162                                       'position',[x y wX yH],'
                                          ForegroundColor','red','callback'
                                          ,@StartStop);
163    y = y + dY;
164    h.clrButton         = uicontrol( h.fig,'style','pushbutton','string','
          Clear Data','position',[x y wX yH],'callback',@Clear);
165    y = y + dY;
166    h.quitButton        = uicontrol( h.fig,'style','pushbutton','string','
          Quit','position',[x y wX yH],'callback',@Quit);
167    y = y + dY;
168    h.calibrateButton = uicontrol( h.fig,'style','pushbutton','string','
          Calibrate','position',[x y wX yH],'callback',@Calibrate);
169    y = y + dY;
170    h.saveButton        = uicontrol( h.fig,'style','pushbutton','string','
          Save',    'position',[x    y wX yH],'callback',@SaveFile);   y =
          y + dY;
171       h.on3D               = uicontrol( h.fig,'style','togglebutton','
             string','3D on/off',...
172          .                            'position',[x y wX yH],'
                                          ForegroundColor','red','callback'
                                          ,@On3D);
173
174    h.matFile           = uicontrol( h.fig,'style','edit',       'string','
          MyDancer','position',[x+wX y wX yH]);
```

uicontrol takes parameter pairs, except for the first argument that can be a figure handle. There are a lot of parameter pairs. The easiest way to explore them is to type:

```
h = uicontrol;
get(h)
```

All types of uicontrol that handle user interaction have "callbacks" that are functions that do something when the button is pushed or menu item is selected. We have five uicontrol with callbacks. The first uses uiwait and uiresume to start and stop data collection.

```
184    function StartStop(hObject, ~, ~ )
185      if( hObject.Value )
186        uiresume;
187      else
188        SaveFile;
189        uiwait
190      end
191    end
```

The second uses questdlg to ask if you want to save the data that has been stored in the GUI. This produces the modal dialog shown in Figure 4.20.

171

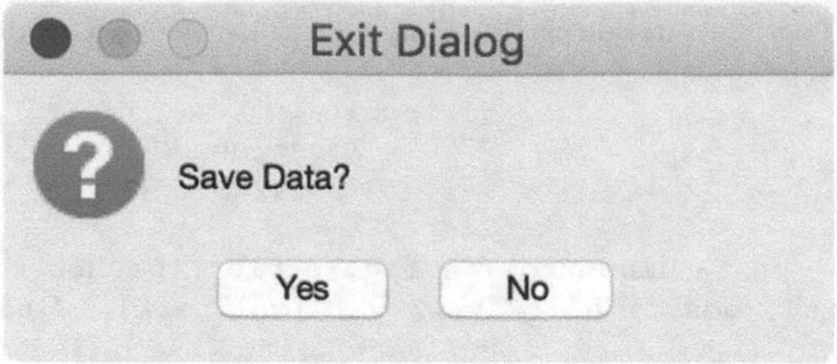

Figure 4.20: *Modal dialog.*

```
194    function Quit(~, ~, ~ )
195      button = questdlg('Save Data?','Exit Dialog','Yes','No','No');
196      switch button
197        case 'Yes'
198          % Save data
199        case 'No'
200      end
201      quitNow = true;
202      uiresume
203    end
```

The third, `Clear`, clears the data storage arrays. It resets the quaternion to a unit quaternion.

```
206    function Clear(~, ~, ~ )
207      kStore      = 1;
208      accelStore  = zeros(3,1000);
209      gyroStore   = zeros(3,1000);
210      quatStore   = zeros(4,1000);
211      timeStore   = zeros(1,1000);
212      time        = 0;
213    end
```

The fourth, `calibrate`, runs the calibration procedure.

```
216    function Calibrate(~, ~, ~ )
217      a      = fread(b,91);
218      dIMU   = DataFromIMU( a );
219      a0     = dIMU.accel;
220      q0     = dIMU.quat;
221      QuaternionVisualization( 'update', q0 )
222    end
```

The fifth, `SaveFile`, saves the recorded data into a mat file for use by the Deep Learning algorithm.

```
225    function SaveFile(~,~,~)
226       cd TestData
227       fileName = get(h.matFile,'string');
228       s = dir;
229       n = length(s);
230       fNames = cell(1,n-2);
231       for kF = 3:n
232          fNames{kF-2} = s(kF).name(1:end-4);
233       end
234       j = contains(fNames,fileName);
235       m = 0;
236       if( ~isempty(j) )
237          for kF = 1:length(j)
238             if( j(kF))
239                f = fNames{kF};
240                i = strfind(f,'_');
241                m = str2double(f(i+1:end));
242             end
243          end
244       end
```

We make it easier for the user to save files by reading the directory and adding a number to the end of the dancer filename that is one greater than the last filename number.

Summary

In this chapter, we introduced figure controls as well as updating graphics in a loop. MATLAB provides a rich GUI building environment including buttons, sliders, listboxes, menus, and plot axes. You can create `uicontrols` directly or use the App Designer tool to build a GUI-based app. It is important to consider performance when designing graphics that update, as some operations have widely disparate performance. Table 4.3 lists the code developed in the chapter.

Table 4.3: *Chapter Code Listing*

File	Description
DancerGUI.m	A GUI for data acquisition
PatchAnimation	Animate a 3D patch in a `for` loop
PatchAnimationStorage	Animate a cube using patch and store as an avi file
QuadAnimator	Create a quad plot page with animation
TimeDisplayDemo	Demonstrate a GUI that shows the time to go in a process
UIControlDemo	Demonstrate the use of a uicontrol button with a callback

CHAPTER 5

■ ■ ■

Testing and Debugging

The MATLAB unit test framework now allows you to incorporate testing into your MATLAB software just as you would your C++ or Java packages. Since entire textbooks have been written on testing methodologies, we will limit ourselves in this chapter to covering the mechanics of using the test framework itself. We also present a couple of recipes that are useful for debugging.

We should, however, say a few words about the goal of software testing. Testing should determine if your software functions as designed. The first step is to have a concrete design against which you are coding. The functionality needs to be carefully described as a set of requirements. The requirements need to specify what inputs the software expects and what outputs it will generate. Testing needs to verify that for all valid inputs, it generates the expected outputs. A second consideration is that the software should handle expected errors and warn the user. For example, a simple function adds two MATLAB variables:

```
c = a + b;
```

You need to verify that it will work for any numeric a and b. You would not generally need to have a warning to the user if a or b is not numeric; that would just fill your code up with unneeded tests. A case where you might want a check is a function containing

```
b = acos(a);
```

If it is supposed to return a real number (perhaps as part of another function), you might want to limit a to have a magnitude less than 1. If you have the code

```
if( abs(a) > 1 )
  a = sign(a);
end
b = acos(a);
```

© Michael Paluszek and Stephanie Thomas 2020

M. Paluszek and S. Thomas, *MATLAB Recipes*,

https://doi.org/10.1007/978-1-4842-6124-8_5

in this case, your test code needs to pass in values of a that are greater than one. This is also a case where you might want to add a custom warning to the user if the magnitude limiting code is exercised, as shown in the following. If you have custom warnings and errors in your code, you also need to test them.

```
if( abs(a) > 1 )
  warning('MyToolbox:MyFunction:OutOfBounds','Input a is out of bounds');
  a = sign(a);
end
b = acos(a);
```

For engineering software, your test code should include known outputs generated by known inputs. In the preceding code, you might include inputs of 1.1, 1, 0.5, 0, -0.5, -1, and -1.1. This would span the range of expected inputs. You might also be very thorough and input linspace(-1.1,1.1) and test against another source of values for the inverse cosine. As shown in the later chapters, we usually include a demo function that tests the function with an interesting set of inputs. Your test code can use the code from the demo function as part of the testing.

All test procedures should employ the MATLAB code coverage tools. The Coverage Report, used in conjunction with the MATLAB Profiler, keeps track of what lines of code are exercised during execution. For a given function or script, it is essential that all code be exercised in its test. Studies have shown that testing done without coverage tools typically exercises only 55% of the code. In reality, it is impossible to actually test every path in anything but the simplest software, and this must be factored into the software development and quality assurance processes. MATLAB does not currently support running the coverage tools on a suite of tests, or during your regression testing, so you should exercise the coverage tools on a per-test basis as you design them.

Once you start using your software, any bug you find should be used to add an additional test case to your software.

5.1   Creating a Unit Test

Problem

Your functions require unit tests.

Solution

Use MATLAB's built-in test capabilities (now available using Java classes) to write and execute unit test functions. Test functions and scripts are identified by using the word "test" as a prefix or suffix to the filename and are run via the runtests function.

How It Works

The `matlab.unittest` package is an xUnit-style, unit testing framework for MATLAB. You can write scripts with test cases separated using cell titles, or functions with test cases in subfunctions, and execute them using the framework. We will show an example of each. There is extensive documentation of the framework and examples in the MATLAB documentation; these lists will get you started.

These are the relevant MATLAB packages implementing the framework:

- matlab.unittest
- matlab.unittest.constraints
- matlab.unittest.fixtures
- matlab.unittest.qualifications

The qualifications package provides all the methods for checking function results, including numerical values, errors, and warnings. The fixtures package allows you to provide setup and teardown code for individual or groups of tests.

Here are the relevant classes you will use when coding tests:

- matlab.unittest.TestCase
- matlab.unittest.TestResult
- matlab.unittest.TestSuite
- matlab.unittest.qualifications.Verifiable

`TestCase` is the superclass for writing test classes.

Here are the relevant functions:

- `assert`
- `runtest`
- `functiontests`
- `localfunctions`

The simplest way to implement some tests for a function is to write a script. Each test case is identified with a cell title, using %%. Use the `assert` function to check the function output. The script can then be run via `runtest`, which will run each test even if a prior test fails, and collate the output into a useful report.

Let's write tests for an example function, `CompleteTriangle`, that computes the remaining data for a triangle given two sides and the interior angle:

CompleteTriangle.m

```
22  function [A,B,c] = CompleteTriangle(a,b,C)
23
24  c = sqrt(a^2 + b^2 - 2*a*b*cosd(C));
25  sinA = sind(C)/c*a;
26  sinB = sind(C)/c*b;
```

```
27  cosA = (c^2+b^2-a^2)/2/b/c;
28  cosB = (c^2+a^2-b^2)/2/a/c;
29  A = atan2(sinA,cosA)*180/pi;
30  B = atan2(sinB,cosB)*180/pi; % insert typo: change a B to A
31
32  end
```

This is similar to the right triangle function used as an example in the MATLAB documentation, but we need the four quadrant inverse tangent as we are allowing obtuse triangles. Since there are very similar lines of code for the two angles A and B, we've made a note that having a typo in one of these lines would be likely, especially if you use copy/paste while writing the function; we'll demonstrate the effect of such a typo via our tests.

Now let's look at a script that defines a few test cases for this function, TriangleTest. We use assert with a logical statement for every check.

TriangleTest.m

```
11  %% Test 1: sum of angles
12  % Test that the angles add up to 180 degrees.
13  C = 30;
14  [A,B] = CompleteTriangle(1,2,C);
15  theSum = A+B+C;
16  assert(theSum == 180,'PSS:Book:triangle','Sum of angles: %f',theSum)
17
18  %% Test 2: isosceles right triangles
19  % Test that if sides a and b are equal, angles A and B are equal.
20  C = 90;
21  [A,B] = CompleteTriangle(2,2,C);
22  assert(A == B,'PSS:Book:triangle','Isoceles Triangle')
23
24  %% Test 3: 3-4-5 right triangle
25  % Test that if side a is 3 and side b is 4, side c (hypotenuse) is 5.
26  C = 90;
27  [~,~,c] = CompleteTriangle(3,4,C);
28  assert(c == 5,'PSS:Book:triangle','3-4-5 Triangle')
29
30  %% Test 4: equilateral triangle
31  % Test that if sides a and b are equal, all angles are 60.
32  [A,B,c] = CompleteTriangle(1,1,60);
33  assert(A == 60,'PSS:Book:triangle','Equilateral Triangle %d',1)
34  assert(B == 60,'PSS:Book:triangle','Equilateral Triangle %d',2)
35  assert(c == 1,'PSS:Book:triangle','Equilateral Triangle %d',3)
```

Note how we have used the additional inputs available to `assert` to add a message ID string and an error message. The error message can take formatted strings with any of the specifiers supported by `sprintf`, such as `%d` and `%f`.

You can simply execute this script, in which case it will exit on the first `assert` that fails. Even better, you can run it with `runtests`, which will automatically distinguish between the test cases and run them independently should one fail.

```
>> runtests('TriangleTest');

Running TriangleTest
...
===========================================================================
Error occurred in TriangleTest/Test4_EquilateralTriangle and it did not
    run to completion.

    --------------
    Error Details:
    --------------
    Equilateral Triangle 1
===========================================================================
.
Done TriangleTest
_____


Failure Summary:

    Name                                Failed  Incomplete  Reason(s)
    ===================================================================
    TriangleTest/Test4_EquilateralTriangle   X        X        Errored.
```

The equilateral triangle test failed, and we know it was the first `assert` in that case due to the index we printed out, `Equilateral Triangle 1`. If you run the code for that test at the command line, you will see that the output does in fact look correct:

```
>> [A,B,c] = CompleteTriangle(1,1,60)
A =
        60
B =
        60
c =
    1
```

179

If we actually subtract the expected value, 60, from A and B, we see why our test has failed.

```
>> A-60
ans =
   7.1054e-15
>> B-60
ans =
   7.1054e-15
```

We are within the tolerances of the trigonometric functions in MATLAB, but our assert did not take that into account. You can add a tolerance like so:

```
1  assert(abs(A-60)<1e-10,'PSS:Book:triangle','Equilateral Triangle %d',1)
2  assert(abs(B-60)<1e-10,'PSS:Book:triangle','Equilateral Triangle %d',2)
```

And now our tests all pass:

```
>> runtests('TriangleTest')
Running TriangleTest
....
Done TriangleTest


ans =
  1x4 TestResult array with properties:

    Name
    Passed
    Failed
    Incomplete
    Duration
Totals:
  4 Passed, 0 Failed, 0 Incomplete.
  0.012243 seconds testing time.
```

Note that we left off the terminating semicolon, so in addition to the brief report, we see that `runtests` returns an array of `TestResult` objects and prints additional total information, including the test duration.

Now let's consider the case of a typo in the function that you have not yet debugged. We will change a B to an A on the last line of the function, so that it reads

```
1  B = atan2(sinB,cosA)*180/pi; % insert typo: change a B to A
```

and run the tests again, using the tolerance check. We use the `table` class with the `TestResult` output to get a nicely formatted version of the test results.

```
>> tr = runtests('TriangleTest');
>> table(tr)
ans =
                Name                       Passed  Failed  Incomplete
```

Duration			
'TriangleTest/Test1_SumOfAngles' 0.0040209	false	true	true
'TriangleTest/Test2_IsoscelesRightTriangles' 0.002971	true	false	false
'TriangleTest/Test3_3_4_5RightTriangle' 0.0027831	true	false	false
'TriangleTest/Test4_EquilateralTriangle' 0.0031556	true	false	false

Despite this being a major error in the code, only one test has failed: the sum of the angles test. The isosceles and equilateral triangle tests still passed because A and B are equal in both cases. You could introduce errors into each line of your code to see if your tests catch them!

Now let's consider the other possibility for the unit tests: a test function, as opposed to the script. In this case, each test case has to be in its own subfunction, and the main function has to return an array of tests. This provides you the opportunity to write setup and teardown functions for the tests. It also makes use of the TestCase class and the qualifications package. Here is what our tests look like in this format:

TriangleFunctionTest.m

```
16  function tests = TriangleFunctionTest
17  % Create an array of local functions
18  tests = functiontests(localfunctions);
19  end
20
21  %%% Test Functions
22  function testAngleSum(testCase)
23  C      = 30;
24  [A,B]  = CompleteTriangle(1,2,C);
25  theSum = A+B+C;
26  testCase.verifyEqual(theSum,180)
27  end
28
29  function testIsosceles(testCase)
30  C      = 90;
31  [A,B]  = CompleteTriangle(2,2,C);
32  testCase.verifyEqual(A,B)
33  end
34
35  function test345(testCase)
36  C      = 90;
37  [~,~,c] = CompleteTriangle(3,4,C);
38  testCase.verifyEqual(c,5)
39  end
40
41  function testEquilateral(testCase)
42  [A,B,c] = CompleteTriangle(1,1,60);
```

```
43  assert(abs(A-60)<testCase.TestData.tol)
44  testCase.verifyEqual(B,60,'absTol',1e-10)
45  testCase.verifyEqual(c,1)
46  end
47
48  %%% Optional file fixtures
49  function setupOnce(testCase)  % do not change function name
50  % set a tolerance that can be used by all tests
51  testCase.TestData.tol = 1e-10;
52  end
53
54  function teardownOnce(testCase)  % do not change function name
55  % change back to original path, for example
56  end
57
58  %%% Optional fresh fixtures
59  function setup(testCase)  % do not change function name
60  % open a figure, for example
61  end
62
63  function teardown(testCase)  % do not change function name
64  % close figure, for example
65  end
```

If you just run this function, you will get an array of the four test methods.

```
>> TriangleFunctionTest
ans =
  1x4 Test array with properties:

    Name
    Parameterization
    SharedTestFixtures
```

We have showed two methods for setting a tolerance for the tests in testEquilateral; in one case, we hard-coded a tolerance in using the absTol parameters, and in the other we used a setup function to pass a tolerance in via TestData. There are two types of setup and teardown functions to choose from: *file* fixtures, which will run just once for the entire set of tests in the file, and *fresh* fixtures, which will run for each test case. The file fixtures are identified with the *Once* suffix. In the case of this tolerance, the setupOnce function is appropriate.

To run the tests, use runtests as for the script. Happily, our tests all pass!

```
>> runtests('TriangleFunctionTest')
Running TriangleFunctionTest
....
Done TriangleFunctionTest
```

```
...
Totals:
   4 Passed, 0 Failed, 0 Incomplete.
   0.043001 seconds testing time.
```

You can run either set of tests in the Profiler (i.e., Run and Time) to verify the coverage of the function being tested. It is a bit easier to navigate to the results for `CompleteTriangle` using the script version of the tests; the results from the test function list many functions from the test framework. The result in the Profiler, showing 100% coverage of our function, is shown in Figure 5.1.

After you have run the Profiler, you can run a Coverage Report. To run the report, you have to use the Current Folder pane of the editor, and select **Reports/Coverage Report** from the context menu. We show an example in Figure 5.2. Our example function runs too quickly to take any measurable time, but generally this report will give you insight into the time taken by your function as well as the coverage you achieved.

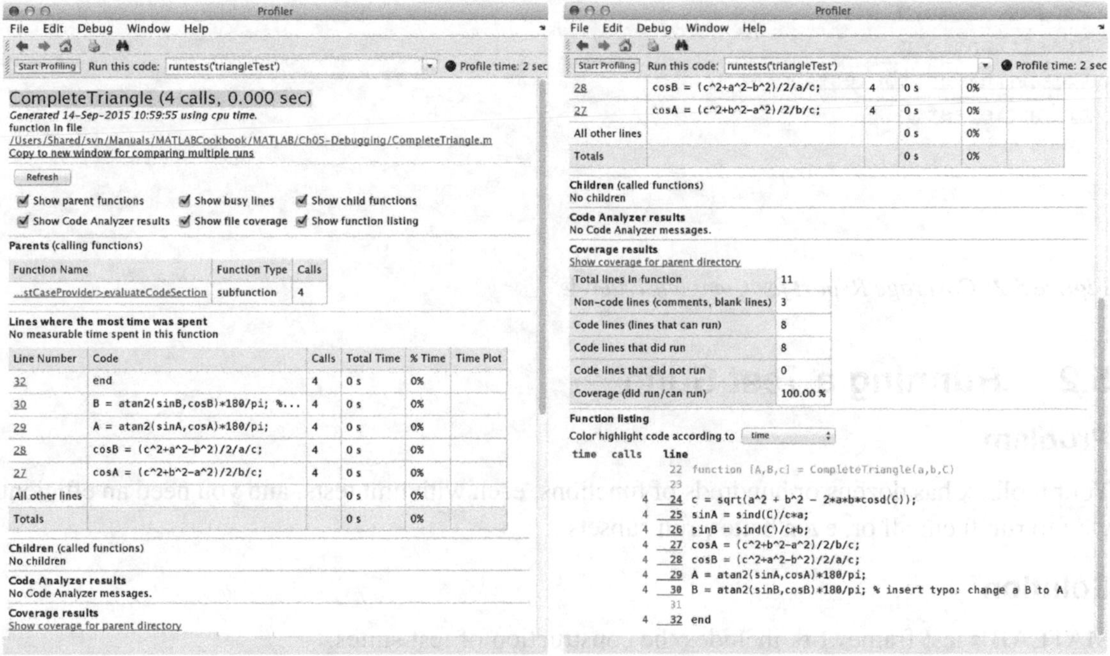

Figure 5.1: *Triangle tests in the Profiler.*

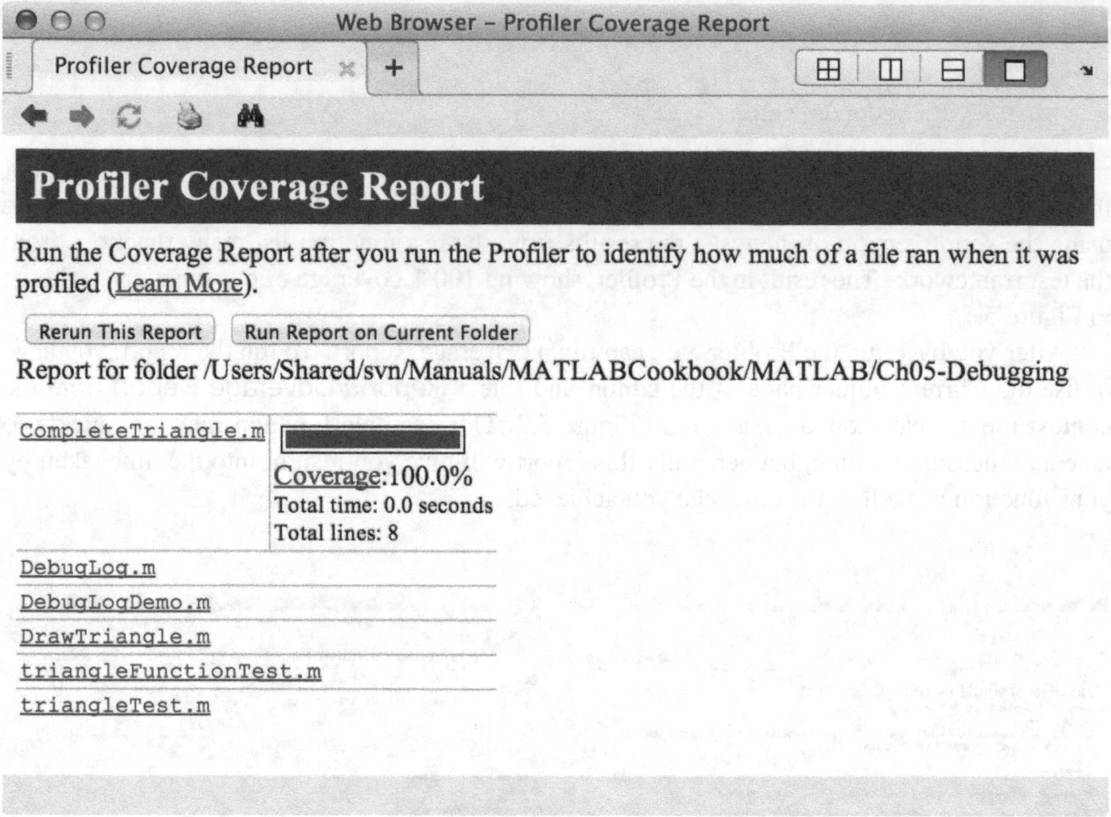

Figure 5.2: *Coverage Report for CompleteTriangle.*

5.2 Running a Test Suite

Problem

Your toolbox has dozens or hundreds of functions, each with unit tests, and you need an efficient way to run them all or, even better, run subsets.

Solution

MATLAB's test framework includes the construction of test suites.

How It Works

After you have generated tests for the functions in your toolbox, you can group them into suites in several ways. The help for the `TestSuite` class lists the options:

```
1    TestSuite methods:
2           fromName    - Create a suite from the name of the test element
3           fromFile    - Create a suite from a TestCase class filename
4           fromFolder  - Create a suite from all tests in a folder
5           fromPackage - Create a suite from all tests in a package
6           fromClass   - Create a suite from a TestCase class
7           fromMethod  - Create a suite from a single test method
```

You can also concatenate test suites made using these methods and pass the array to the test runner. In this way, you can easily generate subsets of your tests to run.

In the previous recipe, we create two test files for `CompleteTriangle`: a test script and a test function. We can create a test suite for the folder containing this code, and it will automatically find both sets of test cases. We assume that the current folder contains the two test files.

```
>> import matlab.unittest.TestSuite
>> testSuite = TestSuite.fromFolder(pwd);
>> result = run(testSuite)

Running TriangleFunctionTest
....
Done TriangleFunctionTest
_____


Running TriangleTest
.......
Done TriangleTest
_____

result =
  1x8 TestResult array with properties:

    Name
    Passed
    Failed
    Incomplete
    Duration
Totals:
   8 Passed, 0 Failed, 0 Incomplete.
   0.04218 seconds testing time.
```

As you can see, test suites are really quite simple. Some advanced features of suites include the ability to apply selectors to a suite to obtain a subset of tests. To see the full documentation of `TestSuite` at the command line, type either

```
>> help matlab.unittest.TestSuite
```

or

```
>> import matlab.unittest.TestSuite
>> help TestSuite
```

The function for performing selections is `selectIf`. Here is an example that selects the two tests of an equilateral triangle from the suite:

```
>> subSuite = testSuite.selectIf('Name', '*Equilateral*');
>> subSuite
subSuite =
  1x2 Test array with properties:

    Name
    Parameterization
    SharedTestFixtures
>> subSuite.Name
ans =
TriangleFunctionTest/testEquilateral
ans =
TriangleTest/Test4_EquilateralTriangle
```

You can run the tests in the resulting suite, or concatenate it with other suites, as before.

5.3   Setting Verbosity Levels in Tests

Problem

The printouts from your tests are getting out of control, but you don't want to just delete or comment out all the information you have needed as you are developing the tests. If a test fails in the future, you may need those messages.

Solution

The test framework includes a logging feature that has four levels of verbosity. To utilize it, you create a test runner using the logging plugin and add `log` calls in your test cases.

How It Works

The four verbosity levels supported are Terse, Concise, Detailed, and Verbose, and they are enumerated as follows:

1	Terse	Minimal amount of information
2	Concise	Typical amount of information
3	Detailed	Supplemental amount of information
4	Verbose	Surplus of information

The default test runner uses the lowest verbosity setting, Terse. The `log` function you use in your test cases is a method of `TestCase`, so to access the help, you need to use the fully qualified name:

```
>> help matlab.unittest.TestCase/log
```

The log method syntax from the help is as follows:

> log(TESTCASE, LEVEL, DIAG) logs the diagnostic at the specified LEVEL. LEVEL can be either a numeric value (1, 2, 3, or 4) or a value from the matlab.unittest.Verbosity enumeration. When level is unspecified, the log method uses level Concise (2).

Logging requires a `TestCase` object. The diagnostic data for `DIAG` can be a string or an instance of `matlab.unittest.diagnostics.Diagnostic`. Let's write an example test for `eig` that demonstrates verbosity.

VerboseEigTest.m

```matlab
1  %% VERBOSEEIGTEST Demonstrate verbosity levels in tests
2  % Run a test of the eig function using log messages. Demonstrates
3  % all four levels of verbosity. To run the tests, at the command line
   use
4  % a TestRunner configured with the LoggingPlugIn:
5  %
6  %    import matlab.unittest.TestRunner;
7  %    import matlab.unittest.plugins.LoggingPlugin;
8  %    runner = TestRunner.withNoPlugins;
9  %    runner.addPlugin(LoggingPlugin.withVerbosity(4));
10 %    results = runner.run(VerboseEigTest);
11 %% Form
12 %    tests = VerboseEigTest
13 %% Inputs
14 % None.
15 %% Outputs
16 %    tests  (:)  Array of test functions
21
22 function tests = VerboseEigTest
23 % Create an array of local functions
24 tests = functiontests(localfunctions);
25 end
26
27 %% Test Functions
28 function eigTest(testCase)
29 log(testCase,'Generating test data'); % default is level 2
30 m = rand(2000);
31 A = m'*m;
32 log(testCase, 1, 'About to call eig.');
33 [V,D,W] = eig(A);
34 log(testCase, 4, 'Eig finished.');
35 assert(norm(W'*A-D*W')<1e-6)
```

```
36  log(testCase, 3, 'Test of eig completed.');
37  end
38
39  % If you want to use the Verbose enumeration in your code instead of
       numbers,
40  % import the class matlab.unittest.Verbosity
41  function eigWithEnumTest(testCase)
42  import matlab.unittest.Verbosity
43  m = rand(1000);
44  A = m'*m;
45  log(testCase, Verbosity.Detailed, 'About to call eig (with enum).');
46  [V,D,W] = eig(A);
47  assert(norm(W'*A-D*W')<1e-6)
48  log(testCase, Verbosity.Terse, 'Test of eig (with enum) completed.');
49  end
```

If you just run this test with `runtests`, you will get the Terse level of output. Note that the system time is displayed along with your log message.

```
>> runtests('VerboseEigTest');
Running VerboseEigTest
    [Terse] Diagnostic logged (2015-09-14T12:15:29): About to call eig.
.   [Terse] Diagnostic logged (2015-09-14T12:15:40): Test of eig (with
    enum) completed.

.

Done VerboseEigTest
```

To get higher levels of verbosity requires a test runner with the logging plugin. This requires a few imports at the command line (or in your script). You need to generate a "plain" runner, with no plugins, then add the logging plugin with the desired level of verbosity. The verbosity level of the message is displayed in the output.

```
>> import matlab.unittest.TestRunner;
>> import matlab.unittest.plugins.LoggingPlugin;
>> runner = TestRunner.withNoPlugins;
>> runner.addPlugin(LoggingPlugin.withVerbosity(4));
>> results = runner.run(VerboseEigTest);
 [Concise] Diagnostic logged (2015-09-14T12:19:57): Generating test data
    [Terse] Diagnostic logged (2015-09-14T12:19:57): About to call eig.
 [Verbose] Diagnostic logged (2015-09-14T12:20:01): Eig finished.
[Detailed] Diagnostic logged (2015-09-14T12:20:07): Test of eig completed
    .
[Detailed] Diagnostic logged (2015-09-14T12:20:07): About to call eig (
    with enum).
    [Terse] Diagnostic logged (2015-09-14T12:20:08): Test of eig (with
       enum) completed.
```

5.4   Create a Logging Function to Display Data

Problem

It is easy and convenient to print out variable values by removing the semicolons from statements, but code left in this state can produce unwanted printouts that are very difficult to track down. Even using `disp` and `fprintf` can make unwanted printouts hard to find as you probably use these functions elsewhere.

Solution

Create a custom logging function to display a variable with a helpful identifying message. You can extend this to a logging mechanism with verbosity settings similar to that described in the previous recipe, as used in the MATLAB testing framework and in most C++ and Java testing frameworks.

How It Works

Our example logging function is implemented in `DebugLog`. `DebugLog` prints out a message, which can be anything, and before that displays the path to where `DebugLog` is called. The backtrace is obtained using `dbstack`.

DebugLog.m

```
1   %% DEBUGLOG Logging function for debugging
2   % Use this function instead of adding disp() statements or leaving out
3   % semicolons.
4   %% Form
5   %   DebugLog( msg, fullPath )
6   %% Decription
7   % Prints out the data in in msg using disp() and shows the path to the
       message.
8   % The full path option will print a complete backtrace.
9   %% Inputs
10  %   msg            (.)      Any message
11  %   fullPath       (1,1)    If entered, print the full backtrace
12  %% Outputs
13  %   None
18
19  function DebugLog( msg, fullPath )
20
21  % Demo
22  if( nargin < 1 )
23    DebugLog(rand(2,2));
24    return;
25  end
26
27  % Get the function that calls this one
28  f = dbstack;
29
30  % The second path is only if called directly from the command line
```

189

```
31  if( length(f) > 1 )
32    f1 = 2;
33  else
34    f1 = 1;
35  end
36
37  if( nargin > 1 && fullPath )
38    f2 = length(f);
39  else
40    f2 = f1;
41  end
42
43  for k = f1:f2
44    disp(['-> ' f(k).name]);
45  end
52  disp(msg);
```

DebugLog is demonstrated in DebugLogDemo. The function has a subfunction to demonstrate the backtrace.

DebugLogDemo.m

```
1  %% Demonstrate DebugLog
2  % Log a variable to the command window using DebugLog.
7
8  function DebugLogDemo
9
10  y = linspace(0,10);
11  i = FindInY(y);
12
13  function i = FindInY(y)
14
15  i = find(y < 0.5);
16  DebugLog( i, true );
```

The output of the demo is shown as follows:

```
>> DebugLogDemo
-> FindInY
-> DebugLogDemo
     1     2     3     4     5
```

One extension of this function is to add the name of the variable being logged, if msg is a variable, using the function inputname. These additional lines of code look like this:

```
47  str = inputname(1);
48  if ~isempty(str)
49    disp(['Variable: ' str]);
50  end
```

The demo output now looks like this:

```
>> DebugLogDemo
-> FindInY
-> DebugLogDemo
Variable: i
      1     2     3     4     5
```

Consistently using your own logging functions for displaying messages to the user and printing debug data will make your code easier to maintain.

5.5   Generating and Tracing MATLAB Errors and Warnings

Problem

You would like to display errors and warnings to the user in an organized fashion.

Solution

Always use the additional inputs to `warning` and `error` to specify a message ID. This allows your message to be traced back to the function in your code that generated it, as well as controlling the display of certain warnings.

How It Works

The `warning` function has several helpful parameters for customizing and controlling warning displays. When you are generating a warning, use the full syntax with a message identifier:

```
1  warning('MSGID', 'MESSAGE', A, B, ...)
```

The `MSGID` is a mnemonic in the form `<component>[:<component>]:<mnemonic>`, such as `PSS:FunctionName:IllegalInput`. The ID is not normally displayed when you give a warning, unless you have turned verbose display on, via `warning on verbose` and `warning off verbose`. This is easy to demonstrate at the command line:

```
>> warning('PSS:Example:DemoWarning', 'This is an example warning')
Warning: This is an example warning
>> warning verbose on
>> warning('PSS:Example:DemoWarning', 'This is an example warning')
Warning: This is an example warning
(Type "warning off PSS:Example:DemoWarning" to suppress this warning.)
```

As displayed, you can turn a given warning off using its message ID by using the command form shown or the functional form, `warning('off', 'msgid')`.

The `lastwarn` function also can return the message ID if passed an additional output, as in

```
>> [lastmsg, lastid] = lastwarn
lastmsg =
This is an example warning
lastid =
PSS:Example:DemoWarning
```

The `error` and `lasterr` functions work the same way. An added benefit of using message identifiers is that you can select them when debugging, as an option when stopping for errors or warnings. The debugger is integrated into the editor window, and the debugger options are grouped under the Breakpoints toolbar button. The button and the "more options" pop-up window are shown in Figure 5.3.

In this case, we entered an example PSS message identifier. Remember, you should always mention any warnings and errors that may be generated by a function in its header!

5.6   Testing Custom Errors and Warnings

Problem

You have code that generates warnings or errors for problematic inputs, and you need to test it.

Solution

You have two possibilities for testing the generation of errors in your code: try/catch blocks with `assert` and the `verifyError` method available to a TestCase. With warnings, you can either use `lastwarn` or `verifyWarning`.

How It Works

A comprehensive set of tests for your code that includes all paths, or as close to all paths as possible, must necessarily exercise all the warnings and errors that can be generated by your code. You can do this manually, using try/catch blocks to catch errors and comparing the error (MException object) to the expected error. For warnings, you can check `lastwarn` to see that a warning was issued, like so:

```
>> lastwarn('');
>> warning('PSS:Book:id','Warning!')
Warning: Warning!
>> [anywarn,anyid] = lastwarn;
>> assert(strcmp(anyid,'PSS:Book:wrongid'))
Assertion failed.
```

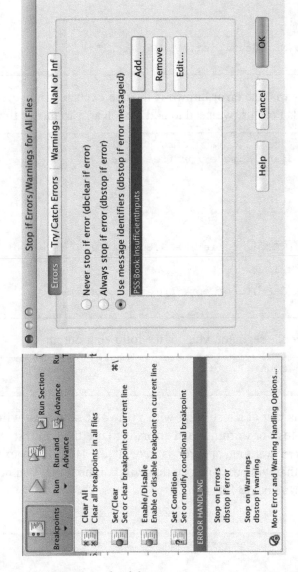

Figure 5.3: *Option to stop on an error in the debugger.*

Here is an example of a try/catch block with `assert` to detect a specific error.

CatchErrorTest.m

```
1   %% Test that we get the expected error, and pass
2   errFun = @() error('PSS:Book:id','Error!');
3   try
4     feval(errFun);
5   catch ME
6     assert(strcmp(ME.identifier,'PSS:Book:id'));
7   end
```

This test will verify that the error thrown is the one expected; however, it will not detect if no error is thrown at all. For this, we need to add a boolean variable to the try block.

```
9    %% This time we don't get any error at all
10   wrongFun = @() disp('Some error-free code.');
11   tf = false;
12   try
13     feval(wrongFun);
14     tf = true;
15   catch ME
16     assert(strcmp(ME.identifier,'PSS:Book:id'));
17   end
18   if (tf)
19     assert(false,'CatchErrorTest: No error thrown');
20   end
```

When you run this code segment, you get the following output:

```
1   Some error-free code.
2   CatchErrorTest: No error thrown
```

If you run the test as part of a test script with `runtests`, the test will fail.

A far better way to test for warnings and errors is to use the unit test framework's qualifiers to check that the desired warning or error is generated. Here is an example of verifying a warning, with one test that will pass and one that will fail; note that you need to pass a function handle to the `verifyWarning` function.

WarningsTest.m

```
1   %% WARNINGSTEST Test generation of warnings.
2   %% Form
3   %    tests = WarningsTest
4   %% Output
5   %    tests    (:)    Array of Tests.
6
7   function tests = WarningsTest
```

```
8   % Create an array of local functions
9   tests = functiontests(localfunctions);
10  end
11
12  %% Test Functions
13  function passTest(testCase)
14  warnFun = @() warning('PSS:Book:id','Warning!');
15  testCase.verifyWarning(warnFun, 'PSS:Book:id');
16  end
17
18  function failTest(testCase)
19  warnFun = @() warning('Wrong:id','Warning!');
20  testCase.verifyWarning(warnFun, 'PSS:id', 'Wrong id');
21  end
```

When we run this test function with runtests, we can see that failTest did in fact fail.

```
>> runtests('WarningsTest')
Running WarningsTest
.Warning: Warning!

================================================================
Verification failed in WarningsTest/failTest.

    ----------------
    Test Diagnostic:
    ----------------
    Wrong id

    --------------------
    Framework Diagnostic:
    --------------------
    verifyWarning failed.
    --> The function handle did not issue the expected warning.

        Actual Warnings:
                Wrong:id
        Expected Warning:
                PSS:id

    Evaluated Function:
            @()warning('Wrong:id','Warning!')

    ------------------
    Stack Information:
    ------------------
    In /Users/Shared/svn/Manuals/MATLABCookbook/MATLAB/Ch05-Debugging/
        WarningsTest.m (failTest) at 12
================================================================
.
```

```
Done WarningsTest
_____

Failure Summary:

    Name                        Failed  Incomplete  Reason(s)
    ================================================================
    WarningsTest/failTest    X                      Failed by verification.

Totals:
    1 Passed, 1 Failed, 0 Incomplete.
    0.047691 seconds testing time.
```

verifyError works the same way. In practice, you will need to make a function handle that includes the inputs to your function that cause the error or warning to be generated.

For advanced programmers, there is a further mechanism for constructing tests using verifyThat with the Constraint class. You can supply your own Diagnostic objects as well. For more information, see the reference pages for these classes along with the Verifiable class.

5.7   Testing Generation of Figures

Problem

Your function generates a figure instead of an output variable. How do you test it?

Solution

While you may need a human to verify that the figure looks correct, you can at least verify that the correct set of figures is generated by your function using findobj.

How It Works

Routinely assigning names to your figures makes it easy to test that they have been generated, even if you don't have access to the handles. You can also assign tags to figures, such as having a single tag for your entire toolbox, which allows you to locate sets of figures.

```
>> figure('Name','Figure 1','Tag','PSS');
>> figure('Name','Figure 2','Tag','PSS')
>> h = findobj('Tag','PSS')
h =
  2x1 Figure array:

  Figure    (PSS)
  Figure    (PSS)
>> h = findobj('Name','Figure 1')
h =
  Figure (PSS) with properties:

      Number: 1
```

196

```
      Name: 'Figure 1'
     Color: [0.94 0.94 0.94]
  Position: [440 378 560 420]
     Units: 'pixels'
```

In your test, you can then check that you have the correct number of figures generated using `length(h)` or that each specific named figure exists using `strcmp`. If you are storing any data in your figures using `UserData`, you can test that as well.

If you are not using tags or need to check for figures that do not have names or tags, you can find all figures currently open using the `type` input to `findobj`:

```
>> findobj('type','figure')
ans =
  2x1 Figure array:

  Figure    (PSS)
  Figure    (PSS)
```

Note that figures will only be returned by `findobj` if they are visible to the command line via their `HandleVisibility` property. This property can have the values `'on'`, `'off'`, and `'callback'`. GUIs generated by the App Designer are generally hidden to prevent users from accidentally altering the GUI using `plot` or similar commands; these figures use the value `'callback'`. Regular figures will have the value `'on'` and can be located as before. A figure with `HandleVisibility` set to `'off'` can only be accessed using its handle.

Summary

This chapter has demonstrated how to use MATLAB's unit test framework and provided some recipes to help you in debugging your functions. Table 5.1 lists the code developed in the chapter.

Table 5.1: *Chapter Code Listing*

File	Description
CatchErrorTest	Script showing how to catch errors in a try block
CompleteTriangle	Example function calculating angles in a triangle
DebugLog	Custom data logging function
DebugLogDemo	Demo of DebugLog showing a backtrace
TriangleFunctionTest	A function with test cases for CompleteTriangle
TriangleTest	A script with test cases for CompleteTriangle
VerboseEigTest	A test function showing all levels of verbosity
WarningsTest	A test function using verifyWarning

CHAPTER 6

■ ■ ■

Classes

MATLAB provides a framework for object-oriented programming. MATLAB created a new framework in 2008a, although the old one is still available. The new framework will be familiar to those of you who program in Python and C++.

Basically, objects contain both data and the operations that work on the data in one package. Classes conceptually derive from the `struct` which only contains data. Combining data with the code that operates on the data can lead to more reliable software. Once you create a class, you can create new classes that inherit from the old class, without changing the old class adding functionality. A new class can inherit from multiple classes. Subclasses get you out of the habit of adding flags to change the functionality of a block of code and data. The subclass usually adds features that are not available in the original class (i.e., the code/data conglomerate).

In this chapter, we will give an example of a class for state space systems. We will create two subclasses, one that is for a continuous system and one that is for discrete systems. This is in line with the many examples in this book.

6.1   Object-Oriented Programming

Object-oriented programming can be thought of as a method for the software designer to impose restrictions on how the software is used by a programmer. In compiled software, the restrictions are imposed at compile time rather than when the software is executed. This, hopefully, catches many errors. In MATLAB, and other scripting languages, the restrictions are imposed whenever you use the function.

Generally, software has moved from unfettered access to memory to more restrictions. In FORTRAN (MATLAB was originally written in FORTRAN), all variables were pointers. You could pass any variable of any size to a function, and the function only saw the first spot in memory. The old linear algebra package, LAPACK, would have you pass the sizes of each variable. Or you could do all sorts of interesting programming, taking advantage of that feature, much to the detriment of anyone wanting to use your code. MATLAB was designed to solve many of the problems of using LAPACK. A really fun thing you could use was the COMMON block. This is perfectly good FORTRAN code.

© Michael Paluszek and Stephanie Thomas 2020
M. Paluszek and S. Thomas, *MATLAB Recipes*,
https://doi.org/10.1007/978-1-4842-6124-8_6

```
1   COMMON i1 i2
2
3   COMMON x
```

Everything starting with an i was an integer. Everything else was float. If you did this, you were mapping i1 and i2 into x. Confusing?

Problems would arise in practice when multiple people used the same code base and changed COMMON blocks without letting other programmers know. Languages like C required all data to be type defined, but even that wasn't enough as software became more complex. Object-oriented programming was devised to catch as many problems as possible at the compile stage.

An object is a conglomeration of data and operations on that data. Classes evolved from structures. A MATLAB structure is

```
1   s = struct('controlInput',[],'state',[],'stateOutput','data',[])
```

This organizes your variables with names that indicate their purpose. Now your function can take as an input s:

```
1   q = ControlFunction(s);
```

rather than

```
1   stateOutput = ControlFunction(controlInput,state,data);
```

However, even with the structure, we still have to know that ControlFunction takes s as an argument. Object-oriented programming helps in this regard.

A class is a definition of the object. An object is an instantiation of the class. The operations are often called methods. At the very least, we want to be able to add data to the object and read data from the object. So the minimal class is

1. Data

2. Input methods

3. Output methods

The input and output methods control access to the data in the object. You might not want the user of the object to be able to change all of the data, and you might not want a user to have access to all of the data. For example, you might have constants in the object that are fundamental to the functioning of the object that you don't want users to ever change.

After this minimal object definition, you then can add methods that operate on the data in the object. These methods are just functions. This leads to the concept of overloading when a function can have one name but operate on different classes. For example, you could create a member function Add for your double class.

200

```
1  a = 1;
2  b = 2;
3  c = Add(a,b);
```

And then create a member function for your image class.

```
1  a = imread('a');
2  b = imread('b');
3  c = Add(a,b);
```

So you don't need names like `AddDouble` and `AddImage`.

6.2 State Space Systems Base Class

Problem

We want to create a class for state space systems.

Solution

Create the base class for state space systems.

How It Works

A state space system consists of four matrices that connect the system inputs to the system outputs. In between are the states of the systems. The states are dynamical quantities that can change with time. The inputs are external inputs and the outputs are what we see outside. Two versions of a state space system are the continuous and discrete. The continuous system is

$$\dot{x}(t) \quad = \quad ax(t) + bu(t) \tag{6.1}$$
$$y \quad = \quad cx(t) + du(t) \tag{6.2}$$

and the discrete system is

$$x_{k+1} \quad = \quad ax_k + bu_k \tag{6.3}$$
$$y_k \quad = \quad cx_k + du_k \tag{6.4}$$

t is the time and k is the step. A step usually means a value taken at a fixed interval of time. Define a time vector:

```
1  t = linspace(0,1000,101);
```

If you use this vector, your step is every 10 time units.

Both systems have the vectors x, y, and u and the matrices a, b, c, and d. Our base class will just involve the vectors and matrices along with their names.

If we were using functional programming, as opposed to object-oriented programming, we would create the structure:

```
1  s = struct('a',[],'b',[],'c',[],'d',[],'x',[],'y',[],'u',[],'xName',{},
       'yName',{},'uName';
```

We'd then create a set of functions to operate on this structure. There are a couple of problems with this approach. The first is that the arrays have specific sizes. If we have n states, m inputs, and p outputs, then a is n by n, b is n by m, c is p by n, and d is p by m. Another problem is that the name fields don't restrict the matrix dimensions which can lead to bugs. Also, there is nothing that says 'xName' is a cell array. Another issue in MATLAB is that anyone can change the structure on the fly, leading to more issues when sharing software, or even using your old software!

To create a class, select "Class" from the New pull-down in the command window.

```
1  classdef untitled
2      %UNTITLED Summary of this class goes here
3      %   Detailed explanation goes here
4
5      properties
6          Property1
7      end
8
9      methods
10         function obj = untitled(inputArg1,inputArg2)
11             %UNTITLED Construct an instance of this class
12             %   Detailed explanation goes here
13             obj.Property1 = inputArg1 + inputArg2;
14         end
15
16         function outputArg = method1(obj,inputArg)
17             %METHOD1 Summary of this method goes here
18             %   Detailed explanation goes here
19             outputArg = obj.Property1 + inputArg;
20         end
21     end
22 end
```

This provides a good starting framework for the class. There is a method to create an instance of the class, the function that is `untitled`. Internally, you see that `obj` is a data structure. `properties` are the data stored in the class. `methods` are operations that work on the data stored in the class. We create the `StateSpace` class to have only data. It does input validation so that once you have created the class, all the matrices are the right sizes. The class definition is

StateSpace.m

```
1  classdef StateSpace
2    % StateSpace Dynamical state space class
3    %  This class contains the matrices and vectors for a state space
4    %  system.
```

The properties, that is, the data, are in the next block of code.

```
6    properties (Access = public)
7      a(:,:) double
8      b(:,:) double
9      c(:,:) double
10     d(:,:) double
11     xN(1,:) cell
12     uN(1,:) cell
13     yN(1,:) cell
14     x(:,:) double
15     u(:,:) double
16     y(:,:) double
17   end
18
19   properties (Access = protected)
20     n(1,1) double
21     m(1,1) double
22     p(1,1) double
23   end
```

We made n,m,p private to restrict its visibility to subclasses. There are many possible properties. Each set goes with its own block. If it were private, subclasses could not see it. You would use private if you didn't want people who are deriving subclasses to have access to that property.

We use property validation by specifying

```
1      a(:,:) double
2      b(:,:) double
3      c(:,:) double
4      d(:,:) double
5      xN(1,:) cell
```

If we don't set the property correctly, we will get

```
>> s.xN = 1
Error setting property 'xN' of class 'StateSpace':
Invalid data type. Value must be cell or be convertible to cell.
```

However, if we do, we will get

```
>> s.a = 'mike'

s =

  StateSpaceDiscrete with properties:

      a: [109 105 107 101]
      b: [2x1 double]
      c: [1 0]
      d: 0
     xN: {'r'  'v'}
     uN: {'u'}
     yN: {'y'}
      x: [2x1 double]
      u: 0
      y: 0
```

It happily makes a a 1-by-4 array. This is because char is numeric. For example, in the char "Mike", each character is an integer, which is of course numeric. A more sophisticated property validation is possible. You should use as much property validation as you deem necessary for your class.

The remaining code is the class constructor.

```
26    function obj = StateSpace(a,b,c,d,xN,uN,yN)
27      % StateSpace Construct an instance of this class
28      %    Checks all of the sizes
29      obj.a    = a;
30      obj.n    = size(a,1);
31      obj.b    = b;
32      [n,m]    = size(b);
33      if(n ~= obj.n)
34        error('b must have as many rows as a');
35      end
36      obj.m = m;
37
38      [p,n]     = size(c);
39      if(n ~= obj.n)
40        error('c must have as many columns as a');
41      end
42      obj.c    = c;
43      obj.p    = p;
44
45      [p,m]     = size(d);
46      if(p ~= obj.p)
47        error('d must have as many rows as c');
48      end
49      if(m ~= obj.m)
50        error('d must have as many columns as b');
51      end
```

```
52        obj.d    = d;
53
54        if( nargin > 4 )
55          n = length(xN);
56          if( n ~= obj.n )
57            error('xN must have as many strings as the rows of a');
58          end
59          obj.xN   = xN;
60
61          m = length(uN);
62          if( m ~= obj.m )
63            error('uN must have as many strings as the columns of b');
64          end
65          obj.uN   = uN;
66
67          p = length(yN);
68          if( n ~= obj.n )
69            error('yN must have as many strings as the rows of c');
70          end
71          obj.yN   = yN;
72        else
73          for k = 1:obj.n
74            obj.xN{k} = sprintf('%d',k);
75          end
76          for k = 1:obj.p
77            obj.yN{k} = sprintf('%d',k);
78          end
79          for k = 1:obj.m
80            obj.uN{k} = sprintf('%d',k);
81          end
82        end
83
84        obj.x = zeros(n,1);
85        obj.u = zeros(m,1);
86        obj.y = zeros(p,1);
```

These methods are all fully implemented in the code. We add one to compute the eigenvalues since this is common to all state space systems.

```
89        function e = Eig(obj)
90        %Eig Get the eigenvalues
91          e = eig(obj.a);
92        end
```

We can then create a double integrator using our class.

```
>> s = StateSpace([0 1;0 0],[0;1],[1 0],1,{'r' 'v'},{'u'},{'y'})

s =

  StateSpace with properties:
```

```
a: [2x2 double]
b: [2x1 double]
c: [1 0]
d: 1
xN: {'r'   'v'}
uN: {'u'}
yN: {'y'}
x: [2x1 double]
u: 0
y: 0
```

`n, m, p` are not listed.

The first argument to every member class is `obj`. You don't pass this as an argument; it is implicit in the member function call.

The eigenvalues are

```
>> s.Eig

ans =

    0
    0
```

which is what we expect.

6.3   State Space Systems Discrete Class

Problem

We want to create a class to propagate discrete time state space systems.

Solution

Create a subclass of `StateSpace` and add a step propagator and a general propagator.

How It Works

We make `StateSpaceDiscrete` a subclass of `StateSpace` in the first line with the `>` operator.

StateSpaceDiscrete.m

```
1  classdef StateSpaceDiscrete<StateSpace
```

If you have multiple super classes, list multiple superclasses `superclass1 & superclass2 & superclass3`, for example:

```
1  classdef automobile>RigidBody>GroundVehicle>FourWheels
```

The constructor just passes the inputs to the super class.

```matlab
 6        function obj = StateSpaceDiscrete(a,b,c,d,xN,uN,yN)
 7          if( nargin == 0 )
 8            a  = [];
 9            b  = [];
10            c  = [];
11            d  = [];
12            xN = {};
13            yN = {};
14            uN = {};
15          end
16          obj@StateSpace(a,b,c,d,xN,uN,yN);
17        end
```

We add two methods to propagate the discrete time class.

```matlab
19        function y = Propagate(obj)
20          %Propagate Propagates the state space system
21          %    Propagates the state space system
22          n    = size(obj.u,2);
23          y    = zeros(size(obj.x,n));
24          y(1) = obj.c*obj.x;
25          for k = 2:n
26            y(k) = obj.c*obj.x + obj.d*obj.u(:,k-1);
27            obj.x = obj.a*obj.x + obj.b*obj.u(:,k-1);
28          end
29        end
30
31        function y = Step(obj,n)
32          %Step Applies a step to the state space system
33          %    Generates the step response. Only the first value of u is
                   used.
34          y    = zeros(obj.p,n);
35          y(1) = obj.c*obj.x + obj.d*obj.u(:,1);
36          for k = 2:n
37            y(k) = obj.c*obj.x + obj.d*obj.u(:,1);
38            obj.x = obj.a*obj.x + obj.b*obj.u(:,1);
39          end
40        end
```

6.4   Using the State Space Class

Problem

We want to create a script to use the state space class.

Solution

Create a script to propagate the continuous subclass of StateSpace.

How It Works

We create a script to use both propagation methods. We assign values to u by using the dot operator, just like any structure. We don't need to write setter methods.

DiscretePropagate.m

```
1  %% Demonstrate using methods of a subclass.
2
3  a    = [0 1;0 0];
4  b    = [0;1];
5  dT   = 0.03;
6
7  % Convert to discrete time
8  [a,b] = C2DZOH(a,b,dT);
9
10 % Step response
11 s    = StateSpaceDiscrete(a,b,[1 0],0,{'r' 'v'},{'u'},{'y'});
12 s.u = 1;
13 y    = s.Step(100);
14
15 PlotSet(1:length(y),y,'x label','Step','y label','y',...
16         'figure title','Sub Class Step');
17
18 % Pulse Response
19 s.u = [zeros(1,20) ones(1,30) zeros(1,50)];
20
21 y    = s.Propagate;
22
23 PlotSet(1:length(y),y,'x label','Step','y label','y',...
24         'figure title','Sub Class Pulse');
```

method uses whatever u is in the object when you used Propagate. You can add help using % just below the method names and at the top.

```
>> help StateSpace
  StateSpace Dynamical state space class
   This class contains the matrices and vectors for a state space
   system.

    Documentation for StateSpace

>> help StateSpace.Eig
 Eig Get the eigenvalues
```

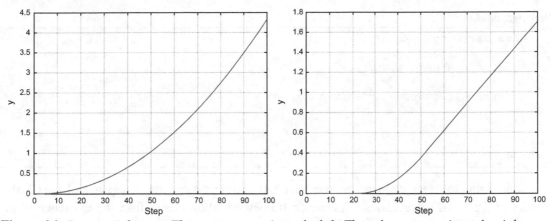

Figure 6.1: *Propagated states. The step response is on the left. The pulse response is on the right.*

The resulting plots are shown in Figure 6.1.

You can see how compact the code is. StateSpaceDiscrete can handle any linear time-invariant (i.e., the state space matrices are constant) discrete time state space system.

6.5 Using a Mocking Framework

Problem

We want to test an incomplete class for which other needed classes are unavailable.

Solution

Use a mocking framework which is a framework that allows us to interface to an incomplete class, that is, a "mock" class.

How It Works

We create a class to test a function that calls a class that does not exist or is unavailable. In this case, we have a function Drag:

Drag.m

```
1  function drag = Drag(DensityModel,h,v,s)
2  drag = 0.5*DensityModel.LookUpDensity(h)*s*v^2;
3  end
```

that uses the density class DensityModel. When you are testing Drag, the density table is not yet available. Create the class for DensityModel with an abstract method.

DensityModel.m

```
1  classdef DensityModel
2
3    methods(Abstract,Static)
4
5       rho = LookUpDensity(altitude);
6
7    end
8
9  end
```

Now write the test class using the `matlab.unittest.TestCase` and `matlab.mock.TestCase` superclasses. The `mock` framework allows us to fake the existence of the density model. The `unittest` framework allows us to evaluate the results of the test.

DragTest.m

```
1  classdef DragTest < matlab.unittest.TestCase & matlab.mock.TestCase
2      methods(Test)
3          function NegativeDensity(testCase)
4              [stubDensity,densityModelBehavior]=createMock(testCase,?
                   DensityModel);
5              testCase.assignOutputsWhen(densityModelBehavior.LookUpDensity
                   (-1),-1);
6              d = Drag(stubDensity,-1,1,2);
7              testCase.verifyLessThan(d,0);
8          end
9      end
10 end
```

The question mark, `?`, is used to get a metaclass of `DensityModel`. In this code snippet, the Drag function is called with an altitude of h= -1, a velocity of v=1, and a surface area of s=2. It is given a "stubDensity" class that is created just for the purpose of this test, using the mock framework `createMock`.

We create a mock object, `stubDensity`. The method is implemented with `density ModelBehavior.LookUpDensity(0)`. When we pass a negative altitude, we get a negative density and negative drag.

Run the test.

```
>> results = runtests('DragTest');
Running DragTest
.
Done DragTest
_____


>> table(results)

ans =
```

```
   1x6 table

   Name                                Passed  Failed  Incomplete Duration
        Details
   _____                _____  _____  _____ _____

{'DragModelTest/NegativeDensity'}  true    false   false      0.04766 {1x1
     struct}
```

This verifies that we get the correct behavior from Drag.

Summary

This chapter has demonstrated how to use MATLAB classes and mocking frameworks in classes. Table 6.1 lists the code developed in the chapter.

Table 6.1: Chapter Code Listing

File	Description
StateSpace	State space dynamical system class
StateSpaceDiscrete	Subclass of StateSpace for discrete systems
DragTest	A test class for Drag and DensityModel
Drag	A test class for Drag
DensityModel	A placeholder class for DensityModel

Part II

Applications

In this part of the book, we will explore the use of MATLAB for dynamical systems and control system design in a number of technologies. In each area, we will derive the equations of motion for the system. A system is defined by its state equations, states, and parameters. The equations of motion are the equations of the states of a system. The state variables are the set of variables that evolve with time that completely define the current state of the system and allow for future prediction of the state without any knowledge of the past. We also need parameters that are independent of the states to fully define the system along with the inputs to the system. The state vector will always be represented by an n-by-1 MATLAB array.

In the equations that we present, we will use the dot notation for derivatives, that is

$$\dot{r} = \frac{dr}{dt} \tag{6.5}$$

State equations are of the form

$$\dot{x} = ax + bu \tag{6.6}$$
$$y = cx + du$$

x is the state and is an $n \times 1$ vector represented by an n-row by 1-column MATLAB array. u is the input matrix and is $n \times m$. y is the measurement. a relates the state to the state derivative and is an $n \times n$ array. b is the input array and is $n \times m$ where the number of inputs, u, is m.

We are not going to delve into control theory in detail. That would require a complete textbook by itself, or many textbooks if you wanted to explore control system design in depth. We will provide an intuitive approach to allow you to get control systems up and running quickly without too much code!

CHAPTER 7

■ ■ ■

The Double Integrator

A double integrator is a dynamical model for a wide variety of physical systems. This includes, for example, a mass moving in one dimension and an object rotating around a shaft. It represents a broad class of systems with two time-varying quantities that we will call states. In this chapter, we will learn how to model a double integrator and how to control it. In the process, we will create some very important functions for implementing control systems, plotting, and performing numerical integration. This will provide a springboard to other more complex systems in later chapters.

7.1   Writing the Equations for the Double Integrator Model

Problem

A double integrator is a second-order system with a second derivative. This model appears in many engineering disciplines.

Solution

We will write the equations for the model dynamics and implement these in a function.

How It Works

One-dimensional linear motion can be modeled with the following differential equations:

$$\dot{r} = v \qquad (7.1)$$

$$m\dot{v} = F \qquad (7.2)$$

r and v are states; m, the mass, is a parameter; and F, the force, is an input. In this case, the states are r, position, and v, velocity. The state vector is

$$x = \begin{bmatrix} r \\ v \end{bmatrix} \qquad (7.3)$$

The variable x is represented in MATLAB with a 2-row by 1-column array. The first element of the array is r and the second is v.

© Michael Paluszek and Stephanie Thomas 2020
M. Paluszek and S. Thomas, *MATLAB Recipes*,
https://doi.org/10.1007/978-1-4842-6124-8_7

■ **TIP**   This equation works equally well for rotational motion. Just replace m with I for inertia, r with θ for angle, v with ω for angular velocity, and F with T for torque.

To write a function for the derivatives, divide by m to isolate the derivatives on the left-hand side of the equations. The terms on the right-hand side are what we will calculate in our so-called *RHS* function. The state equations are of the form

$$\dot{x} = f(x, F) \tag{7.4}$$

$f(x, F)$ is our right-hand side.

$$\dot{r} = v \tag{7.5}$$
$$\dot{v} = \frac{F}{m} \equiv a \tag{7.6}$$

Writing these equations in vector notation, we have

$$\dot{x} = \begin{bmatrix} v \\ a \end{bmatrix} \tag{7.7}$$

Now we can write the function, `RHSDoubleIntegrator`, for the derivative of the state vector x. Note the prefix of `RHS` in the name, which we use to identify all functions that are to be integrated. The velocity term is the second element of our state x and is the derivative of the position state r. The derivative of the velocity state is the acceleration a. The RHS function has a placeholder ~ for the first argument where the integrator will pass the time t, which this function doesn't require.

RHSDoubleIntegrator.m

```
 1  %% RHSDOUBLEINTEGRATOR Right hand side of a double integrator.
31  function xDot = RHSDoubleIntegrator( ~, x, a )
32
33  xDot = [x(2);a];
```

7.2   Creating a Fixed-Step Numerical Integrator

Problem

We need to use numerical integration in our simulations to evolve the state of the systems.

Solution

We will present the equations for a fourth-order Runge-Kutta integrator and develop a function to perform a fixed-step integration.

How It Works

Mathematical Modeling

Let's look at a simple model for linear motion. We need to put the equations in a state equation form, which is a set of first-order differential equations with the derivatives on the left-hand side. Take, for example, our one-dimensional motion model from Recipe 7.1, with the derivative terms for r and v on the left-hand side of the equations.

$$\frac{dr}{dt} = v \tag{7.8}$$

$$\frac{dv}{dt} = \frac{F}{m} \tag{7.9}$$

Multiply both sides by dt.

$$dr = vdt \tag{7.10}$$

$$dv = \frac{F}{m}dt \tag{7.11}$$

Replace dt with the fixed time interval h. We can now write our simplest type of numerical integrator for computing the state at step $k + 1$ from the state at step k.

$$r_{k+1} = r_k + v_k h \tag{7.12}$$

$$v_{k+1} = v_k + \frac{F_k}{m}h \tag{7.13}$$

Step k is at time t_k and step $k + 1$ is at time t_{k+1}, where $t_{k+1} = t_k + h$.

This simple integrator is called Euler integration. It assumes that the force F_k is constant over the time interval h. Euler integration works fairly well for simple systems, like the one given earlier. For example, for a spring system

$$\ddot{r} + \omega^2 r = a \tag{7.14}$$

the natural frequency of oscillation is ω. If the time step is greater than half the period of the oscillation $2\pi/\omega$, the numerical integration cannot capture the dynamics. In practice, the time step, h, must be much lower than half of the period of the oscillation.

Euler integration is rarely used for engineering due to its limited accuracy. One of the most popular methods used for control system simulation is the fourth-order Runge-Kutta method. The equations for Runge-Kutta integration are

$$
\begin{aligned}
k_1 &= f(x, u(t), t) \\
k_2 &= f(x + \frac{h}{2}k_1, u(t) + \frac{h}{2}, t + \frac{h}{2}) \\
k_3 &= f(x + \frac{h}{2}k_2, u(t) + \frac{h}{2}, t + \frac{h}{2}) \\
k_4 &= (x + hk_3, u(t + h - \epsilon), t + h) \\
x &= x + \frac{h}{6}(k_1 + 2(k_2 + k_3) + k_4) + O(h^4)
\end{aligned}
\tag{7.15}
$$

$f(x, u, t)$ is the right-hand side of the differential equations. $O(h^4)$ means the truncation error due to the order of the integration goes as the fourth power of the time step. This means that if we halve our time step, the error drops to 0.0625 of the error with the bigger time step. For the preceding system

$$
f(x, F, t) = \begin{bmatrix} v \\ \frac{F}{m} \end{bmatrix}
\tag{7.16}
$$

The right-hand sides are computed at four different points, twice at $h/2$ and once at t and $t + h$.

■ **NOTE**   There are other fourth-order Runge-Kutta methods with different coefficients.

MATLAB has many numerical integrators. They are designed to integrate over any interval. Some, such as `ode113`, adjust the step size (this would be the time step in this example) and the order of the integration (Euler is first order, the preceding Runge-Kutta is fourth order), between the desired interval. For digital control, we need to integrate with a step size that is, at a minimum, the sample time of the digital controller. You could use `ode113` for this, but usually the fourth-order Runge-Kutta is sufficient. We will use this method for all of our examples.

MATLAB Code

We next want to look at the `RungeKutta` function that implements Equation 7.15. Note the use of `feval` to evaluate the right-hand-side function in the following code:

RungeKutta.m

```
27  function x = RungeKutta( Fun, t, x, h, varargin )
28
29  hO2    = 0.5*h;
30  tPHO2  = t + hO2;
31
32  k1     = feval( Fun,    t,       x,              varargin{:} );
33  k2     = feval( Fun,    tPHO2,   x + hO2*k1,     varargin{:} );
34  k3     = feval( Fun,    tPHO2,   x + hO2*k2,     varargin{:} );
```

```
35   k4      = feval( Fun,      t+h,     x + h*k3,   varargin{:} );
36
37   x       = x + h*(k1 + 2*(k2+k3) + k4)/6;
```

Fun is a pointer to the right-hand-side function. varargin is passed to that function, which enables the dynamics model to have any number of parameters. This RungeKutta function will be used in all of the examples in this book. We precompute all values that are used multiple times, such as $t + \frac{h}{2}$. This is particularly important in functions that are called repeatedly.

■ **TIP**   Compute values, such as $h/2$, once and store in a variable.

We pass RungeKutta a handle to the right-hand-side function, RHSDoubleInteg rator, that implements Equation 7.1.

■ **TIP**   Replace unused variables in function calls with the tilde.

To integrate the model one step, we call

```
xNew = RungeKutta( @RHSDoubleIntegrator, ~, x, h, a )
```

Our RungeKutta function and all MATLAB integrators have the dependent variable first, which in this case is t. Since it isn't used in this case, we replace it with the tilde. MATLAB's code analyzer will suggest this for efficiency for all unused function inputs and outputs in your code.

7.3   Implement a Discrete Proportional-Derivative Controller

Problem

We want digital control software that can control a double integrator system and other dynamical systems.

Solution

We will derive the equations for a damped second-order system in the time domain and then create a sampled data version and implement it in a MATLAB function.

How It Works

If a constant force F is applied to the system, the mass m will accelerate and its position will change with the square of time. The analytical solution for the two states, r and v, is

$$r(t) \;=\; r(0) + v(0)\,(t - t(0)) + \frac{1}{2}\frac{F}{m}\,(t - t(0))^2 \tag{7.17}$$

$$v(t) \;=\; v(0) + \frac{F}{m}\,(t - t(0)) \tag{7.18}$$

If we wish to have the mass stay close to zero, we can use a control system known as a regulator. We will use a proportional-derivative regulator. This regulator measures the position and applies a control force proportional to the position error and to the derivative of the position error. Let's look at a particularly simple form of this controller. Our control would be

$$F_c = -k_r r - k_v v \tag{7.19}$$

We don't have to be able to measure the disturbance force, F, for this to work. Picking the gains, k_r and k_v, is easy if we write the dynamical system as a second-order system.

$$m\ddot{r} = F_c + F \tag{7.20}$$

$$m\ddot{r} = F - k_r r - k_v v \tag{7.21}$$

$$m\ddot{r} + k_v \dot{r} + k_r r = F \tag{7.22}$$

Our controlled system is a damped second-order oscillator. We can write the desired differential equation as

$$\ddot{r} + 2\zeta\sigma\dot{r} + \sigma^2 r = \frac{F}{m} \tag{7.23}$$

where the gains are

$$k_r = m\sigma^2 \tag{7.24}$$

$$k_v = 2m\zeta\sigma \tag{7.25}$$

σ is the undamped natural frequency, and ζ is the damping ratio. This system will always be stable as long as $\zeta > 0$. If F is constant, the position will settle to an offset. This will be

$$r = \frac{F}{m\sigma^2} \tag{7.26}$$

This method of control design is called "pole placement."

Virtually, all control systems are implemented on digital computers so we must transform this controller into a digital form. We assume that we only measure the position. The first step is to assemble the control in a continuous time state space form. For this implementation, we

220

will add a rate filter to our PD controller.

$$\omega = \omega_r + 2\zeta\omega_n \tag{7.27}$$

$$k = \omega_r\omega_n^2 * m/w \tag{7.28}$$

$$\tau = (\frac{m}{k}\omega_n(\omega_n + 2\zeta\omega_r) - 1)/\omega \tag{7.29}$$

$$a = -\omega \tag{7.30}$$

$$b = \omega \tag{7.31}$$

$$c = -k\omega\tau \tag{7.32}$$

$$d = k * (\tau\omega + 1) \tag{7.33}$$

where ω_r is the cutoff frequency for the first-order filter on the rate term, ω_n is the undamped natural frequency for the controller, and ζ is the controller damping ratio. m is the "mass" or inertia. You can always set this to 1 and scale the control output. The undamped natural frequency gives the bandwidth of the controller. The higher the bandwidth, the faster it responds to errors. The higher the bandwidth, the smaller the offset error will be due to a constant input. Higher bandwidth requires more control force. In addition, measurement noise will be passed into the controller and "controlled." Generally, you want the bandwidth to be no higher than the frequency of the expected disturbances.

$$\dot{x} = ax + bu \tag{7.34}$$

$$y = -cx - du \tag{7.35}$$

x is the controller state, and u is the position measurement. The state space form is convenient for computation but still assumes that we are sampling continuously. There are many ways to convert this to digital form. We will use a zero-order hold, meaning we will compute the control of each sample and hold the value over that sample period. We convert this using the matrix exponential function in MATLAB, expm. If T is the sample period, we assemble the matrix:

$$\sigma = \begin{bmatrix} aT & bT \\ 0 & 0 \end{bmatrix} \tag{7.36}$$

The sampled time versions of a and b are

$$\begin{bmatrix} a_d \\ b_d \end{bmatrix} = e^\sigma \tag{7.37}$$

Our digital controller is now

$$x_{k+1} = a_dx_k + b_du_k \tag{7.38}$$

$$y_k = -cx_k - du_k \tag{7.39}$$

Let's now look at PDControl. This function designs and implements the control system derived in Equation 7.27. The name stands for "Proportional-Derivative Control." It has several child functions. First, we review the header. It has a link to the help for one of the subfunctions which will be active when the header is displayed at the command line. We list each data structure field for the input and output. In the header, d is used as the feedthrough matrix but is used as a data structure in the function.

PDControl.m

```
1  %% PDCONTROL Design and implement a PD Controller in sampled time.
2  %% Forms
3  %   d = PDControl( 'struct' )
4  %   d = PDControl( 'initialize', d )
5  %   [y, d] = PDControl( 'update', u, d )
6  %
7  %% Description
8  % Designs a PD controller and implements it in discrete form.
9  %
10 %   y = -c*x - d*u
11 %   x = a*x + b*u
12 %
13 % where u is the input and y is the output. This controller has a first
14 % order rate filter on the derivative term.
15 %
16 % Set the mode to initialize to create the state space matrices for the
17 % controller. Set the mode to update to update the controller and get a
18 % new output.
19 %
20 % Utilizes the subfunction C2DZOH to discretize, see <a href="matlab:
       help PDControl>CToDZOH">CToDZOH help</a>
21 %
22 %% Inputs
23 %   mode     (1,1) 'initialize' or 'update'
24 %   u        (1,1) Measurement
25 %   d        (.)   Data structure
26 %                  .m       (1,1) Mass
27 %                  .zeta    (1,1) Damping ratio
28 %                  .wN      (1,1) Undamped natural frequency
29 %                  .wD      (1,1) Derivative term filter cutoff
30 %                  .tSamp   (1,1) Sampling period*
31 %                  .x       (1,1) Controller state
32 %
33 %% Outputs
34 %   y        (1,1) Control
35 %   d        (.)   Data structure additions
36 %                  .a       (1,1) State transition matrix
37 %                  .b       (1,1) Input matrix
38 %                  .c       (1,1) State output matrix
39 %                  .d       (1,1) Feedthrough matrix
40 %                  .x       (1,1) Updated controller state
```

Next, let us look at the body of the function. Note the switch statement and the two child functions, CToDZOH and DefaultStruct, at the bottom.

```matlab
43  function [y, d] = PDControl( mode, u, d )
44
45  % Demo
46  if( nargin < 1 )
47    disp('Demo of PDControl using the default struct')
48    d = PDControl('struct');
49    d = PDControl('initialize',d);
50    disp(d)
51    return
52  end
53
54  switch lower(mode)
55    case 'initialize'
56      d              = u;
57      w              = d.wD + 2*d.zeta*d.wN;
58      k              = d.wD*d.wN^2*d.m/w;
59      tau            = ((d.m/k)*d.wN*(d.wN + 2*d.zeta*d.wD) - 1 )/w;
60      d.a            = -w;
61      d.b            = w;
62      d.c            = -k*w*tau;
63      d.d            = k*(tau*w + 1);
64
65      [d.a, d.b]     = CToDZOH(d.a,d.b,d.tSamp);
66      y              = d;
67
68    case 'update'
69      y   = -d.c*d.x - d.d*u;
70      d.x = d.a*d.x + d.b*u;
71
72    case 'struct'
73      y = DefaultStruct;
74
75    otherwise
76      error('%s is not a valid mode',mode);
77  end
79
80  function [f, g] = CToDZOH( a, b, T )
81  %% PDControl>CToDZOH
82  % Continuous to discrete transformation using a zero order hold.
       Discretize
83
84  q = expm([a*T b*T;zeros(1,2)]);
85
86  f = q(1,1);
87  g = q(1,2);
89
90  function d = DefaultStruct
91  %% PDControl> DefaultStruct
```

```
92
93   d = struct('m',1,'zeta',0.7,'wN',0.1,'wD',0.5,'tSamp',1.0,'x',0,'a'
        ,[],...
94               'b',[],'c',[],'d',[]);
```

This is the standard format for an engineering function. Here are its important features:

- It combines design and implementation in one function.
- It returns the default data structure that it uses.
- It has modes.
- It has a built-in demo.
- It has nested functions.

The built-in demo uses the default values. The default values give the user an idea of reasonable parameters for the function. This built-in demo just generates the state space matrices, which are scalars in this case.

```
>> PDControl
Demo of PDControl using the default struct
        m: 1
     zeta: 0.7000
       wN: 0.1000
       wD: 0.5000
    tSamp: 1
        x: 0
        a: 0.5273
        b: 0.4727
        c: -0.0722
        d: 0.0800
```

A more elaborate demo, with a simulation, could have been added. The built-in demos are very useful because they show the user a simple example of how to use the function. It also is helpful in developing the function because you can test the function by just typing the function name in the command line.

The first argument is the mode variable that indicates which case in the `switch` statement the function should execute. The `'initialize'` mode must always be run first. The initialization modifies the data structure which is used as the function's memory. You could also use persistent variables for the function memory. Using an output makes it easier to programmatically inspect the contents of the memory. The `'update'` is used to update the controller as new inputs arrive. The switch statement has an `'otherwise'` case to warn the user of mode errors. This throws an error stopping execution of the script. You may not always want to do this and may just use a `warning` to warn the user.

The nested function `CToDZOH` converts the continuous control system to a sampled data control systems using Equation 7.36. The name stands for "Continuous to Discrete Zero-Order Hold." A non-control expert wouldn't immediately understand the acronym, but the expanded name would be too long for a useful function name.

■ **TIP**   Make function names consistent in form and use terms that are standard for your field. Remember that not all readers of your code will be English language native speakers!

If you were building a toolbox, the CToDZOH function would likely be a separate file. For this book, it is only used by PDControl so we put it into that function file.

7.4   Simulate the Double Integrator with Digital Control

Problem

We want to simulate digital control of the double integrator model.

Solution

We will write a script that calls the control function and integrator sequentially in a loop and plots the results.

How It Works

Here are the nominal values for the control parameters we will use for the double integrator simulation.

Table 7.1: *Control Parameters*

zeta	Damping ratio	1.0
wN	Undamped natural frequency	0.1 rad/sec
wD	Derivative term filter cutoff	1.0 rad/sec
dT	Time step	0.1 sec

The simulation script is implemented in DoubleIntegratorSim.m. Note the use of cell breaks to divide the script into sections that can be run independently. The See also section lists the functions used, which will be links when the header is displayed via the command-line help.

DoubleIntegratorSim.m

```
1  %% Double Integrator Demo
2  % Demonstrate control of a double integrator.
3  %% See also
4  % PDControl, RungeKutta, RHSDoubleIntegrator, TimeLabel
9
10 %% Initialize
11 tEnd        = 100; % Simulation end time (sec)
12 dT          = 0.1; % Time step (sec)
13 aD          = 1.0; % Disturbance acceleration (m/s^2)
14 controlIsOn = false;  % True if the controller is to be used
15 x           = [0;0]; % [position;velocity]
16
17 % Controller parameters
```

```
18  d          = PDControl( 'struct' );
19  d.zeta     = 1.0;
20  d.wN       = 0.1;
21  d.wD       = 1.0;
22  d.tSamp    = dT;
23  d          = PDControl( 'initialize', d );
24
25  %% Simulation
26  nSim   = tEnd/dT+1;
27  xPlot = zeros(3,nSim);
28
29  for k = 1:nSim
30    if( controlIsOn )
31      [u, d] = PDControl('update',x(1),d);
32    else
33      u = 0;
34    end
35    xPlot(:,k)  = [x;u];
36    x           = RungeKutta( @RHSDoubleIntegrator, 0, x, dT, aD+u );
37  end
38
39  %% Plot the results
40  yL     = {'r (m)' 'v (m/s)' 'u (m/s^2)'};
41  [t,tL] = TimeLabel(dT*(0:(nSim-1)));
42
43  PlotSet( t, xPlot, 'x label', tL, 'y label', yL );
```

The first code block sets up simulation parameters that the user can change. The `control IsOn` variable is set to true if the controller is to be used. This makes it easy to test the script without the controller. When the controller is disabled, you get the "open loop" response. It is a good idea to make sure that the open loop response makes sense before testing the controller.

■ **TIP**   Put all parameters that the user can change at the beginning of the script.

The second block sets up the controller. Recall that `PDControl` has three arguments, `'struct'`, `'initialize'`, and `'update'`. The first returns the data structure required by the function. We fill the structure fields with the values selected for our problem in the lines that follow that statement. At the end of this block, we initialize the controller. This sets the controller state to zero and creates the sampled time state space matrices, which in this case are four scalars.

The third block is the simulation with the check to see if the controller is on. Note the sequential use of the control function followed by the integrator. This is discrete control because the control, `u`, is constant over the integration time step. The final block plots the results.

Figure 7.1 shows the open loop response obtained by setting the `controlIsOn` flag to `false` and executing `DoubleIntegratorSim`. The velocity increases linearly, and the position increases with the square of time as it should. The output agrees with the analytical

226

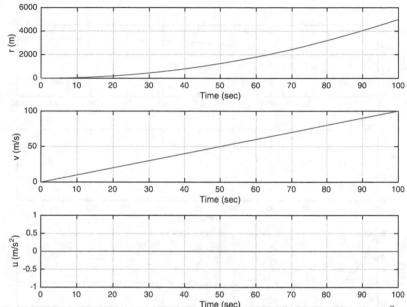

Figure 7.1: *The open loop response with a constant disturbance acceleration of 1 m/s².*

solution in Equation 7.17. Figure 7.2 shows the closed loop response, with `controlIsOn` set to `true`. The velocity goes to zero, and the position reaches a constant, though not zero. The control acceleration u exactly matches the disturbance acceleration a. We could have eliminated the position offset by using a proportional-integral differential (PID) controller.

The two examples show that the simulation works without the control and that the control performs as expected. This script is an integrated test of all of the functions listed in the script. It does a good job of testing their functionality. However, one test isn't sufficient to understand the controller. Let's make the controller underdamped by setting ζ, in the field `d.zeta`, to 0.2. Now the response oscillates; see Figure 7.3. We set `tEnd` to 300 to show that it damps.

Another thing to try is setting the bandwidth really high. Set ω_n, in field `d.wN`, to 8 and the rate filter bandwidth `d.wD` to 50. The result is shown in Figure 7.4. The controller is unstable because the bandwidth is much higher than that allowed by the sampling rate. Your bandwidth has to be less than half the sampling bandwidth which is

$$\omega_s = \frac{2\pi}{T} \tag{7.40}$$

All the results are expected behavior. The last case is a corner case that shows that the expected instability does happen. These four cases are a minimalist set of tests for this admittedly simple control system example.

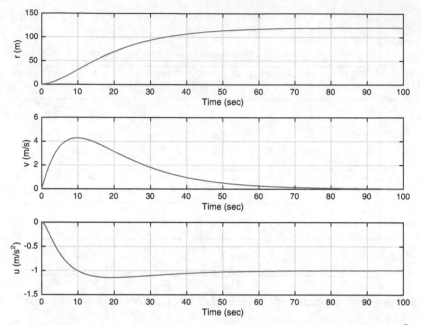

Figure 7.2: *The closed loop response with a constant disturbance acceleration of 1 m/s².*

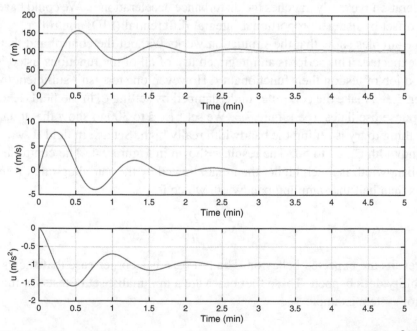

Figure 7.3: *The closed loop response with a constant disturbance acceleration of 1 m/s² and ζ equal to 0.2.*

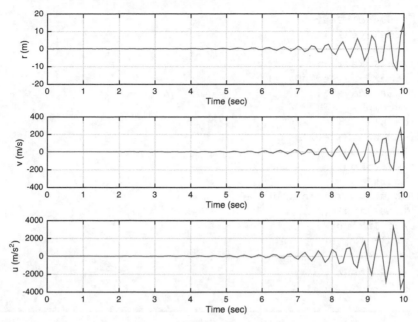

Figure 7.4: *The closed loop response using a bandwidth too high for the sampling.*

7.5 Create Time Axes with Reasonable Time Units

Problem

We want our time axes to have easy to read units, not just seconds.

Solution

We will create a function that checks the duration and converts from seconds to minutes, hours, days, or years.

How It Works

We check the maximum time in the array of times and scale it to larger time units. The time units implemented are seconds, minutes, hours, days, and years. We return the scaled time vector, the label string, and the units string as it might be useful.

TimeLabel.m

```
21  function [t, c, s] = TimeLabel( t )
22
23  secInYear    = 365.25*86400;
24  secInDay     = 86400;
25  secInHour    =  3600;
26  secInMinute  =    60;
27
28  tMax         = max(t);
29
```

```
30  if( tMax > secInYear )
31    c = 'Time (years)';
32    s = 'years';
33    t = t/secInYear;
34  elseif( tMax > 3*secInDay )
35    c = 'Time (days)';
36    t = t/secInDay;
37    s = 'days';
38  elseif( tMax > 3*secInHour )
39    c = 'Time (hours)';
40    t = t/secInHour;
41    s = 'hours';
42  elseif( tMax > 3*secInMinute )
43    c = 'Time (min)';
44    t = t/secInMinute;
45    s = 'min';
46  else
47    c = 'Time (sec)';
48    s = 'sec';
49  end
```

The rules for changing the time scale are reasonable, but you could pick any breakpoints you wished.

7.6   Create Figures with Multiple Subplots

Problem

We frequently generate figures with multiple subplots during control analysis, and this results in large blocks of repetitive code at the bottom of every script.

Solution

We make a function that can easily generate subplots with a single line.

How It Works

We will write a function that uses parameter pairs to flexibly create subplot figures in a single function call. The y input can have multiple rows, and the x input can have one row or the same number of rows as y. We supply default labels so that the function can be called with just two inputs.

The only parameters supported in this version are various labels and the plot type – standard plot, semilogx, semilogy, and loglog – but you can easily imagine adding handling for line thickness, plot markers, shading, and so on. We use a for loop to check every other component of varargin in a switch statement. Remember that varargin provides a cell array of arguments.

Note the use of a built-in demo showing both main branches of the function, for one x series and for two.

The plotting code creates subplots in the figure for each plot based on the number of rows in x and y. This function assumes the subplots are in a single column, but you could extend the logic to create multiple columns or any arrangement of subplots that suits your application. The grid is turned on.

PlotSet.m

```matlab
28  function PlotSet( x, y, varargin )
29
30  % Demo
31  %-----
32  if( nargin < 1 )
33    x = linspace(1,1000);
34    y = [sin(0.01*x);cos(0.01*x)];
35    disp('PlotSet: One x and two y rows')
36    PlotSet( x, y, 'figure title', 'PlotSet Demo' )
37        disp('PlotSet: Two x and two y rows')
38    PlotSet( [x;y(1,:)], y )
39
40    return;
41  end
42
43  % Defaults
44  nCol      = 1;
45  n         = size(x,1);
46  m         = size(y,1);
47
48  yLabel    = cell(1,m);
49  xLabel    = cell(1,n);
50  plotTitle = cell(1,n);
51  for k = 1:m
52    yLabel{k} = 'y';
53  end
54  for k = 1:n
55    xLabel{k}    = 'x';
56    plotTitle{k} = 'y vs. x';
57  end
58  figTitle = 'PlotSet';
59  plotType = 'plot';
60
61  % Handle input parameters
62  for k = 1:2:length(varargin)
63    switch lower(varargin{k} )
64      case 'x label'
65        for j = 1:n
66          xLabel{j} = varargin{k+1};
67        end
68      case 'y label'
69        temp = varargin{k+1};
```

```
70        if( ischar(temp) )
71          yLabel{1} = temp;
72        else
73          yLabel      = temp;
74        end
75      case 'plot title'
76        plotTitle{1}  = varargin{k+1};
77      case 'figure title'
78        figTitle      = varargin{k+1};
79      case 'plot type'
80        plotType      = varargin{k+1};
81      otherwise
82        fprintf(1,'%s is not an allowable parameter\n',varargin{k});
83    end
84  end
85
86  h = figure;
87  set(h,'Name',figTitle);
88
89  % First path is for just one row in x
90  if( n == 1 )
91    for k = 1:m
92      subplot(m,nCol,k);
93      plotXY(x,y(k,:),plotType);
94      xlabel(xLabel{1});
95      ylabel(yLabel{k});
96      if( k == 1 )
97        title(plotTitle{1})
98      end
99      grid on
100   end
101 else
102   for k = 1:n
103     subplot(n,nCol,k);
104     plotXY(x(k,:),y(k,:),plotType);
105     xlabel(xLabel{k});
106     ylabel(yLabel{k});
107     title(plotTitle{k})
108     grid on
109   end
110 end
112
113 %%% PlotSet>plotXY Implement different plot types
114 % log and semilog types are supported.
115 %
116 %   plotXY(x,y,type)
117 function plotXY(x,y,type)
118
119 switch type
120   case 'plot'
121     plot(x,y);
122   case {'log' 'loglog' 'log log'}
```

232

```
123      loglog(x,y);
124    case {'xlog' 'semilogx' 'x log'}
125      semilogx(x,y);
126    case {'ylog' 'semilogy' 'y log'}
127      semilogy(x,y);
128    otherwise
129      error('%s is not an available plot type',type);
130  end
```

This fairly long function results in the plotting in `DoubleIntegratorSim.m` taking place in one line.

Summary

The double integrator is a very useful model for developing control systems, as it represents an ideal version of many systems, such as a spring. In this chapter, we developed the mathematical model for the double integrator and wrote the dynamics in a *right-hand-side* function. We introduced numerical integration and wrote the `Runge-Kutta` integrator which will be used throughout the remaining applications in this book. Our recipe for the control function combines design and implementation, contains a built-in demo, and defines a data structure that is used for memory between calls. Our first demo script showed how to initialize a controller for a double integrator, simulate it, and plot the results. This is the basis for almost any mathematical or control analysis you will do in MATLAB. Table 7.2 lists the code developed in the chapter.

Table 7.2: *Chapter Code Listing*

File	Description
RHSDoubleIntegrator	Dynamical model for the double integrator.
RungeKutta	Fourth order Runge-Kutta integrator.
PDControl	Proportional-derivative controller.
DoubleIntegratorSim	Simulation of the double integrator with discrete control.
PlotSet	Create two-dimensional plots from a data set.
TimeLabel	Produce time labels and scaled time vectors.

CHAPTER 8

■ ■ ■

Robotics

The SCARA robot (Selective Compliance Articulated Robot Arm) is a simple industrial robot that can be used for placing components in a two-dimensional space. We will derive the equations of motion for a SCARA robot arm. It has two rotational joints and one prismatic joint. A prismatic joint allows only linear motion. Each joint has a single degree of freedom. A SCARA robot is broadly applicable to work where a part needs to be inserted in a two-dimensional space or where drilling needs to be done.

Our input to the robot system will be a new location for the arm effector. We have to solve two control problems. One is to determine what joint angles we need to place the arm at a particular xy coordinate. This is the inverse kinematics problem. The second is to control the two joints so that we get a smooth response to commands to move the arm. We will also develop a custom visualization function that can be used to create animations of the robot motion.

For more information on the dynamics used in this chapter, see Example 9.8.2 (p. 405) in Lung-Wen Tsai's book *Robot Analysis: The Mechanics of Serial and Parallel Manipulators*, John Wiley & Sons, New York, 1999.

8.1 Creating a Dynamical Model of the SCARA Robot

Problem

The robot has two rotational joints and one prismatic or linear "joint." We need to write the dynamics that link the forces and torques applied to the arm to its motion so that we can simulate the arm.

Solution

The equations of motion are derived using the Lagrangian formulation. We will need to solve a set of coupled linear equations in MATLAB.

© Michael Paluszek and Stephanie Thomas 2020
M. Paluszek and S. Thomas, *MATLAB Recipes*,
https://doi.org/10.1007/978-1-4842-6124-8_8

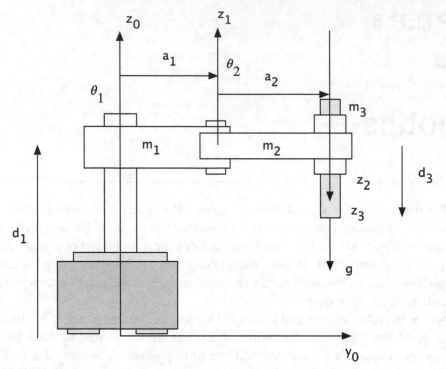

Figure 8.1: *SCARA robot. The two arms move in a plane. The plunger moves perpendicular to the plane.*

How It Works

The SCARA robot is shown in Figure 8.1. It has two arms that move in the xy plane and a plunger that moves in the z direction. The angles θ_1 and θ_2 are measured around the z_0 and z_1 axes.

The equations of motion for the SCARA robot are

$$I \begin{bmatrix} \ddot{\theta}_1 \\ \ddot{\theta}_2 \\ \ddot{d}_3 \end{bmatrix} + \begin{bmatrix} -(m_2 + 2m_3)a_1a_2 \sin\theta_2(\dot{\theta}_1\dot{\theta}_2 + \frac{1}{2}\dot{\theta}_2^2) \\ \left(\frac{1}{2}m_2 + m_3\right)a_1a_2 \sin\theta_2\dot{\theta}_1^2 \\ -m_3g \end{bmatrix} = \begin{bmatrix} T_1 \\ T_2 \\ F_3 \end{bmatrix} \tag{8.1}$$

The first term is the product of the generalized inertia matrix and the acceleration vector. The second array contains the rotational coupling terms. The final array is the control vector. The generalized inertia matrix, I, is

$$\begin{bmatrix} I_{11} & I_{21} & 0 \\ I_{21} & I_{22} & 0 \\ 0 & 0 & I_{33} \end{bmatrix} \tag{8.2}$$

where

$$I_{11} = \left(\frac{1}{3}m_1 + m_2 + m_3\right) a_1^2 + (m_2 + 2m_3) a_1 a_2 \cos\theta_2 + \left(\frac{1}{3}m_2 + m_3\right) a_2^2 \tag{8.3}$$

$$I_{21} = \left(\frac{1}{2}m_2 + m_3\right) a_1 a_2 \cos\theta_2 + \left(\frac{1}{3}m_2 + m_3\right) a_2^2 \tag{8.4}$$

$$I_{22} = \left(\frac{1}{3}m_2 + m_3\right) a_2^2 \tag{8.5}$$

$$I_{33} = m_3 \tag{8.6}$$

Note that in developing this inertia matrix, the author is treating the links as point masses and not solid bodies.

The inertia matrix is symmetric as it should be. There is coupling between the two rotational degrees of freedom but no coupling between the plunger and the rotational hinges. The inertia matrix is not constant, so it cannot be precomputed.

First, we will define a data structure for the robot, in SCARADataStructure.m, defining the length and mass of the links, and with fields for the forces and torques that can be applied. The function can supply a default structure to be filled in, or the fields can be specified a priori.

SCARADataStructure.m

```
11  %% Inputs
12  %    a1  (1,1)   Link 1 length
13  %    a2  (1,1)   Link 2 length
14  %    d1  (1,1)   Distance of link 1 from ground
15  %    m1  (1,1)   Link 1 mass
16  %    m2  (1,1)   Link 2 mass
17  %    m3  (1,1)   Link 3 mass
29  function d = SCARADataStructure( a1, a2, d1, m1, m2, m3 )
30
31  if( nargin < 1 )
32    d = struct('a1',0.1,'a2',0.1,'d1',0.05,'m1',1,'m2',1,'m3',1,'t1',0,'
        t2',0,'f3',0);
33  else
34    d = struct('a1',a1,'a2',a2,'d1',d1,'m1',m1,'m2',m2,'m3',m3,'t1',0,'t2
        ',0,'f3',0);
35  end
```

Then we write the right-hand-side (RHS) function from our equations. We need to solve for the state derivatives $\ddot{\theta}_1, \ddot{\theta}_2, \ddot{d}_3$ which we will do with a left matrix divide. This is easily done in MATLAB with a backslash, which uses a QR, triangular, LDL, Cholesky, Hessenberg, or LU solver, as appropriate for the inputs. The function does not have a built-in demo as this impacts performance in RHS functions, which are called repeatedly by integrators. Note the definition of the constant for gravity at the top of the file. The inertia matrix is returned as an additional output. This is handy for debugging.

RHSSCARA.m

```
26  function [xDot, i] = RHSSCARA( ~, x, d )
27
28  g    = 9.806; % The acceleration of gravity (m/s^2)
29
30  c2   = cos(x(2));
31  s2   = sin(x(2));
32
33  theta1Dot = x(4);
34  theta2Dot = x(5);
35
36  % Inertia matrix
37  i        = zeros(3,3);
38  a1Sq     = d.a1^2;
39  a2Sq     = d.a2^2;
40  a12      = d.a1*d.a2;
41  m23      = 0.5*d.m2 + d.m3;
42  i(1,1)   = (d.m1/3 + d.m2 + d.m3)*a1Sq  + 0.5*m23*a12*c2 + (d.m2/3 + d.
          m3)*a2Sq;
43  i(2,2)   = (d.m2/3 + d.m3)*a2Sq;
44  i(3,3)   = d.m3;
45  i(1,2)   = m23*a12*c2 + (d.m2/3 + d.m3);
46  i(2,1)   = i(1,2);
47
48  % Right hand side
49  u = [d.t1;d.t2;d.f3];
50  f = [-(d.m2 + 2*d.m3)*a12*s2*(theta1Dot*theta2Dot + 0.5*theta2Dot^2)
          ;...
51          0.5*m23*a12*s2*theta1Dot^2;...
52          -d.m3*g];
53
54  xDot = [x(4:6);i\(f-u)];
```

8.2   Customize a Visualization Function for the Robot

Problem

We would like to be able to visualize the motion of the robot arm, without relying on simple time histories or 3D lines.

Solution

We will write a function to draw a 3D SCARA robot arm. This will allow us to easily visualize the movement of the robot arm.

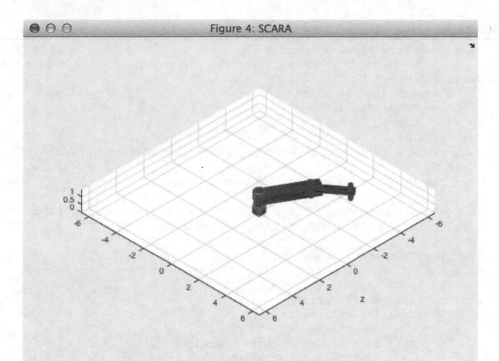

Figure 8.2: *SCARA robot visualization using* `patch`.

How It Works

This function demonstrates the use of the low-level plotting functions `patch` and `light`. We will create box and cylinder shapes for the components of the robot arm. It also demonstrates how to produce MATLAB movies of the robot motion using `getframe`. The resulting visualization is shown in Figure 8.2.

This function's first argument is an action. We define an initialization action to generate all the patch objects, which are stored in a persistent variable. Then, during the update action, we only need to update the patch vertices. The function has one output, which is movie frames from the animation. Note that the input x is vectorized, meaning we can pass a set of states to the function and not just one at a time. The header of the function is as follows.

■ **TIP**   "patch" is a computer graphics term for a part of a surface. It consists of vertices organized into faces.

DrawSCARA.m

```
1  %% DRAWSCARA Draw a SCARA robot arm.
2  %
3  %% Forms
4  %      DrawSCARA( 'initialize', d )
5  %   m = DrawSCARA( 'update', x )
6  %
7  %% Description
8  % Draws a SCARA robot using patch objects. A persistent variable is
        used to
9  % store the graphics handles in between update calls.
10 %
11 % The SCARA acronym stands for Selective Compliance Assembly Robot Arm
12 % or Selective Compliance Articulated Robot Arm.
13 %
14 % Type DrawSCARA for a demo.
15 %
16 %% Inputs
17 %   action   (1,:)  Action string
18 %   x        (3,:)  [theta1;theta2;d3]
19 %     or
20 %   d          (.)  Data structure for dimensions
21 %                     .a1 (1,1) Link arm 1 joint to joint
22 %                     .a2 (1,1) Link arm 2 joint to joint
23 %                     .d1 (1,1) Height of link 1 and link2
24 %
25 %% Outputs
26 %   m         (1,:) If there is an output it makes a movie using getframe
```

Next, we show the body of the main function. Note that the function has a built-in demo demonstrating a vector input with 100 states. The function will initialize itself with default data if the data structure d is omitted.

```
32  function m = DrawSCARA( action, x )
33
34  persistent p
35
36  % Demo
37  %-----
38  if( nargin < 1 )
39    disp('Demo of DrawSCARA using the default data:');
40    DrawSCARA( 'initialize' );
41    t        = linspace(0,100);
42    omega1   = 0.1;
43    omega2   = 0.2;
44    omega3   = 0.3;
45    x        = [sin(omega1*t);sin(omega2*t);0.01*sin(omega3*t)];
46    m        = DrawSCARA( 'update', x );
47    snapnow;  % publishing
48    if( nargout < 1 )
```

240

```
49     clear m;
50   end
51   return
52 end
53
54 switch( lower(action) )
55   case 'initialize'
56     if( nargin < 2 )
57       d = SCARADataStructure;
58     else
59       d = x;
60     end
61
62     p = Initialize( d );
63
64   case 'update'
65     if( nargout == 1 )
66       m = Update( p, x );
67     else
68       Update( p, x );
69     end
70 end
```

Note that we used subfunctions for the `Initialize` and `Update` actions. This keeps the switch statement clean and easy to read. In the `Initialize` function, we define additional parameters for creating the box and cylinder objects we use to visualize the entire robot arm. Then we use `patch` to create the graphics objects, using parameter pairs instead of the (x, y, z) input. Specifying the unique vertices this way can reduce the size of the data needed to define the patch and is simple conceptually. See the *Introduction to Patch Objects* and *Specifying Patch Object Shapes* sections of the MATLAB help for more information. The handles are stored in the persistent variable p. We only show the creation of the base. The `patch` function creates the 3D object from vertices and faces. Phong lighting is defined.

```
87 function p = Initialize( d )
88
89 p.fig   = figure( 'name','SCARA' );
90
91 % Create parts
92 c = [0.5 0.5 0.5]; % Color
93 r = [1.0 0.0 0.0];
94
95 % Store for use in updating
96 p.a1 = d.a1;
97 p.a2 = d.a2;
98
99 % Physical parameters for drawing
100 d.b  = [1 1 1]*d.d1/2;
101 d.l1 = [0.12 0.02 0.02 0.005 0.03]*10*d.a1;
102 d.l2 = [0.12 0.02 0.01]*10*d.a2;
```

```
103   d.c1   = [0.1 0.4]*d.a1;
104   d.c2   = [0.06 0.3]*d.a1;
105   d.c3   = [0.06 0.5]*d.a2;
106   d.c4   = [0.05 0.6]*d.a2;
107
108   % Base
109   [vB, fB] = Box( d.b(1), d.b(2), d.b(3) );
110   [vC, fC] = Cylinder( d.c1(1), d.c1(2) );
111   f        = [fB;fC+size(vB,1)];
112   vB(:,3)  = vB(:,3) + d.b(3)/2;
113   vC(:,3)  = vC(:,3) + d.b(3);
114   v        = [vB;vC];
115   p.base   = patch('vertices', v, 'faces', f,...
116                    'facecolor', c, 'edgecolor', c,...
117                    'facelighting', 'phong' );
```

The Update function updates the vertices for each patch object. The nominal vertices are stored in the persistent variable p and are rotated using a transformation matrix calculated from the sine and cosine of the link angles. If there is an output argument, the function uses getframe to grab the figure as a movie frame.

```
181   function m = Update( p, x )
182
183   if( nargout > 0 )
184     % Allocate movie frame array
185     n = getframe(p.fig);
186     m(1:size(x,2)) = n;
187   end
188
189   for k = 1:size(x,2)
190
191     % Link 1
192     c       = cos(x(1,k));
193     s       = sin(x(1,k));
194     b1      = [c -s 0;s c 0;0 0 1];
195     v       = (b1*p.v1')';
196     set(p.link1,'vertices',v);
197
198     % Link 2
199     r2      = b1*[p.a1;0;0];
200     c       = cos(x(2,k));
201     s       = sin(x(2,k));
202     b2      = [c -s 0;s c 0;0 0 1];
203     v       = (b2*b1*p.v2')';
204     v(:,1)  = v(:,1) + r2(1);
205     v(:,2)  = v(:,2) + r2(2);
206     set(p.link2,'vertices',v);
207
208     % Link 3
209     r3      = b2*b1*[p.r3;0;0] + r2;
210     v       = p.v3;
```

```
211    v(:,1)   = v(:,1) + r3(1);
212    v(:,2)   = v(:,2) + r3(2);
213    v(:,3)   = v(:,3) + x(3,k);
214    set(p.link3,'vertices',v);
215
216    if( nargout > 0 )
217      m(k) = getframe(p.fig);
218    else
219      drawnow;
220    end
221
222  end
```

The subfunctions Box, Cylinder, and UChannel create the vertices and faces for each type of 3D object. Faces are defined using indices of the vertices, in this case, triangles. We will show the Box function here to demonstrate how the vertices and faces matrices are created.

```
240  function [v, f] = Box( x, y, z )
241
242  f    = [2 3 6;3 7 6;3 4 8;3 8 7;4 5 8;4 1 5;2 6 5;2 5 1;1 3 2;1 4 3;5 6
         7;5 7 8];
243  x    = x/2;
244  y    = y/2;
245  z    = z/2;
246
247  v = [-x  x   x  -x  -x   x   x  -x;...
248       -y -y   y   y  -y  -y   y   y;...
249       -z -z  -z  -z   z   z   z   z]';
```

8.3   Using Numerical Search for Robot Inverse Kinematics

Problem

The goal of the robot controller is to place the end effector at a desired position. We need to know the link states corresponding to this position.

Solution

We will use a numerical solver to compute the robot states. The MATLAB solver is fminsearch, which implements a Nelder-Mead minimizer.

■ **TIP**   Nelder-Mead is also known as downhill simplex. This is not to be confused with the simplex algorithm.

How It Works

The goal of our control system is to position the end effector as close as possible to the desired position $[x; y; z]$. z is determined by $d_1 - d_3$ from Figure 8.1 in Recipe 8.1. x and y are found from the two angles a_1 and a_2. The position vector for the arm end effector is

$$\begin{bmatrix} x \\ y \\ z \end{bmatrix} = \begin{bmatrix} a_1 \sin\theta_1 + a_2 \sin(\theta_1 + \theta_2) \\ a_1 \cos\theta_1 + a_2 \cos(\theta_1 + \theta_2) \\ d_1 - d_3 \end{bmatrix} \tag{8.7}$$

While these equations don't seem complicated, they can't be used to solve for x and y directly. First of all, if a_2 is less than a_1, there will be a region around the origin that cannot be reached. In addition, there may be more than one solution for each x, y pair.

We will use this equation for the position to create a cost function that can be passed to `fminsearch`. This will compute the state which results in the desired position. The resulting function demonstrates a nested cost function, a built-in demo, and a default plot output. Note that we use a data structure as returned by `optimset` to pass parameters to `fminsearch`.

SCARAIK.m

```
1   %% SCARAIK Generate SCARA states for desired end effector position and
        angle.
2   %
3   %% Form
4   %   x = SCARAIK( r, d )
5   %
6   %% Description
7   % SCARA inverse kinematics. Uses fminsearch to find the link states
        given the
8   % effector location. The cost function is embedded. Type SCARAIK for a
        demo
9   % which creates a plot and a video.
10  %
11  %% Inputs
12  %   r             (3,:) End effector position [x;y;z]
13  %   d             (.)   Robot data structure
14  %                       .a1 (1,1) Link 1 length
15  %                       .a2 (1,1) Link 2 length
16  %                       .d1 (1,1) Distance of link 1 from ground
17  %
18  %% Outputs
19  %   x             (3,:) SCARA states [theta1;theta2;d3]
26
27  if( nargin < 1 )
28    disp('Demo of SCARAIK...');
29    r = [linspace(0,0.2);zeros(2,100)];
30    d = SCARADataStructure;
31
32    SCARAIK( r, d );
33    return;
```

```
34   end
35
36   n   = size(r,2);
37   xY = zeros(2,n);
38
39   TolX        = 1e-5;
40   TolFun      = 1e-9;
41   MaxFunEvals = 1500;
42   Options = optimset('TolX',TolX,'TolFun',TolFun,'MaxFunEvals',
         MaxFunEvals);
43
44   x0 = [0;0];
45   for k = 1:n
46     d.xT      = r(1:2,k);
47     xY(:,k) = fminsearch(@Cost, x0, Options, d );
48     x0        = xY(:,k);
49   end
50
51   x = [xY;d.d1-r(3,:)];
52
53   % Default output is to create a plot
54   %-----------------------------------
55   if( nargout == 0 )
56     DrawSCARA( 'initialize', d );
57     m = DrawSCARA( 'update', x );
58     disp('Saving movie...')
59     vidObj = VideoWriter('SCARAIK.avi');
60     open(vidObj);
61     writeVideo(vidObj,m);
62   end
63
64   end
65
66   %%% SCARAIK>Cost
67   % Cost function. The cost is the difference between the position as
         computed from the
68   % states and the target position xT in d.
69   %
70   %   y = Cost( x, d )
71   function y = Cost( x, d )
72
73   xE = d.a1*cos(x(1)) + d.a2*cos(x(1)+x(2));
74   yE = d.a1*sin(x(1)) + d.a2*sin(x(1)+x(2));
75   y  = sqrt((xE-d.xT(1))^2+(yE-d.xT(2))^2);
76
77   end
```

The function creates a video using a VideoWriter object and the frame data returned by DrawSCARA. Before VideoWriter was introduced, this could be done with movie2avi.

8.4   Developing a Control System for the Robot

Problem

Robot arm control is a critical technology in robotics. We need to be able to smoothly and reliably change the location of the end effector. The speed of the control will determine how many operations we can do in a given amount of time, thus determining the productivity of the robot.

Solution

We will solve this problem using the inverse kinematics function discussed earlier and then feeding the desired angles into two PD controllers as developed in the Chapter 7.

How It Works

We apply our PD controller described in the Chapter 7 using the c array as the desired angle and position vector. We will compute control accelerations, not torques, and then multiply by the inertia matrix to get control torques:

$$T = Ia \tag{8.8}$$

where T is the control torque, I is the inertia matrix, and a is the computed control acceleration. We need to do this because there are cross-coupling terms in the inertia matrix and I_{11} changes as the position of the outer arm changes. We are neglecting the nonlinear terms in the equations of motion. These terms are functions of the angles and the angular rates. If we move slowly, this should not pose a problem. If we move quickly, we could feedforward the nonlinear torques and cancel them.

The first step is to specify a desired position for the end effector and use the inverse kinematics function to compute the target states corresponding to this location.

SCARARobotSim.m

```
32  % Pick the location to place the end effector, [x;y;z]
33  r = [4;2;0];
34
35  % Find the two angles for the joints
36  setPoint = SCARAIK( r, d );
```

Next is the code that designs the controllers, one for each joint, using `PDControl`. Note that we use identical parameters for both controllers. We set the damping ratio, `zeta`, to 1.0 to avoid overshoot. Recall that wN, the undamped natural frequency, is the bandwidth of the controller; the higher this frequency, the faster the response. wD, the derivative term filter cutoff, is set to 5–10 times wN so that the filter doesn't cause lag below wN. The dT variable is the time step of the simulation.

```
38  %% Control Design
39  % We will use two PD controllers, one for each rotational joint.
```

```
40
41  % Controller parameters
42  dC1          = PDControl( 'struct' );
43  dC1.zeta     = 1.0;
44  dC1.wN       = 0.6;
45  dC1.wD       = 60.0;
46  dC1.tSamp    = dT;
47  dC2          = dC1;
48
49  % Create the two controllers
50  dC1          = PDControl( 'initialize', dC1 );
51  dC2          = PDControl( 'initialize', dC2 );
```

This is the portion that computes and applies the control. We eliminate the inertia coupling by computing joint accelerations and multiplying by the inertia matrix, which is computed each time step, to get the desired control torques. We use the feature of the RHS that computes the inertia matrix from the current state.

```
69  [acc(1,1), dC1] = PDControl('update',thetaError(1),dC1);
70  [acc(2,1), dC2] = PDControl('update',thetaError(2),dC2);
71  torque          = inertia(1:2,1:2)*acc;
```

We can run these lines at the command line to see what the acceleration and torque magnitude look like for an example robot. Assuming Meters-Kilogram-Second (MKS) units, we have links of 1 meter in length and masses of 1 kg.

```
>> dC1          = PDControl( 'struct' );
>> dC1.zeta     = 1.0;
>> dC1.wN       = 0.6;
>> dC1.wD       = 60.0;
>> dC1.tSamp    = 0.025;
>> dC2          = dC1;
>> dC1          = PDControl( 'initialize', dC1 );
>> dC2          = PDControl( 'initialize', dC2 );
>> d = SCARADataStructure(1,1,1,1,1,1);
>> x = zeros(6,1);
>> [~,inertia] = RHSSCARA( 0, x, d );
>> inertia
inertia =

    4.4167          2.8333          0
    2.8333          1.3333          0
         0               0          1

>> thetaError = [0.1;0.1];
>> [acc(1,1), dC1] = PDControl('update',thetaError(1),dC1);
>> [acc(2,1), dC2] = PDControl('update',thetaError(2),dC2);
>> acc
acc =
```

```
      -7.236
      -7.236

>> torque  = inertia(1:2,1:2)*acc
torque =

      -52.461
      -30.15
```

8.5   Simulating the Controlled Robot

Problem

We want to test our robot arm under control. Our input will be the desired xy coordinates of the end effector.

Solution

The solution is to build a MATLAB script in which we design the PD controller matrices as before and then simulate the controller in a loop, applying the calculated torques until the states match the desired angles. We will not control the vertical position of the end effector, leaving this as an exercise for the reader.

How It Works

This is a discrete simulation, with a fixed time step and the control torque calculated separately from the dynamics. The simulation runs in a loop calling first the controller code from Recipe 8.4 and the right-hand side from the fourth-order Runge-Kutta function. When the simulation ends, the angles and angle errors are plotted and a 3D animation is displayed. We could plot more variables, but all the essential information is in the angles and errors.

With a very small time step of 0.025 seconds, we could have increased the bandwidth of the controller to speed the response. Remember that the cutoff frequency of the filter must also be below the Nyquist frequency.

Notice that we do not handle large angle errors, that is, errors greater than 2π. In addition, if the desired angle is $2\pi - \epsilon$ and the current position is $2\pi + \epsilon$, it will not necessarily go the shortest way. This can be handled by adding code that computes the smallest error between two points on the unit circle. The reader can add code for this to make the controller more robust.

The script is as follows, skipping the control design lines from Recipe 8.4. First, we initialize the simulation data including the time parameters and the robot geometry. We initialize our plotting arrays using `zeros` before entering the simulation loop. There is a control flag which allows the simulation to be run open loop or closed loop. The integration occurs in the last line of the loop.

```
16  %% Initialize
17  % Specify the time, robot geometry and the control target.
18
19  % Simulation time settings
20  tEnd        = 20.0;      % sec
21  dT          = 0.025;
22  nSim        = tEnd/dT+1;
23  controlIsOn = true;
24
25  % Robot parameters
26  d = SCARADataStructure(3,2,1,4,6,1);
27
28  % Set the initial arm states
29  x0      = zeros(6,1);
30  %x0(5) = 0.05;
31
32  % Pick the location to place the end effector, [x;y;z]
33  r = [4;2;0];
34
35  % Find the two angles for the joints
36  setPoint = SCARAIK( r, d );
52
53  %% Simulation
54  % The simulation can be run with or without control, i.e. closed or
        open
55  % loop.
56  x        = x0;
57  xPlot    = zeros(4,nSim);
58  tqPlot   = zeros(2,nSim);
59  inrPlot  = zeros(2,nSim);
60
61  for k = 1:nSim
62    % Control error
63    thetaError   = setPoint(1:2) - x(1:2);
64    [~,inertia] = RHSSCARA( 0, x, d );
65    acc          = zeros(2,1);
66
67    % Apply the control
68    if( controlIsOn )
69      [acc(1,1), dC1] = PDControl('update',thetaError(1),dC1);
70      [acc(2,1), dC2] = PDControl('update',thetaError(2),dC2);
71      torque            = inertia(1:2,1:2)*acc;
72    else
73      torque = zeros(2,1);
74    end
75    d.t1 = torque(1);
76    d.t2 = torque(2);
77
78    % Plotting array
79    xPlot(:,k)   = [x(1:2);thetaError];
80    tqPlot(:,k)  = torque;
```

```
81    inrPlot(:,k) = [inertia(1,1);inertia(2,2)];
82
83    % Enter the motor torques into the dynamics model
84    x = RungeKutta( @RHSSCARA, 0, x, dT, d );
85  end
86
87  %% Plot the results
88  % Plot a time history and perform an animation.
89
90  % Plot labels
91  yL = {'\theta_1 (rad)' '\theta_2 (rad)' 'Error \theta_1 (rad)' 'Error \
        theta_2 (rad)'};
92
93  % Time histories
94  [t,tL] = TimeLabel(dT*(0:(nSim-1)));
95  PlotSet( t, xPlot, 'y label', yL, 'x label', tL );
96  PlotSet( t, tqPlot, 'y label', {'T_x','T_y'}, 'x label', tL );
97  PlotSet( t, inrPlot, 'y label', {'I_{11}','I_{22}'}, 'x label', tL );
98
99  % Animation
100 DrawSCARA( 'initialize', d );
```

Figure 8.3 shows the transient response of the two joints. Both converge to their set points but look different than the double integrator response that we saw in the previous chapter. This system is nonlinear due to the coupling between the links. For instance, in a double integrator, we would expect no overshoot of the target angle for a damping ratio of 1.0. However, we do see some in the second subplot of θ_2, and otherwise the shape is similar to a double integrator

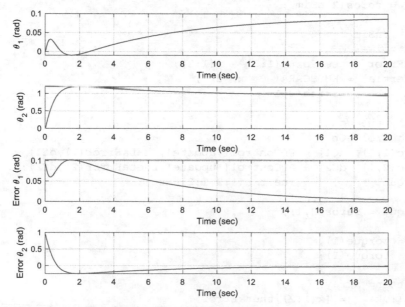

Figure 8.3: *SCARA robot angles showing the transient response.*

250

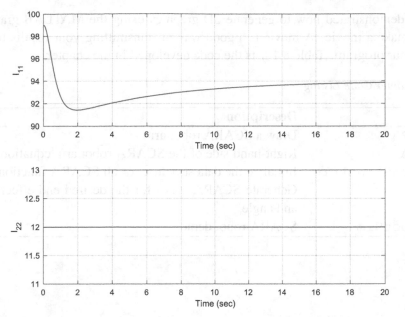

Figure 8.4: *SCARA robot inertia as the arm moves.*

response. We see that θ_1, in contrast, reverses direction as it reacts to the motion of the outer joint; after about 2 seconds when the θ_2 has peaked, θ_1 also resembles a double integrator. Keep in mind that the two controllers are independent and are, in some ways, working at cross-purposes.

Figure 8.4 shows the resulting inertia components. We expected I_{11} to change and I_{22} to remain constant, which is in fact the case.

After the simulation is done, the script runs an animation of the arm motion. Both 2D plots and the 3D animation are needed to debug the controller and for production runs.

The same script could be extended to show a sequence of commands to the arm.

Summary

This chapter has demonstrated how to write the dynamics and implement a simple control law for a two-link manipulator, the SCARA robot. We have implemented coupled nonlinear equations in the right-hand side with the simple controller developed in the previous chapter. The format of the simulation is very similar to the double integrator. There are more sophisticated ways of performing this control that would take into account the coupling between the links, which can be added to this framework. We have not implemented any constraints on the motion or the control torque.

We also demonstrated how to generate 3D graphics using the MATLAB graphics engine and how to make a movie. A movie is a good way of transmitting your results to people and debugging your program. Table 8.1 lists the code developed in the chapter.

Table 8.1: *Chapter Code Listing*

File	Description
DrawSCARA	Draw a SCARA robot arm
RHSSCARA	Right-hand side of the SCARA robot arm equations
SCARADataStructure	Initialize the data structure for all SCARA functions
SCARAIK	Generate SCARA states for the desired end effector position and angle
SCARARobotSim	SCARA robot demo

CHAPTER 9

■ ■ ■

Electric Motors

We will model a three-phase permanent magnet motor driven by a direct current (DC) power source. This has three coils on the stator and permanent magnets on the rotor. This type of motor is driven by a DC power source with six semiconductor switches that are connected to the three coils, known as the A, B, and C coils. Two or more coils can be used to drive a brushless DC motor, but three coils are particularly easy to implement. This type of motor is used in many industrial applications today, including electric cars and robotics. It is sometimes called a brushless DC motor (BLDC) or a permanent magnet synchronous motor (PMSM).

Pulsewidth modulation is used for the switching because it is efficient; the switches are off when not needed. Coding the model for the motor and the pulsewidth modulation is relatively straightforward. In the simulation, we will demonstrate using two time steps, one for the simulation to handle the pulsewidths and one for the outer control loop. The simulation script will have multiple control flags to allow for debugging this complex system.

Figure 9.1 shows the big picture in this chapter. We will look at the motor model first (the AC motor block), then at the pulsewidth modulation (the SVPWM and three-phase inverter block). The controller is covered last. Each of these major functions is in a separated gray block.

9.1 Modeling a Three-Phase Brushless Permanent Magnet Motor

Problem

We want to model a three-phase permanent magnet synchronous motor in a form suitable for control system design. A conceptual drawing is shown in Figure 9.2. The motor has three stator windings and one permanent magnet on the rotor. The magnet has two poles or one pole pair. The coordinate axes are the a, b, and c on the stator, one axis at the center of each coil following the right-hand rule, and the (d,q) coordinates fixed to the magnet in the rotating frame. In motor applications, the axes represent currents or voltages, not positions like in mechanical engineering.

© Michael Paluszek and Stephanie Thomas 2020
M. Paluszek and S. Thomas, *MATLAB Recipes*,
https://doi.org/10.1007/978-1-4842-6124-8_9

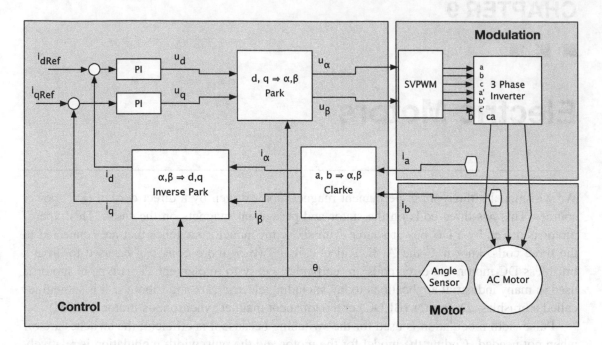

Figure 9.1: *Motor controller. PI is the proportional-integral controller. PWM the is pulsewidth modulation. There are two current sensors measuring i_a and i_b and one angle sensor measuring θ.*

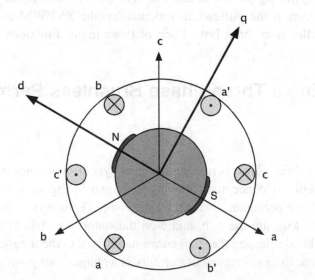

Figure 9.2: *Motor diagram showing the three-phase coils a, b, and c on the stator and the two-pole magnet (N,S) on the rotor. The × means the current is going into the paper; the dot means it is coming out of the paper.*

Solution

The solution is to model a motor with three stator coils and permanent magnets on the rotor. We have to model the coil currents and the physical state of the rotor.

How It Works

Permanent magnet synchronous motors use two or more windings in the stator and permanent magnets in the rotor. The rotor can have any even number of magnet poles. The phasing of the currents in the stator coils must be synchronized with the position of the rotor. Define the inductance matrix L, which gives the coupling between currents in different loops[1]:

$$
L = \frac{1}{d}
\begin{bmatrix}
2L_{ss} - L_m & L_m & L_m \\
L_m & 2L_{ss} - L_m & L_m \\
L_m & L_m & 2L_{ss} - L_m
\end{bmatrix}
\tag{9.1}
$$

where

$$
d = 2L_{ss}^2 - L_{ss}L_m - L_m^2
\tag{9.2}
$$

L_m is the mutual inductance of the phase windings and L_{ss} is the self-inductance. Self-inductance is the effect of a current in a loop on itself. Mutual inductance is the effect of the current in one loop on another loop. The three-phase current array, i, is

$$
i =
\begin{bmatrix}
i_a \\
i_b \\
i_c
\end{bmatrix}
\tag{9.3}
$$

where i_a is the phase A stator winding current, i_b is the phase B current, and i_c is the phase C current.

The phase voltage vector, u, is

$$
u =
\begin{bmatrix}
u_a \\
u_b \\
u_c
\end{bmatrix}
\tag{9.4}
$$

where u_a is the phase A stator winding voltage. The dynamical equations are a set of first-order differential equations and are

$$
\begin{bmatrix}
\dot{i} \\
\dot{\omega}_e \\
\dot{\theta}_e
\end{bmatrix}
=
\begin{bmatrix}
-r_s L & 0 & 0 \\
0 & -\frac{b}{J} & 0 \\
0 & 1 & 0
\end{bmatrix}
\begin{bmatrix}
i \\
\omega_e \\
\theta_e
\end{bmatrix}
+ \psi
\begin{bmatrix}
-L\omega_e \\
\frac{p^2 i^T}{4J} \\
0\ 0\ 0
\end{bmatrix}
\begin{bmatrix}
\cos\theta_e \\
\cos(\theta_e + \frac{2\pi}{3}) \\
\cos(\theta_e - \frac{2\pi}{3})
\end{bmatrix}
+
\begin{bmatrix}
L \\
0\ 0\ 0 \\
0\ 0\ 0
\end{bmatrix}
u
$$

$$
+
\begin{bmatrix}
0 \\
\frac{p}{2J} \\
0
\end{bmatrix}
T_L
\tag{9.5}
$$

---

[1]Lyshevski, S. E. "Electromechanical Systems, Electric Machines, and Applied Mechatronics," CRC Press, 2000, pp. 589-627.

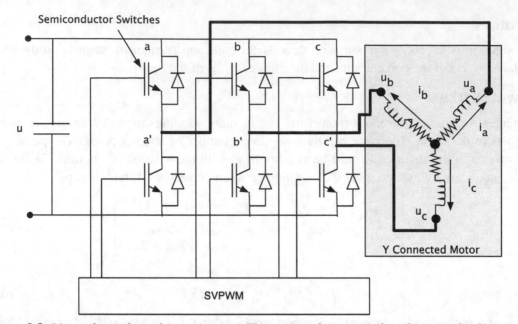

Figure 9.3: *Motor three-phase driver circuitry. The semiconductor switches shown in the diagram are IGBT (integrated gate bipolar transistors). The pulsewidth modulation block, SVPWM, is discussed in Recipe 9.3.*

where ω is the rotor angular rate, θ is the rotor angle, p is the number of rotor poles, b is the viscous damping coefficient, r_s is the stator resistance, ψ is the magnetic flux, T_L is the load torque, and J is the rotor inertia. i and u are the phase winding 3-vectors shown earlier, and L is the 3-by-3 inductance matrix also shown earlier. Equation 9.5 is actually five equations in matrix form. The first three equations, for the current array i, are the electrical dynamics. The last two for ω_e and θ_e are the mechanical dynamics represented in electrical coordinates.

The driver circuitry is shown in Figure 9.3. It has six semiconductor switches. In this model, they are considered ideal, meaning they can switch instantaneously at any frequency we desire. In practice, switches will have a maximum switching speed and will have some transient response. Note that the motor is Y connected, meaning that the ends of the three-phase windings are tied together.

The right-hand-side code is shown in the following. The first output is the state derivative, as needed for integration. The second output is the electrical torque needed for the control. The first block of code defines the motor model data structure with the parameters needed by our dynamics equation. This structure can be retrieved by calling the function with no inputs. The remaining code implements Equation 9.5. Note the suffix M used for ω and θ, to reinforce that these are mechanical quantities; this distinguishes them from the electrical quantities which are related by $p/2$, where p is the number of poles. The use of M and E subscripts is typical when writing software for motors.

The function returns a default data structure if no input arguments are passed to it. This is a convenient way for the designer of the code to give users a working starting point for the model.

This way, the user only has to change parameters that are different from the default. It lets the user get up and running quickly.

The electrical torque is a second output argument. It is not used during numerical integration but is helpful when debugging the function. It is useful to output quantities that a user might want to plot too. MATLAB is helpful in allowing multiple outputs for a function.

RHSPMMachine.m

```
1   %% RHSPMMACHINE Permanent magnet machine model in ABC coordinates.
2   % Assumes a 3 phase machine in a Y connection. The permanent magnet
3   % flux
4   % distribution is assumed sinusoidal.
5   %% Forms
6   %   d = RHSPMMachine
7   %   [xDot,tE] = RHSPMMachine( ~, x, d )
8   %
9   %% Inputs
10  %   t    (1,1)    Time, unused
11  %   x    (5,1)    The state vector [iA;iB;iC;omegaE;thetaE]
12  %   d    (.)      Data structure
13  %                       .lM    (1,1) Mutual inductance
14  %                       .psiM  (1,1) Permanent magnet flux
15  %                       .lSS   (1,1) Stator self inductance
16  %                       .rS    (1,1) Stator resistance
17  %                       .p     (1,1) Number of poles (1/2 pole pairs)
18  %                       .u     (3,1) [uA;uB;uC]
19  %                       .tL    (1,1) Load torque
20  %                       .bM    (1,1) Viscous damping (Nm/rad/s)
21  %                       .j     (1,1) Inertia
22  %                       .u     (3,1) Phase voltages [uA;uB;uC]
23  %
24  %% Outputs
25  %   x    (5,1)    The state vector derivative
26  %   tE   (1,1)    Electrical torque
27  %
28  %% Reference
29  % Lyshevski, S. E., "Electromechanical Systems, Electric Machines, and
30  % Applied Mechatronics," CRC Press, 2000.
35
36  function [xDot, tE] = RHSPMMachine( ~, x, d )
37
38  if( nargin == 0 )
39    xDot = struct('lM',0.0009,'psiM',0.069, 'lSS',0.0011,'rS',0.5,'p'
          ,2,...
40                  'bM',0.000015,'j',0.000017,'tL',0,'u',[0;0;0]);
41    return
42  end
43
44  % Pole pairs
45  pP = d.p/2;
46
```

257

```
47  % States
48  i        = x(1:3);
49  omegaE = x(4);
50  thetaE = x(5);
51
52  % Inductance matrix
53  denom = 2*d.lSS^2 - d.lSS*d.lM - d.lM^2;
54  l2      = d.lM;
55  l1      = 2*d.lSS - l2;
56  l       = [l1 l2 l2;l2 l1 l2;l2 l2 l1]/denom;
57
58  % Right hand side
59  tP3      = 2*pi/3;
60  c        = cos(thetaE + [0;-tP3;tP3]);
61  iDot     = l*(d.u - d.psiM*omegaE*c - d.rS*i);
62  tE       = pP^2*d.psiM*i'*c;
63  omegaDot = (tE - d.bM*omegaE - 0.5*pP*d.tL)/d.j;
64  xDot     = [iDot;omegaDot;omegaE];
```

9.2   Controlling the Motor

Problem

We want to control the motor to produce a desired torque. Specifically, we need to compute the voltages to apply to the stator coils.

Solution

We will use *field-oriented control* with a proportional-integral controller to control the motor. Field-oriented control is a control method where the stator currents are transformed into two orthogonal components. One component defines the magnetic flux of the motor and the other defines the torque. The control voltages we calculate will be implemented using pulsewidth modulation of the semiconductor switches as developed in the previous recipe. Torque control is only one type of motor control. Speed control is often the goal. Robots often have position control as the goal. One could use torque control as an inner loop for either a speed controller or position controller.

How It Works

The motor controller is shown in Figure 9.1. This implements field-oriented control (FOC). FOC effectively turns the brushless three-phase motor into a commutated DC motor.

There are three electrical frames of reference in this problem. The first is the (a,b,c) frame which is the frame of the three-phase stator as in Figure 9.2. This is a time-varying frame. We next want to transform into a two-axis time-varying frame, the (α,β) frame, and then into a two-axis time-invariant frame, the (d, q) frame, which is also known as the direct-quadrature axes and is fixed to the permanent magnet. In our frames, each axis is a current. Since with a

Y-connected motor the sum of the currents is zero:

$$0 = i_a + i_b + i_c \tag{9.6}$$

we need only work with two currents, i_a and i_b.

The (d, q) to (α, β) transformation is known as the Forward Park transformation:

$$\begin{bmatrix} u_\alpha \\ u_\beta \end{bmatrix} = \begin{bmatrix} \cos\theta_e & -\sin\theta_e \\ \sin\theta_e & \cos\theta_e \end{bmatrix} \begin{bmatrix} u_d \\ u_q \end{bmatrix} \tag{9.7}$$

This transforms from the stationary d, q frame to the rotating (α, β) frame. θ_e is in electrical axes and equals $\frac{1}{2}p\,\theta_M$ where p is the number of magnet poles. The Forward Clarke transformation for a Y-connected motor is

$$\begin{bmatrix} u_\alpha \\ u_\beta \end{bmatrix} = \begin{bmatrix} 1 & 0 \\ \frac{1}{\sqrt{3}} & \frac{2}{\sqrt{3}} \end{bmatrix} \begin{bmatrix} u_a \\ u_b \end{bmatrix} \tag{9.8}$$

These two transformations are implemented in the functions `ClarkeTransformationMatrix` and `ParkTransformationMatrix`. They allow us to go from the time-varying (a,b,c) frame to the time-invariant, but rotating, (d,q) frame.

The equations for a general permanent magnet machine in the direct-quadrature frame are

$$u_q = r_s i_q + \omega_e(L_d i_d + \psi) + \frac{dL_q i_q}{dt} \tag{9.9}$$

$$u_d = r_s i_d - \omega_e L_q i_q + \frac{d(L_d i_d + \psi)}{dt} \tag{9.10}$$

where u are the voltages, i are the currents, r_s is the stator resistance, L_q and L_d are the d and q phase inductances, ω_e is the electrical angular rate, and ψ is the flux due to the permanent magnets. The electrical torque produced is

$$T_e = \frac{3}{2}p((L_d i_d + \psi)i_q - L_q i_q i_d) \tag{9.11}$$

where p is the number of pole pairs.

The torque equation is

$$T_e = T_L + b\omega_m + J\frac{d\omega_m}{dt} \tag{9.12}$$

where b is the mechanical damping coefficient, T_L is the external load torque, and J is the inertia, and the relationship between the mechanical and the electrical angular rate is

$$\omega_e = p\omega_m \tag{9.13}$$

The more pole pairs you have, the higher the electrical frequency. In a magnet surface mount machine with coils in slots, $L_d = L_q \equiv L$, and ψ and the inductances are not functions of time. The equations simplify to

$$u_q = r_s i_q + \omega_e L i_d + \omega_e \psi + L\frac{di_q}{dt} \tag{9.14}$$

$$u_d = r_s i_d - \omega_e L i_q + L\frac{di_d}{dt} \tag{9.15}$$

We control direct current i_d to zero. If i_d is zero, control is linear in i_q. The torque is now

$$T_e = \frac{3}{2}p\psi i_q \tag{9.16}$$

Thus, the torque is a function of the quadrature current i_q only. We can therefore control the electrical torque by controlling the quadrature current. The quadrature current is in turn controlled by the direct and quadrature phase voltages. The desired current i_q^s can now be computed from the torque set point T_e^s.

$$i_q^s = \frac{2}{3}T_e^s/(p\psi) \tag{9.17}$$

We will use a proportional-integral controller to compute the (d,q) voltages. The proportional part of the control drives errors to zero. However, if there is a steady disturbance, there will be an offset. The integral part can drive an error due to such a steady disturbance to zero. Without the integral term, a steady disturbance will result in a steady error. A proportional-integral controller is of the form

$$u = K\left(1 + \frac{1}{\tau}\int\right)y \tag{9.18}$$

where u is the control, y is the measurement, τ is the integrator time constant, and K is the forward (proportional) gain. Our control u will be the phase voltages, and our measurement y is the current error in the (d,q) frame.

$$u_{(d,q)} = -k_F\left(i_{err} + \frac{1}{\tau}\int i_{err}\right) \tag{9.19}$$

where

$$\begin{bmatrix} i_d \\ i_q \end{bmatrix}_{err} = \begin{bmatrix} i_d \\ i_q \end{bmatrix} - \begin{bmatrix} 0 \\ i_q^s \end{bmatrix} \tag{9.20}$$

We now write a function, TorqueControl, that calculates the control voltages $u_{(\alpha,\beta)}$ given the current state x. The state vector is the same as Recipe 9.1, that is, current i in the (a,b,c) frame plus the angle states θ and ω. We use the Park and Clarke transformations to compute the current in the (d,q) frame. We can then implement the proportional-integral controller with Euler integration. The function uses its data structure as memory – the updated structure d is passed back as an output. TorqueControl is shown as follows. This function will return a default data structure if no inputs are passed into the function.

TorqueControl.m

```
1  %% TORQUECONTROL Compute torque control of an AC machine
2  % Determines the quadrature current needed to produce a torque and uses
   a
3  % proportional integral controller to control the motor. We control the
```

```
 4  % direct current to zero since we want to use just the magnet flux to
       react
 5  % with the quadrature current. We could control the direct current to
 6  % another value to implement field-weakening control but this would
       result
 7  % in a nonlinear control system.
 8  %% Forms
 9  %   d = TorqueControl
10  %   [u, d, iAB] = TorqueControl( torqueSet, x, d )
11  %
12  %% Inputs
13  %   torqueSet (1,1)     Set point torque
14  %   x         (5,1)     State [ia;ib;ic;omega;theta]
15  %   d         (.)        Control data structure
16  %                       .kF      (1,1) Forward gain
17  %                       .tauI    (1,1) Integral time constant
18  %                       .iDQInt  (2,1) Integral of current errors
19  %                       .dT      (1,1) Time step
20  %                       .psiM    (1,1) Magnetic flux
21  %                       .p       (1,1) Number of magnet poles
22  %
23  %% Outputs
24  %   u        (2,1)      Control voltage [alpha;beta]
25  %   d        (.)        Control data structure
26  %   iAB      (2,1)      Steady state currents [alpha;beta]
27  %
32
33  function [u, d, iAB] = TorqueControl( torqueSet, x, d )
34
35  % Default data structure
36  if( nargin == 0 )
37    u = struct('kF',0.003,'tauI',0.001, 'iDQInt',[0;0], 'dT', 0.01,...
38                'psiM',0.0690,'p',2);
39    if( nargout == 0 )
40      disp('TorqueControl struct:');
41    end
42    return
43  end
44
45  % Clarke and Park transforms
46  thetaE = 0.5*d.p*x(5);
47  park   = ParkTransformationMatrix( thetaE );
48  iPark  = park';
49  clarke = ClarkeTransformationMatrix;
50  iDQ    = iPark*clarke*x(1:2);
51
52  % Set point to produce the desired torque [iD;iQ]
53  iDQSet = [0;(2/3)*torqueSet/(d.psiM*d.p)];
54
55  % Error
56  iDQErr = iDQ - iDQSet;
57
```

261

```
58  % Integral term
59  d.iDQInt = d.iDQInt + d.dT*iDQErr;
60
61  % Control
62  uDQ = -d.kF*(iDQErr + d.iDQInt/d.tauI);
63  u   = park*uDQ;
64
65  % Steady state currents
66  if( nargout > 2 )
67    iAB = park*iDQSet;
68  end
```

9.3 Pulsewidth Modulation of the Switches

Problem

In the previous recipe, we calculate the control voltages to apply to the stator. Now we want to take those control voltages as an input and drive the switches via pulsewidth modulation.

Solution

We will use the Space Vector Modulation to go from a rotating two-dimensional (α,β) frame to the rotating three-dimensional (a,b,c) stator frame, which is more computationally efficient than modulating in (a,b,c) directly.

How It Works

We will use Space Vector Modulation to drive the switches for pulsewidth modulation.[2] This goes from (α,β) coordinates to switch states (a,b,c). Each node of each phase is either connected to ground or to $+u$. These values are shown in Figure 9.4. The six spokes in the diagram, as well as the origin, correspond to the eight discrete switch states.

Table 9.1 delineates each of these eight discrete switch states, the corresponding vector in the (α,β) coordinates, and the resulting voltages. Note that the O vectors are at the origin of the Space Vector Modulation, while the U vectors are at 60-degree increments. The states are indexed from 0 to 7 with 0 being all open states and 7 being all closed.

In order to produce the desired torque, we must use a combination of the vectors or switch states so that we achieve the desired voltage on average. We select the two vectors O or U bracketing the desired angle in the (α,β) plane; these are designated k and $k+1$ where k refers to the number of the vector in Table 9.1. We must then calculate the amount of time to spend in each switch state, for each pulsewidth period. The durations of these two segments, T_k and T_{k+1}, are found from this equation:

$$\begin{bmatrix} T_k \\ T_{k+1} \end{bmatrix} = \frac{\sqrt{3}}{2}\frac{T_s}{u_d}\begin{bmatrix} \sin\frac{k\pi}{3} & -\cos\frac{k\pi}{3} \\ -\sin\frac{(k-1)\pi}{3} & \cos\frac{(k-1)\pi}{3} \end{bmatrix}\begin{bmatrix} u_\alpha \\ u_\beta \end{bmatrix} \tag{9.21}$$

---

[2] Analog Devices, "Implementing Space Vector Modulation with the ADMCF32X," ANF32X-17, January 2000.

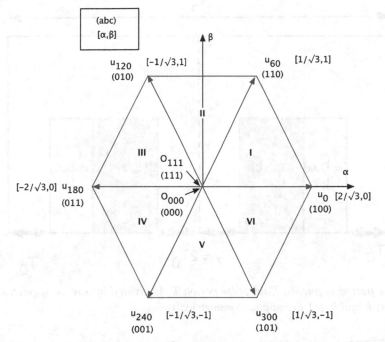

Figure 9.4: *Space Vector Modulation in (α, β) coordinates. We determine which sector (in Roman numerals) we are in and then pick the appropriate vectors to apply so that they on average attain the desired voltage. The numbers in brackets are the normalized $[\alpha, \beta]$ voltages.*

Table 9.1: *Space Vector Modulation. In the vector names, O means open and U means a voltage is applied, while the subscripts denote the angle in the α-β plane. The switch states are a, b, c as shown in Figure 9.3, where 1 means a switch is closed and 0 means it is open.*

k	abc	Vector	u_a/u	u_b/u	u_c/u	u_{ab}/u	u_{bc}/u	u_{ac}/u
0	000	O_{000}	0	0	0	0	0	0
1	110	U_{60}	2/3	1/3	-1/3	1	0	-1
2	010	U_{120}	1/3	1/3	-2/3	0	1	-1
3	011	U_{180}	-1/3	2/3	-1/3	-1	1	0
4	001	U_{240}	-2/3	1/3	1/3	-1	0	1
5	101	U_{300}	-1/3	-1/3	2/3	0	-1	1
6	100	U_{360}	1/3	-2/3	1/3	1	-1	0
7	111	O_{111}	0	0	0	0	0	0

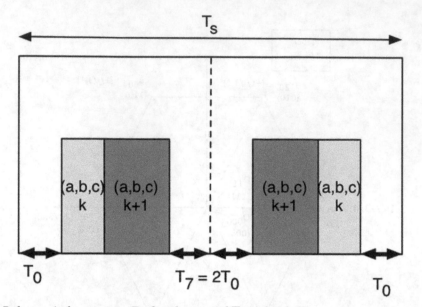

Figure 9.5: *Pulse period segments. Each pulse period T_s is divided into seven segments so that the two switching patterns k and $k + 1$ are applied symmetrically.*

The corresponding (a,b,c) switch patterns are each used for the calculated time, averaging to the designated voltage.

The time spent in each pattern, T_k or T_{k+1}, is then split into two equal portions so that the total pulse pattern is symmetric. The zero time T_0, when no switching is required, is split evenly between the endpoints and the middle of the pulse T_s – so that the time in the middle pattern (O_{111}) is twice the time in each end pattern (O_{000}). This results in a total of seven segments depicted in Figure 9.5. The total middle time is designated T_7.

$$T_0 = \frac{1}{4}\left(T_s - (T_k + T_{k+1})\right) \tag{9.22}$$

The implementation of the pulse segments is slightly different for the even and odd sectors in Figure 9.4. Both are symmetric about the midpoint of the pulse as described, but we reverse the implementation of patterns k and $k + 1$. This is shown for the resulting voltages u in the following equations. We use the first in even sectors and the second in odd sectors.

$$\begin{bmatrix} u_0 & u_k & u_{k+1} & u_7 & u_{k+1} & u_k & u_0 \end{bmatrix} \tag{9.23}$$

and

$$\begin{bmatrix} u_0 & u_{k+1} & u_k & u_7 & u_k & u_{k+1} & u_0 \end{bmatrix} \tag{9.24}$$

Using the different patterns for odd and even vectors minimizes the number of commutations per cycle.

We determine the sector from the angle Θ formed by the commanded voltages u_α and u_β:

$$\Theta = \text{atan}\frac{u_\beta}{u_\alpha} \tag{9.25}$$

The pulsewidth modulation routine, SVPWM, does not actually perform an arctangent. Rather, it looks at the unit u_α and u_β vectors and determines first their quadrant and then their sector without any need for trigonometric operations.

The first section of SVPWM implements the timing for the pulses. Just as in the previous recipe for the controller, the function uses its data structure as memory – the updated structure is passed back as an output. This is an alternative to persistent variables.

SVPWM.m

```
44   function [s, d] = SVPWM( t, d )
45
46   % Default data structure
47   if( nargin < 1 && nargout == 1 )
48     s = struct( 'dT',1e-6,'tLast',-0.0001,'tUpdate',0.001,'u',[0;0],...
49                 'uM',10,'tP',zeros(1,7),'sP',zeros(3,7));
50     return;
51   end
52
53   % Run the demo
54   if( nargin < 1 )
55     disp('Demo of SVPWM:');
56     Demo;
57     return;
58   end
59
60   % Update the pulsewidths at update time
61   if( t >= d.tLast + d.tUpdate || t == 0 )
62     [d.sP, d.tP] = SVPW( d.u, d.tUpdate, d.uM );
63     d.tLast      = t;
64   end
65
66   % Time since initialization of the pulse period
67   dT = t - d.tLast;
68   s  = zeros(3,1);
69
70   for k = 1:7
71     if( dT < d.tP(k) )
72       s = d.sP(:,k);
73       break;
74     end
75   end
```

The pulsewidth vectors are computed in the subfunction SVPW. We first compute the quadrant and then the sector without using any trigonometric functions. This is done using simple if/else statements and a switch statement. Note that the modulation index k is simply designated k and $k+1$ is designated kP1. We then compute the times for the two space vectors that bound the sector. We then assemble the seven subperiods.

```
89   function [sP, tP] = SVPW( u, tS, uD )
90
91   % Make u easier to interpret
92   alpha = 1;
93   beta  = 2;
94
95   % Determine the quadrant
96   if( u(alpha) >= 0 )
97     if( u(beta) > 0 )
98       q = 1;
99     else
100      q = 4;
101    end
102  else
103    if( u(beta) > 0 )
104      q = 2;
105    else
106      q = 3;
107    end
108  end
109
110  sqr3 = sqrt(3);
111
112  % Find the sector. k1 and k2 define the edge vectors
113  switch q
114    case 1 % [+,+]
115      if( u(beta) < sqr3*u(alpha) )
116        k     = 1;
117        kP1   = 2;
118        oddS  = 1;
119      else
120        k     = 2;
121        kP1   = 3;
122        oddS  = 0;
123      end
124    case 2  % [-,+]
125      if( u(beta) < -sqr3*u(alpha) )
126        k     = 3;
127        kP1   = 4;
128        oddS  = 1;
129      else
130        k     = 2;
131        kP1   = 3;
132        oddS  = 0;
133      end
134    case 3 % [-,-]
135      if( u(beta) < sqr3*u(alpha) )
136        k     = 5;
137        kP1   = 6;
138        oddS  = 1;
139      else
```

266

```
140        k        = 4;
141        kP1      = 5;
142        oddS     = 0;
143     end
144   case 4 % [+,-]
145     if( u(beta) < -sqr3*u(alpha) )
146        k        = 5;
147        kP1      = 6;
148        oddS     = 1;
149     else
150        k        = 6;
151        kP1      = 1;
152        oddS     = 0;
153     end
154 end
155
156 % Switching sequence
157 piO3    = pi/3;
158 kPiO3   = k*pi/3;
159 kM1PiO3 = kPiO3-piO3;
160
161 % Space vector pulsewidths
162 t = 0.5*sqr3*(tS/uD)*[ sin(kPiO3)      -cos(kPiO3);...
163                       -sin(kM1PiO3)  cos(kM1PiO3)]*u;
164
165 % Total zero vector time
166 t0 = tS - sum(t);
167
168 t  = t/2;
169
170 % Different order for odd and even sectors
171 if( oddS )
172   sS  = [0 k kP1 7 kP1 k 0];
173   tPW = [t0/4 t(1) t(2) t0/2 t(2) t(1) t0/4];
174 else
175   sS  = [0 kP1 k 7 k kP1 0];
176   tPW = [t0/4 t(2) t(1) t0/2 t(1) t(2) t0/4];
177 end
178 tP  = [tPW(1) zeros(1,6)];
179
180 for k = 2:7
181   tP(k) = tP(k-1) + tPW(k);
182 end
183
184 % The switches corresponding to each voltage vector
185 % From 0 to 7
186 %               a b c
187 s             = [ 0 0 0;...
188                   1 0 0;...
189                   1 1 0;...
190                   0 1 0;...
191                   0 1 1;...
```

```
192                    0  0  1;...
193                    1  0  1;
194                    1  1  1]';
195
196    sP = zeros(3,7);
197    for k = 1:7
198      sP(:,k) = s(:,sS(k)+1);
199    end
```

The built-in demo is fairly complex so it is in a separate subfunction. We simply specify an example input u using trigonometric functions.

```
201    function Demo
202    %%% SVPWM>Demo Function demo
203    % Calls SVPWM with a sinusoidal input u.
204    % This demo will run through an array of times and create a plot of the
205    % resulting voltages.
206
207    d       = SVPWM;
208    tEnd    = 0.003;
209    n       = tEnd/d.dT;
210    a       = linspace(0,pi/4,n);
211    tP3     = 2*pi/3;
212    uABC    = 0.5*[cos(a);cos(a-tP3);cos(a+tP3)];
213    uAB     = ClarkeTransformationMatrix*uABC(1:2,:); % a-b to alpha-beta
214    tSamp   = 0;
215    t       = 0;
216    tPP     = 1;
217    x       = zeros(4,n);
218    for k = 1:n
219      if( t >= tSamp )
220        tSamp = tSamp + d.tUpdate;
221        tPP   = ~tPP;
222      end
223      d.u     = uAB(:,k);
224      [s, d] = SVPWM( t, d );
225      t       = t + d.dT;
226      x(:,k) = [SwitchToVoltage(s,d.uM);tPP];
227    end
```

Figure 9.6 shows the state vector pulsewidth modulation from the built-in demo. There are three pulses in the plot, each 0.001 seconds long. Each pulse period has seven subperiods.

The function `SwitchToVoltage` converts switch states to voltages. It assumes instantaneous switching and no switch dynamics.

SwitchToVoltage.m

```
24
25    % Switch states [a;b;c]
26    sA      = [1  1  0  0  0  1;...
```

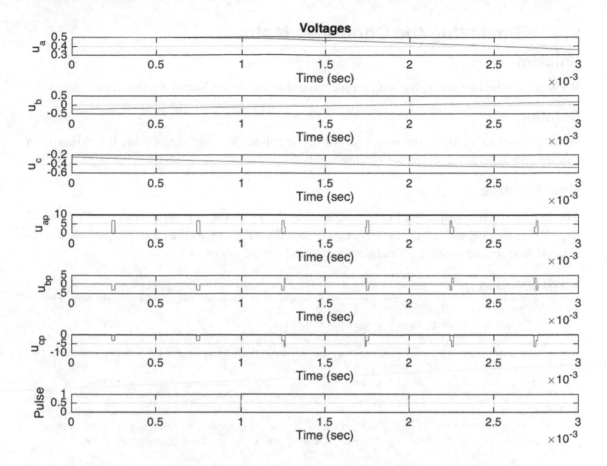

Figure 9.6: *The desired voltage vector and the Space Vector Modulation pulses and pulsewidth. The bottom plot shows the pulse periods. Note that the pulse sequences are symmetric within each pulse period.*

```
27              0   1   1   1   0   0;...
28              0   0   0   1   1   1];
29
30   % Array of voltages
31   uA    = [ 2   1  -1  -2  -1   1;...
32             -1   1   2   1  -1  -2;...
33             -1  -2  -1   1   2   1];
34
35   % Find the correct switch state
36   u    = [0;0;0];
37   for k = 1:6
38     if( sum(sA(:,k) - s) == 0 )
39       u = uA(:,k)*uDC/3;
40       break;
41     end
42   end
```

9.4   Simulating the Controlled Motor

Problem

We want to simulate the motor with torque control using Space Vector Modulation.

Solution

Write a script to simulate the motor with the controller. We include options for closed loop control and balanced three-phase voltage inputs.

How It Works

The header for the script, PMMachineDemo, is shown in the following listing. The control flags bypassPWM and torqueControlOn are described as well as the two periods implemented, one for the simulation and a longer period for the control.

PMMachineDemo.m

```
 1  %% Simulation of a permanent magnet AC motor
 2  % Simulates a permanent magnet AC motor with torque control. The
        simulation has
 3  % two options. The first is torqueControlOn. This turns torque control
        on and
 4  % off. If it is off the phase voltages are a balanced three phase
        voltage set.
 5  %
 6  % bypassPWM allows you to feed the phase voltages directly to the motor
 7  % bypassing the pulsewidth modulation switching function. This is
        useful for
 8  % debugging your control system and other testing.
 9  %
10  % There are two time constants for this simulation. One is the control
        period
11  % and the second is the simulation period. The latter is much shorter
        because it
12  % needs to simulate the pulsewidth modulation.
13  %
14  % For control testing the load torque and setpoint torque should be the
        same.
```

The body of the script follows. Three different data structures are initialized from their corresponding functions as described in the previous recipes, that is, from SVPWM, TorqueControl, and RHSPMMachine. Note that we are only simulating the motor for a small fraction of a second, 0.05 seconds, and the time step is just 1e-6 seconds. The controller time step is set to 100 times the simulation time step.

```
20  %% Initialize all data structures
21  dS      = SVPWM;
22  dC      = TorqueControl;
23  d       = RHSPMMachine;
```

```
24  dC.psiM = d.psiM;
25  dC.p    = d.p;
26  d.tL    = 1.0; % Load torque (Nm)
27
28  %% User inputs
29  tEnd            = 0.05;      % sec
30  torqueControlOn = false;
31  bypassPWM       = false;
32  torqueSet       = 1.0;       % Set point (Nm)
33  dC.dT           = 100*dS.dT; % 100x larger than simulation dT
34  dS.uM           = 1.0;       % DC Voltage at the input to the switches
35  magUABC         = 0.1;       % Voltage for the balanced 3 phase
        voltages
36
37  if (torqueControlOn && bypassPWM)
38    error('The control requires PWM to be on.');
39  end
40
41  %% Run the simulation
42  nSim = ceil(tEnd/dS.dT);
43  xP   = zeros(10,nSim);
44  x    = zeros(5,1);
45
46  % We require two timers as the control period is larger than the
        simulation
47  % period
48  t    = 0.0; % simulation timer
49  tC   = 0.0; % control timer
50
51  for k = 1:nSim
52    % Electrical degrees
53    thetaE = x(5);
54    park   = ParkTransformationMatrix( thetaE );
55    clarke = ClarkeTransformationMatrix;
56
57    % Compute the voltage control
58    if( torqueControlOn && t >= tC )
59      tC          = tC + dC.dT;
60      [dS.u, dC] = TorqueControl( torqueSet, x, dC );
61    elseif( ~torqueControlOn )
62      tP3  = 2*pi/3;
63      uABC = magUABC*dS.uM*[cos(thetaE);cos(thetaE-tP3);cos(thetaE+tP3)];
64      if( bypassPWM )
65        d.u = uABC;
66      elseif( t >= tC )
67        tC   = tC + dC.dT;
68        dS.u = park*clarke*uABC(1:2,:);
69      end
70    end
71
72    % Space Vector Pulsewidth Modulation
73    if( ~bypassPWM )
```

```
74      dS.u    = park'*dS.u;
75      [s,dS]  = SVPWM( t, dS );
76      d.u     = SwitchToVoltage(s,dS.uM);
77    end
78
79    % Get the torque output for plotting
80    [~,tE]  = RHSPMMachine( 0, x, d );
81    xP(:,k) = [x;d.u;torqueSet;tE];
82
83    % Propagate one simulation step
84    x = RungeKutta( @RHSPMMachine, 0, x, dS.dT, d );
85    t = t + dS.dT;
86  end
87
88  %% Generate the time history plots
89  [t, tL]   = TimeLabel( (0:(nSim-1))*dS.dT );
90
91  figure('name','3 Phase Currents');
92  plot(t, xP(1:3,:));
93  grid on;
94  ylabel('Currents');
95  xlabel(tL);
96  legend('i_a','i_b','i_c')
97
98  PlotSet( t, xP([4 10],:), 'x label', tL, 'y label', {'\omega_e' 'T_e (
        Nm)'}, ...
99    'plot title','Electrical', 'figure title','Electrical');
100
101 thisTitle = 'Phase Voltages';
102 if ~bypassPWM
103   thisTitle = [thisTitle ' - PWM'];
104 end
105
106 PlotSet( t, xP(6:8,:), 'x label', tL, 'y label', {'u_a' 'u_b' 'u_c'},
        ...
107   'plot title',thisTitle, 'figure title',thisTitle);
108
109 thisTitle = 'Torque/Speed';
110 if ~bypassPWM
111   thisTitle = [thisTitle ' - PWM'];
112 end
```

We turn off torque control to test the motor simulation with the results shown in Figure 9.7. The two plots show the torque speed curves. The first is with direct three-phase excitation, that is, bypassing the pulsewidth modulation, by setting bypassPWM to false. Directly controlling the phase voltages this way, while creating the smoothest response, would require linear amplifiers which are less efficient than switches. This would make the motor much less efficient overall and would generate unwanted heat. The second plot is with Space Vector Pulsewidth Modulation. The plots are nearly identical, indicating that the pulsewidth modulation is working.

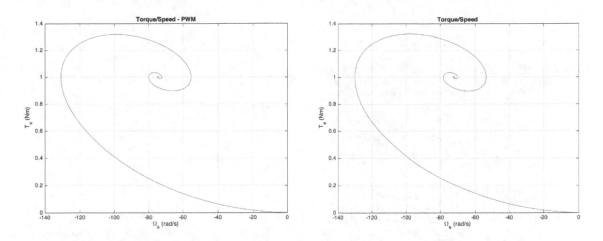

Figure 9.7: *Torque speed curves for a balanced three-phase voltage excitation and a load torque of 1.0 Nm. The left figure shows the curve for the direct three-phase input, and the right shows the curve for the Space Vector Pulsewidth Modulation input. They are nearly identical.*

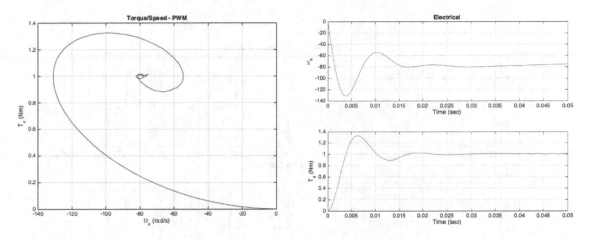

Figure 9.8: *PI torque control of the motor.*

We now turn on torque control, via the `torqueControlOn` flag, and get the results shown in Figure 9.8. The overshoot is typical for torque control. Note that the load torque is set equal to the torque set point of 1 Nm. There is limit cycling near the end point.

The pulsewidths and resulting coil currents are shown in Figure 9.9. A zoomed view of the end of the pulsewidth plot with shading added to alternate pulsewidths is in Figure 9.10. This makes it easier to see the segments of the pulsewidths and verify that they are symmetric.

The code which adds the shading uses `fill` with transparency via the `alpha` parameter. In this case, we hard-code the function to show the last five pulsewidths, but this could be generalized to a time window, or to shade the entire plot. We did take the time to add an input for the pulsewidth length, so that this could be changed in the main script and the function

273

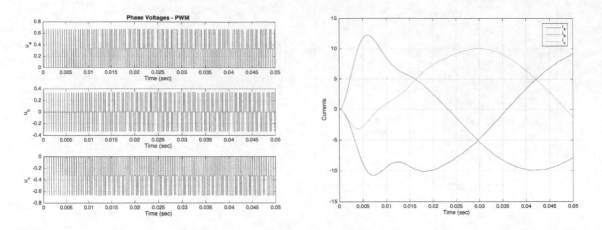

Figure 9.9: *Voltage pulsewidths and resulting currents for PI torque control.*

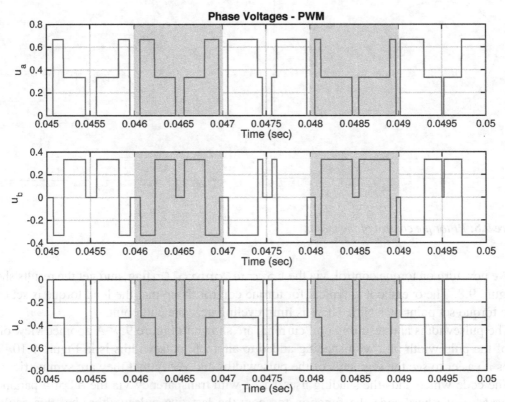

Figure 9.10: *Pulsewidths with shading.*

would still work. Note that we reorder the axes children as the last step, to keep the shading from obscuring the plot lines.

AddFillToPWM.m

```matlab
17   function AddFillToPWM( dT )
18
19   if nargin == 0
20     dT = 0.001;
21   end
22
23   hAxes = get(gcf,'children');
24   nAxes = length(hAxes);
25
26   for j = 1:nAxes
27     if strcmp(hAxes(j).type,'axes')
28       axes(hAxes(j));
29       AddFillToAxes;
30     end
31   end
32
33     function AddFillToAxes
34
35     hold on;
36     y = axis;
37     xMin = y(2) - 5*dT;
38     xMax = y(2);
39     axis([xMin xMax y(3:4)])
40     x0   = xMin;
41     yMin = y(3) + 0.01*(y(4)-y(3));
42     yMax = y(4) - 0.01*(y(4)-y(3));
43     for k = [2 4]
44       xMinK = x0 + (k-1)*dT;
45       xMaxK = x0 + k*dT;
46
47       fill([xMinK xMaxK xMaxK xMinK],...
48            [yMin,yMin,yMax,yMax],...
49            [0.8 0.8 0.8],'edgecolor','none','facealpha',0.5);
50
51     end
52     babes = get(gca,'children');
53     set(gca,'children',[babes(end); babes(1:end-1)])
54     hold off;
55
56     end
57
58   end
```

Summary

This chapter has demonstrated how to write the dynamics and implement a field-oriented control law for a three-phase motor. We use a proportional-integral controller with Space Vector Pulsewidth Modulation to drive the six switches. This produces a low-cost controller for a motor. Table 9.2 lists the code developed in the chapter.

Table 9.2: *Chapter Code Listing*

File	Description
AddFillToPWM	Add shading to the motor pulsewidth plot
ClarkeTransformationMatrix	Clarke transformation matrix
ParkTransformationMatrix	Park transformation matrix
PMMachineDemo	Permanent magnet motor demonstration
RHSPMMachine	Right-hand side of a permanent magnet brushless three-phase electrical machine
SVPWM	Implements Space Vector Pulsewidth Modulation
SwitchToVoltage	Converts switch states to voltages
TorqueControl	Proportional-integral torque controller

CHAPTER 10

■ ■ ■

Fault Detection

Introduction

Fault detection is the process of detecting failures, also known as faults, in a dynamical system. It is an important area for systems that are supposed to operate without human supervision. There are many ways of detecting failures. The simplest is using boolean logic to check against fixed thresholds. For example, you might check an automobile's speed against a speed limit. Other methods include fuzzy logic, parameter estimation, expert systems, statistical analysis, and parity space methods. In this section, we will implement one type of fault detection system, a detection filter. This is based on linear filtering. The detection filter is a state estimator tuned to detect specific failures. We will design a detection filter system for an air turbine. We will also show how to build a graphical user interface (GUI) as a front end to the fault detection simulation.

10.1   Modeling an Air Turbine

Problem

We need to make a numerical model of an air turbine to demonstrate detection filters.

Solution

Write the equations of motion for an air turbine. We will use a linear model of the air turbine to simplify the detection filter design. This will allow us to model the system with a linear state space model.

© Michael Paluszek and Stephanie Thomas 2020
M. Paluszek and S. Thomas, *MATLAB Recipes*,
https://doi.org/10.1007/978-1-4842-6124-8_10

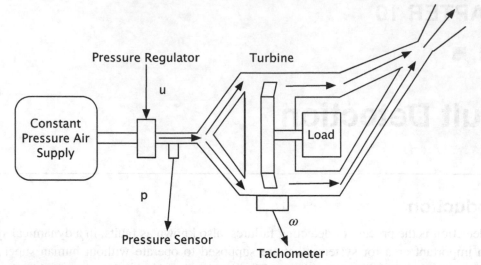

Figure 10.1: *Air turbine. The arrows show the airflow. The air flows through the turbine blade tips causing it to turn.*

How It Works

Figure 10.1 shows an air turbine.[1] It has a constant pressure air supply. We can control the valve from the air supply, the pressure regulator, to control the speed of the turbine. The air flows past the turbine blades causing it to turn. The control needs to adjust the air pressure to handle variations in the load. We measure the air pressure p downstream from the valve, and we also measure the rotational speed of the turbine ω with a tachometer.

The dynamical model for the air turbine is

$$\begin{bmatrix} \dot{p} \\ \dot{\omega} \end{bmatrix} = \begin{bmatrix} -\frac{1}{\tau_p} & 0 \\ \frac{K_t}{\tau_t} & -\frac{1}{\tau_t} \end{bmatrix} \begin{bmatrix} p \\ \omega \end{bmatrix} + \begin{bmatrix} \frac{K_p}{\tau_p} \\ 0 \end{bmatrix} u \qquad (10.1)$$

This is a state space system:

$$\dot{x} = ax + bu \qquad (10.2)$$

where

$$a = \begin{bmatrix} -\frac{1}{\tau_p} & 0 \\ \frac{K_t}{\tau_t} & -\frac{1}{\tau_t} \end{bmatrix} \qquad (10.3)$$

$$b = \begin{bmatrix} \frac{K_p}{\tau_p} \\ 0 \end{bmatrix} \qquad (10.4)$$

---

[1]PhD thesis of Jere Schenck Meserole, "Detection Filters for Fault-Tolerant Control of Turbofan Engines," Massachusetts Institute of Technology, Department of Aeronautics and Astronautics, 1981.

The state vector is

$$\begin{bmatrix} p \\ \omega \end{bmatrix} \tag{10.5}$$

The pressure downstream from the regulator is equal to K_{pu} when the system is in equilibrium. τ_p is the regulator time constant, and τ_t is the turbine time constant. The turbine speed is K_{tp} when the system is in equilibrium. The tachometer measures ω, and the pressure sensor measures p. The load is folded into the time constant for the turbine.

The code for the right-hand side of the dynamical equations is shown in the following. Only one line of code is needed. The rest returns the default data structure. The simplicity of the model is due to its being a state space model. The number of states could be large, yet the code would not change.

RHSAirTurbine.m

```
27  function xDot = RHSAirTurbine( ~, x, d )
28
29  % Default data structure
30  if( nargin < 1 )
31     kP    = 1;
32     kT    = 2;
33     tauP = 10;
34     tauT = 40;
35     c     = eye(2);
36     b     = [kP/tauP;0];
37     a     = [-1/tauP 0; kT/tauT -1/tauT];
38
39     xDot = struct('a',a,'b',b,'c',c,'u',0);
40     if( nargout == 0 )
41       disp('RHSAirTurbine struct:');
42     end
43     return
44  end
```

The response to a step input for u is shown in Figure 10.2. The pressure settles faster than the turbine speed. This is due to the turbine time constant and the lag in the pressure change. The residuals are very small because there are no failures.

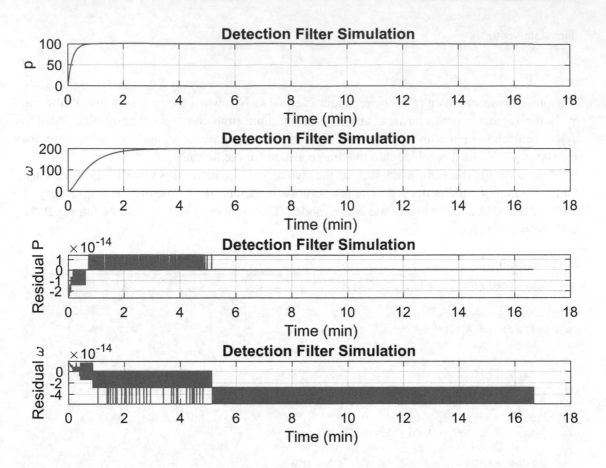

Figure 10.2: *Air turbine response to a step pressure regulator input. The residuals are zero as expected.*

10.2 Building a Detection Filter

Problem

We want to build a system to detect failures in our air turbine using the linear model developed in the previous recipe.

Solution

We will build a detection filter that detects pressure regulator failures and tachometer failures. Our plant model (continuous a, b, and c state space matrices) will be an input to the filter building function.

How It Works

The detection filter is an estimator with a specific gain matrix that multiplies the residuals. The residuals are the difference between the estimated outputs and the outputs:

$$\begin{bmatrix} \dot{\hat{p}} \\ \dot{\hat{\omega}} \end{bmatrix} = \begin{bmatrix} -\frac{1}{\tau_p} & 0 \\ \frac{K_t}{\tau_t} & -\frac{1}{\tau_t} \end{bmatrix} \begin{bmatrix} \hat{p} \\ \hat{\omega} \end{bmatrix} + \begin{bmatrix} \frac{K_p}{\tau_p} \\ 0 \end{bmatrix} u + \begin{bmatrix} d_{11} & d_{12} \\ d_{21} & d_{22} \end{bmatrix} \begin{bmatrix} p - \hat{p} \\ \omega - \hat{\omega} \end{bmatrix} \tag{10.6}$$

where $\hat{p}$ is the estimated pressure and $\hat{\omega}$ is the estimated angular rate of the turbine. The D matrix is the matrix of detection filter gains. These feedback the residuals, the difference between the measured and estimated states, into the detection filter. The residual vector is

$$r = \begin{bmatrix} p - \hat{p} \\ \omega - \hat{\omega} \end{bmatrix} \tag{10.7}$$

The residuals are the difference between the measured values and the estimated values. The D matrix needs to be selected so that this vector tells us the nature of the failure. The gains should be selected so that

1. The filter is stable.

2. If the pressure regulator fails, the first residual $p - \hat{p}$ is nonzero, but the second remains zero.

3. If the turbine fails, the second residual $\omega - \hat{\omega}$ is nonzero, but the first remains zero.

A gain matrix is

$$D = a + \begin{bmatrix} \frac{1}{\tau_1} & 0 \\ 0 & \frac{1}{\tau_2} \end{bmatrix} \tag{10.8}$$

The time constant τ_1 is the pressure residual time constant. The time constant τ_2 is the tachometer residual time constant. In effect, we cancel out the dynamics of the plant and replace them with decoupled detection filter dynamics. These time constants should be shorter than the time constants in the dynamical model so that we detect failures quickly. However, they need to be at least twice as long as the sampling period to prevent numerical instabilities.

We will write a function with three actions, an initialize case, an update case, and a reset case. `varargin` is used to allow the three cases to have different input lists. The function signature is

DetectionFilter.m

```
49   function d = DetectionFilter( action, varargin )
```

The header and syntax for `DetectionFilter` are shown as follows. We used LaTeX equations to describe the function.

```
1   %% DETECTIONFILTER Builds and updates a linear detection filter.
2   %% Forms
3   %    d = DetectionFilter( 'initialize', d, tau, dT )
4   %    d = DetectionFilter( 'update', u, y, d )
5   %    d = DetectionFilter( 'reset', d )
6   %
7   %% Description
8   % The detection filter gain matrix d is designed during the initialize
9   % action. The continuous matrices are then discretized using the
         internal
10  % function CToDZOH. The esimated state and residual vectors are
         initialized
11  % to the size dictated by a. During the update action, the residuals
         and
12  % new estimated state are calculated and stored in the data structure d
13  %
14  % The residuals calculation is
15  %
16  % $$r   = y - c\hat{x}$$
17  %
18  % The estimated state calculated with the detection filter gains is
19  %
20  % $$\hat{x}_{k+1} = a*\hat{x} + +b*u + d*r$$
21  %
22  %% Inputs
23  %    action        (1,:) 'initialize' or 'update'
24  %    d             (.)   Data structure
25  %                        .a (:,:) State space continuous a matrix
26  %                        .b (:,1) State space continuous b matrix
27  %                        .c (:,:) State space continuous c matrix
28  %    tau           (:,1) Vector of time constants
29  %    dT            (1,1) Time step
30  %    u             (:,1) Actuation input
31  %    y             (:,1) Measurement vector
32  %
33  %% Outputs
34  %    d             (.)   Updated data structure
35  %                        .a (:,:) State space discrete a matrix
36  %                        .b (:,1) State space discrete b matrix
37  %                        .c (:,:) State space discrete c matrix
38  %                        .d (:,:) Detection filter gain matrix
39  %                        .x (:,1) Estimated states
40  %                        .r (:,1) Residual vector
```

282

The filter is built and initialized in the following code in `DetectionFilter`. The continuous state space model of the plant, in this case, our linear air turbine model, is an input. The selected time constants τ are also an input, and they are added to the plant model as in Equation 10.8. The function discretizes the plant a and b matrices and the computed detection filter gain matrix d.

```
48
49   function d = DetectionFilter( action, varargin )
50
51   switch lower(action)
52     case 'initialize'
53       d   = varargin{1};
54       tau = varargin{2};
55       dT  = varargin{3};
56
57       % Design the detection filter
58       d.d = d.a + diag(1./tau);
59
60       % Discretize both
61       d.d        = CToDZOH( d.d, d.b, dT );
62       [d.a, d.b] = CToDZOH( d.a, d.b, dT );
63
64       % Initialize the state
66       d.x = zeros(m,1);
67       d.r = zeros(m,1);
```

The update for the detection filter is in the same function, as the next action in the `switch` statement. Note the equations implemented as described in the header.

```
69     case 'update'
70       u = varargin{1};
71       y = varargin{2};
72       d = varargin{3};
73       r = y - d.c*d.x;
74       d.x = d.a*d.x + +d.b*u + d.d*r;
75       d.r = r;
```

Finally, we create a reset action to allow us to reset the residual and state values for the filter in between simulations. After this action, we end the `switch` statement.

```
77     case 'reset'
78       d = varargin{1};
79       m = size(d.a,1);
80       d.x = zeros(m,1);
81       d.r = zeros(m,1);
82   end
```

10.3   Simulating the Fault Detection System

Problem

We want to simulate a failure in the plant and demonstrate the performance of the failure detection.

Solution

We will build a MATLAB script that designs the detection filter using the function from the previous recipe and then simulates it with a user selectable pressure regulator or tachometer failure. The failure can be total or partial.

How It Works

The script designs a detection filter using `DetectionFilter` from the previous recipe and implements it in a loop. Runge-Kutta integration propagates the continuous domain right-hand side of the air turbine, `RHSAirTurbine`. The detection filter is discrete time.

The script has two scale factors `uF` and `tachF` that multiply the regulator input and the tachometer output to simulate failures. Setting a scale factor to zero is a total failure, and setting it to one indicates that the device is working perfectly. If we fail one, we expect the associated residual to be nonzero and the other to stay at zero. Failures can be any number between zero and one. Partial failures are not necessarily related to a specific mechanical failure but are useful for testing the system.

DetectionFilterSim.m

```
1   %% Simulation of a detection filter
2   % Simulates detecting failures of an air turbine.
3   % An air turbine has a constant pressure air source that sends air
4   % through a duct that drives the turbine blades. The turbine is
5   % attached to a load.
6   %
7   % The air turbine model is linear. Failures are modeled by multiplying
8   % the regulator input and tachometer output by a constant. A constant
9   % of 0 is a total failure and 1 is perfect operation.
14
15  %% User inputs
16
17  % Failures. Set to any number betweem 0 and 1 is 0 is total failure. 1
        is working perfectly.
18  % uF scales the actuation u. tachF scales the rate measurement.
19  uF    = 0;
20  tachF = 1;
21
22  % Time constants for failure detection
23  tau1 = 0.3; % sec
```

```
24   tau2 = 0.3; % sec
25
26   % End time
27   tEnd = 1000; % sec
28
29   % State space system
30   d = RHSAirTurbine;
31
32   %% Initialization
33   dT = 0.02; % sec
34   n  = ceil(tEnd/dT);
35
36   % Initial state
37   x = [0;0];
38
39   %% Detection Filter design
40   dF = DetectionFilter('initialize',d,[tau1;tau2],dT);
41
42   %% Run the simulation
43
44   % Control. This is the regulator input.
45   u = 100;
46
47   % Plotting array
48   xP = zeros(4,n);
49   t  = (0:n-1)*dT;
50
51   for k = 1:n
52     % Measurement vector including measurement failure
53     y       = [x(1);tachF*x(2)]; % Sensor failure
54     xP(:,k) = [x;dF.r];
55
56     % Update the detection filter
57     dF = DetectionFilter('update',u,y,dF);
58
59     % Integrate one step
60     d.u = uF*u; % Actuator failure
61     x   = RungeKutta( @RHSAirTurbine, t(k), x, dT, d );
62   end
63
64   %% Plot the states and residuals
65   [t,tL] = TimeLabel(t);
66   yL     = {'p' '\omega' 'Residual P' 'Residual \omega' };
67   tTL    = 'Detection Filter Simulation';
68   PlotSet( t, xP,'x label',tL,'y label',yL,'plot title',tTL,'figure title
          ',tTL)
```

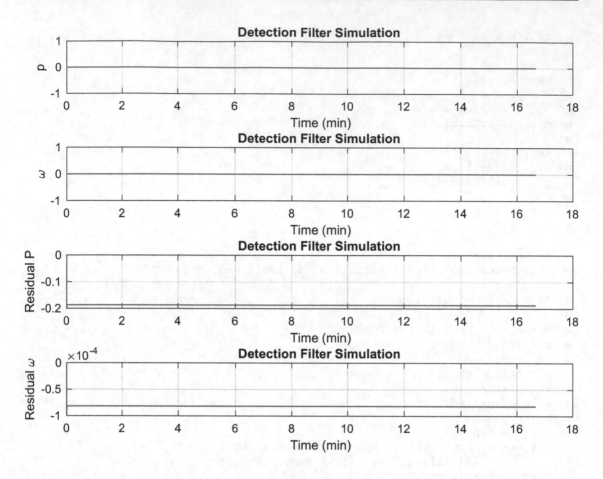

Figure 10.3: *Air turbine response to a failed regulator.*

In Figure 10.3, the regulator fails and its residual is nonzero. In Figure 10.4, the tachometer fails and its residual is nonzero. The residuals show what has failed clearly. Simple boolean logic (i.e., if end statements) are all that is needed.

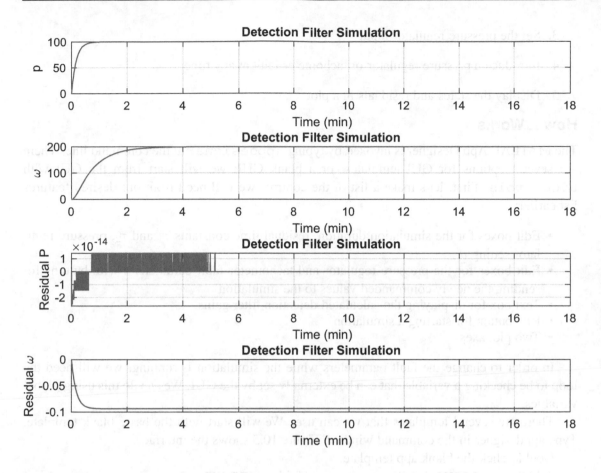

Figure 10.4: *Air turbine response to a failed tachometer.*

10.4   Building a GUI for the Detection Filter Simulation

Problem

We want a GUI to provide a graphical interface to the fault detection simulation that will allow us to evaluate the filter's performance.

Solution

We will use the MATLAB App Designer to build a GUI that will allow us to

1. Set the residual time constants

2. Set the end time for the simulation

287

3. Set the pressure regulator input

4. Introduce a pressure regulator or tachometer fault at any time

5. Display the states and residuals in a plot

How It Works

The MATLAB App Designer is invoked by typing `appdesigner` at the command line. There are several options for GUI templates, or a blank GUI; we will start from the GUI with `uicontrols`. First, let's make a list of the controls we will need from our desired features list earlier:

- Edit boxes for the simulation duration, residual time constants τ_1 and τ_2, pressure regulator setting u
- Edit boxes for the pressure regulator and tachometer fault parameters, with buttons for sending the newly commanded values to the simulation
- Text box for displaying the calculated detection filter gains
- Run button for starting a simulation
- Two plot axes

In order to change the fault parameters while the simulation is running, we will need the loop to be checking a variable that can be externally set by the GUI. We can do this using global variables.

There are several templates that we can use. We will start with the basic blank template. Type appdesigner in the command window. Figure 10.5 shows the interface.

Double-click the blank app template.

Add the app `DFGUI`. It will appear in your folder as `DFGUI.mlapp`.

Add the following to the blank template:

1. Parameter input boxes

    (a) Duration

    (b) Input

    (c) Tau 1

    (d) Tau 2

    (e) Gains (2-by-2 matrix)

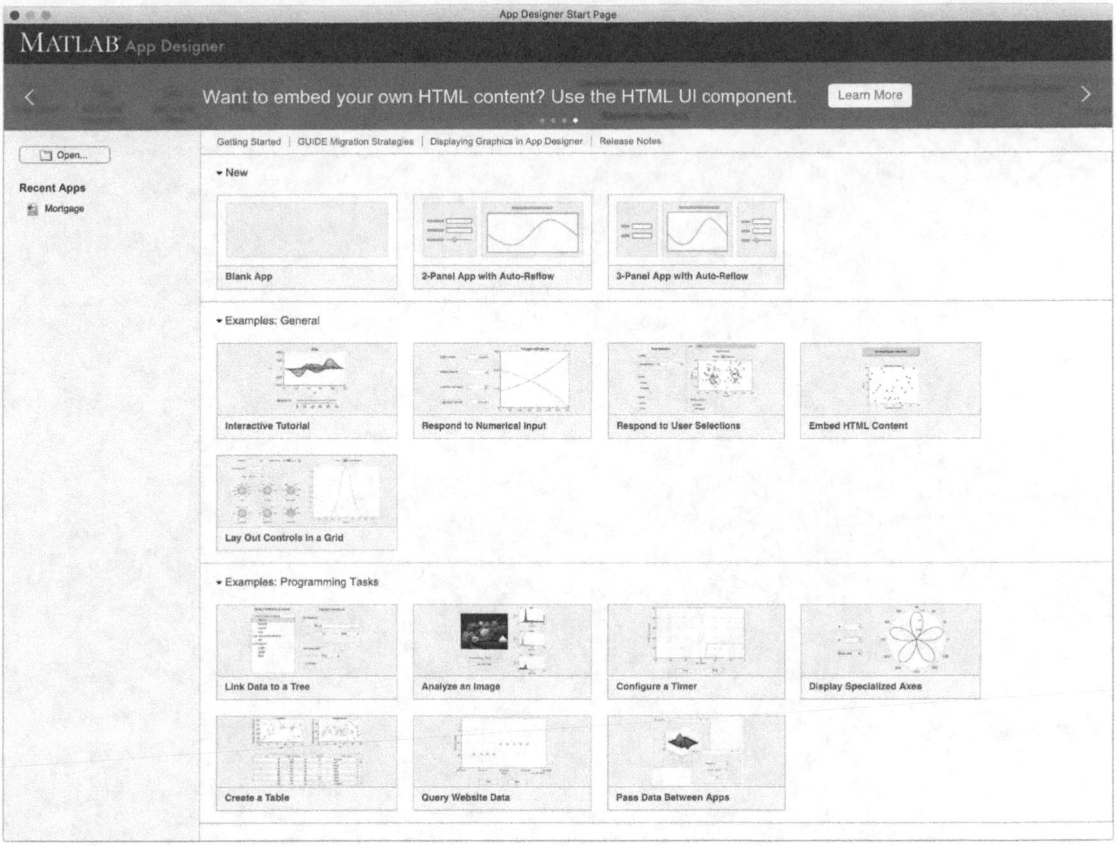

Figure 10.5: *The interface to appdesigner.*

2. Failure input boxes

   (a) Tachometer

   (b) Input

   (c) Send button for tachometer

   (d) Send button for input

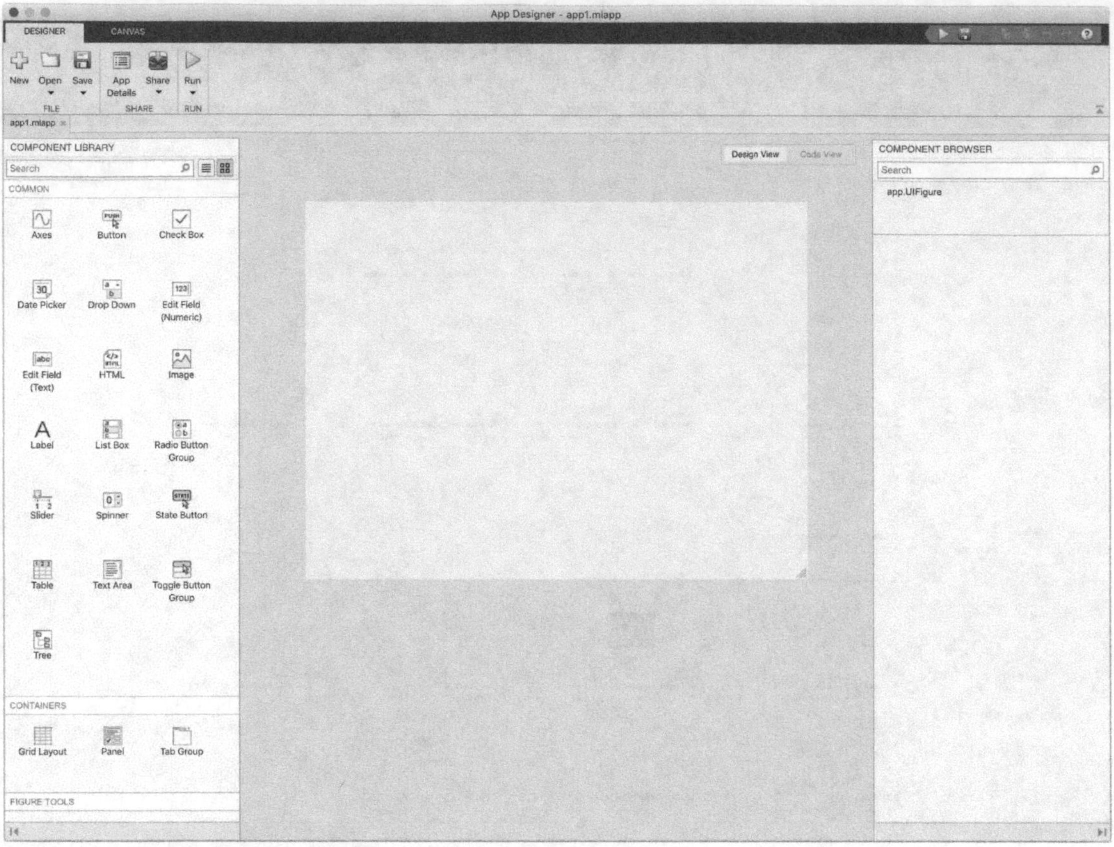

Figure 10.6: *Snapshot of the blank app.*

3. Calculate button

4. Reset button

5. State plot

6. Residual plot

You add items by dragging and dropping them on the window from the items on the left-hand side. We use numeric for the input text boxes. Figure 10.7 shows the completed interface. There are four push buttons.

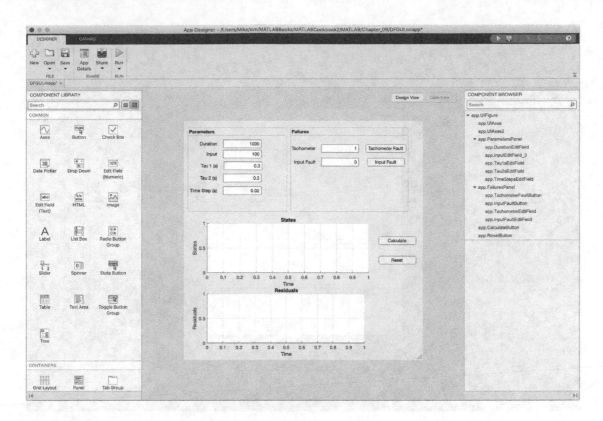

Figure 10.7: *Snapshot of the app after the interface is done.*

You can add information about the app. Figure 10.8 shows the window for app information. The app appears in the app menu as shown in Figure 10.9.

We select the callback for calculate. The App Designer highlights where the code should go. We copy relevant code from the simulation script. We get the inputs from the text boxes.

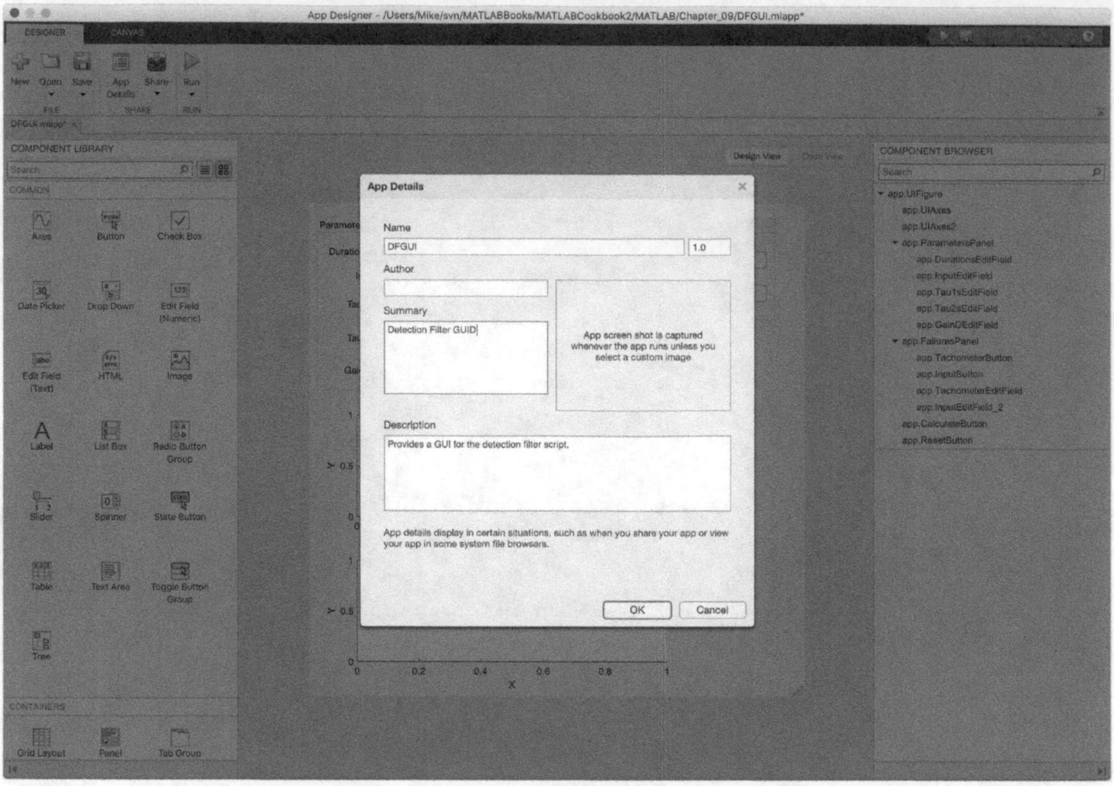

Figure 10.8: *App Details let you add information about the app for users.*

Figure 10.10 shows the code. You access parameters from the text boxes using `app.xxx.Value`. For all plot-related functions, you need to add the axes handle using `app.UAxes.` or `app.UAxe2s.`

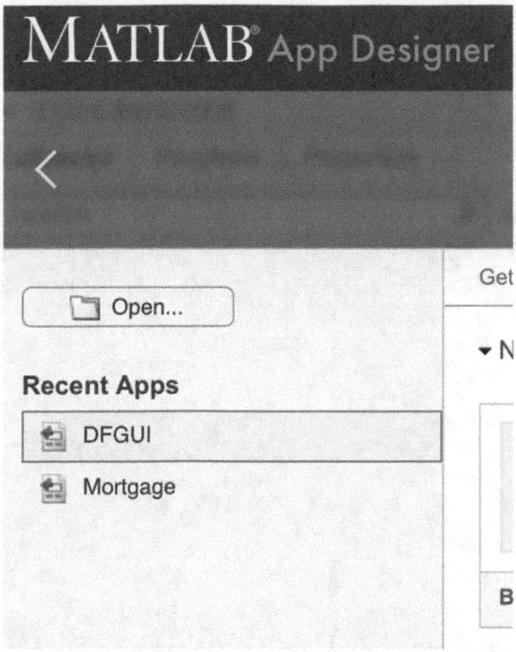

Figure 10.9: *The app appears in the app menu. You get to this window by hitting the design app button.*

Figure 10.11 shows a debugger breakpoint. You have full access to the debugger in App Designer. You will also see MATLAB warnings on the right.

Figure 10.12 shows the app after a run.

App Designer – /Users/Mike/svn/MATLABBooks/MATLABCookbook2/MATLAB/Chapter_09/DFGUI.mlapp*

```matlab
% Button pushed function: CalculateButton
function CalculateButtonPushed(app, event)
% get the data from the app
u        = app.InputEditField_3.Value;
duration = app.DurationEditField.Value;
tachF    = app.TachometerEditField.Value;
uF       = app.InputFaultEditField.Value;
dT       = app.TimeStepsEditField.Value;
tau1     = app.Tau1sEditField.Value;
tau2     = app.Tau2sEditField.Value;

% Set up the model
d        = RHSAirTurbine;
dF       = DetectionFilter('initialize',d,[tau1;tau2],dT);

% Plotting array
n        = ceil(duration/dT);
xP       = zeros(4,n);
t        = (0:n-1)*dT;
x        = [0;0];

for k = 1:n
% Measurement vector including measurement failure
  y      = [x(1);tachF*x(2)]; % Sensor failure
  xP(:,k) = [x;dF.r];

% Update the detection filter
  dF     = DetectionFilter('update',u,y,dF);

% Integrate one step
  d.u    = uF*u; % Actuator failure
  x      = RungeKutta( @RHSAirTurbine, t(k), x, dT, d );
end

[t,tL] = TimeLabel(t);

% Plot the states and residuals
axes(app.UIAxes)
plot(app.UIAxes,t, xP(1:2,:))
xlabel(app.UIAxes,tL);
legend(app.UIAxes,'p','\omega')

axes(app.UIAxes2)
plot(app.UIAxes2,t, xP(3:4,:))
xlabel(app.UIAxes2,tL);
legend(app.UIAxes2,'r_p','r_{\omega}')

end
```

Figure 10.10: *The light area is where the code goes. The code is from the simulation script.*

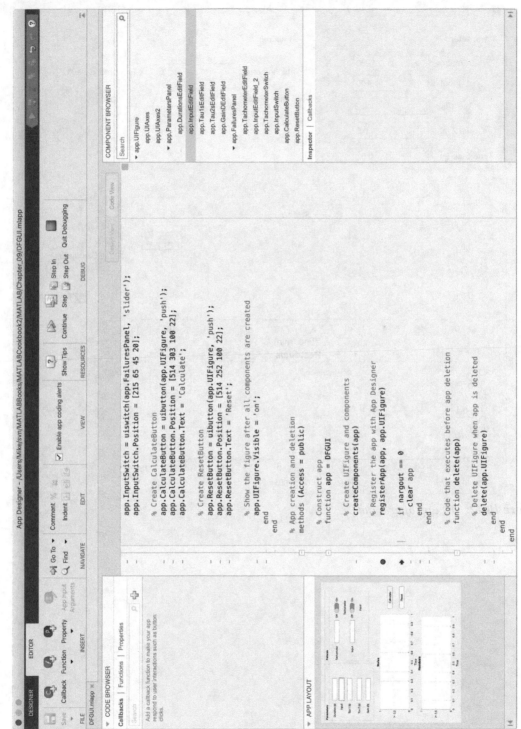

Figure 10.11: *You can use the debugger in the App Designer.*

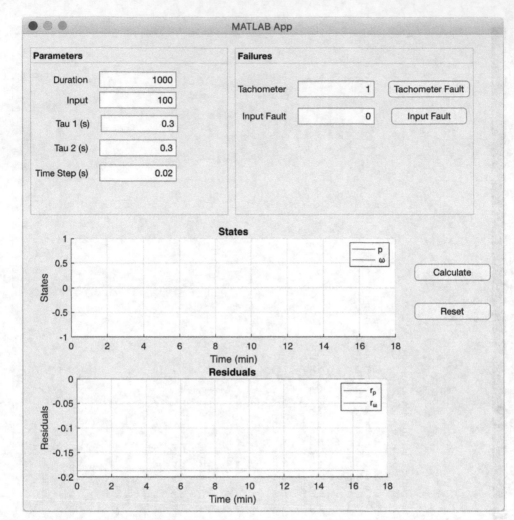

Figure 10.12: *The app after a run.*

Summary

This chapter has demonstrated how to design a detection filter for detecting faults in a dynamical system. The system is demonstrated with an air turbine that can experience a pressure regulator failure and a tachometer failure. In addition, we used App Designer to design a GUI to automate filter simulations. The GUI demonstrates real-time plotting and injecting failures into an ongoing simulation loop. Table 10.1 lists the code developed in the chapter.

Table 10.1: *Chapter Code Listing*

File	Description
RHSAirTurbine	Air turbine dynamical model in continuous state space form
DetectionFilter	Builds and updates a linear detection filter
DetectionFilterSim	Simulation of a detection filter
DetectionFilterGUI	Run the detection filter simulation from a GUI
DFGUI.m1App	App Designer app
DFGUI.mlappinstall	DFGUI app installer

CHAPTER 11

■ ■ ■

Chemical Processes

In chemical engineering, the production of chemicals needs to be modeled and the production process controlled. Our example will be a simple process in which the pH of a mixed solution needs to be controlled. This problem is interesting because the process is highly nonlinear and the sensor model does not have an explicit solution for the pH output. Modeling the sensor will require the use of the numerical solver `fzero`.

The specific chemical process we will study consists of an acid (HNO_3) stream, a buffer ($NaHCO_3$) stream, and a base ($NaOH$) stream that are mixed in a stirred tank.[1] This is based on a bench-scale experiment developed at the University of California, Santa Barbara (UCSB), to study chemical process control.[2] Figure 11.1 shows a diagram of the system, with three incoming streams $q1$, $q2$, and $q3$, and a valve to the output stream $q4$, where we will measure the pH. The goal will be to achieve a neutral pH.

11.1 Modeling the Chemical Mixing Process

Problem

We want to model the chemical process consisting of an acid stream, a buffer stream, and a base stream that are mixed in a stirred tank.

Solution

We model the chemical equilibria by adding two reaction invariants for each inlet stream and write the dynamical equations using the invariants. These are coded in a right-hand-side function that also defines the model data structure.

---

[1]Henson, M. A. and D. E. Seborg, "Nonlinear Process Control," Chapter 4: Feedback Linearizing Control, Prentice-Hall, 1997.

[2]Henson, M. and Seborg, D. "Adaptive Nonlinear Control of a pH Neutralization Process," IEEE Transactions on Control Systems Technology, Vol. 2, No. 3, August 1994.

© Michael Paluszek and Stephanie Thomas 2020

M. Paluszek and S. Thomas, *MATLAB Recipes*,

https://doi.org/10.1007/978-1-4842-6124-8_11

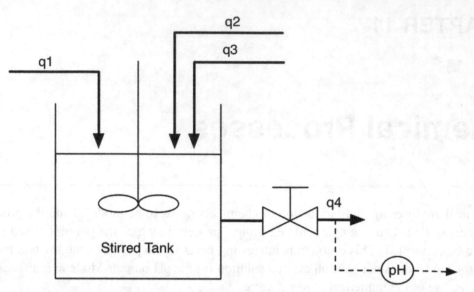

Figure 11.1: *Chemical mixing problem.*

How It Works

Reaction invariants are quantities whose values do not change during a reaction. Each is a combination of chemicals that do not vary. Our inputs are nitric acid (HNO₃), sodium bicarbonate or baking soda (NaHCO₃), and sodium hydroxide or lye (NaOH). There is a pair of invariants for each input stream i. The two reaction invariants W_a and W_b are

$$W_{ai} = [H^+]_i - [OH^-]_i - [HCO_3^-]_i - 2[CO_3^=]_i \qquad (11.1)$$
$$W_{bi} = [H_2CO_3]_i + [HCO_3^-]_i + [CO_3^=]_i \qquad (11.2)$$

$i = 1$ is for the acid stream, $i = 2$ for the buffer stream, $i = 3$ for the base stream, and $i = 4$ is for the mixed effluent. These combinations of chemicals do not change. The amounts of the combinations may change. The dynamical equations for the effluent invariants are derived via mass balances to be

$$\dot{W}_{a4} = \frac{1}{Ah}(W_{a1} - W_{a4})q_1 + \frac{1}{Ah}(W_{a2} - W_{a4})q_2 + \frac{1}{Ah}(W_{a3} - W_{a4})q_3 \quad (11.3)$$
$$\dot{W}_{b4} = \frac{1}{Ah}(W_{b1} - W_{b4})q_1 + \frac{1}{Ah}(W_{b2} - W_{b4})q_2 + \frac{1}{Ah}(W_{b3} - W_{b4})q_3 \quad (11.4)$$

where q_i is the volumetric flow rates for the ith stream, A is the cross-sectional area of the mixing tank, and h is the liquid level. The volumetric flow rates are inputs. The liquid level is governed by a differential equation:

$$\dot{h} = \frac{1}{A}\left[q_1 + q_2 + q_3 - C_v(h + z)^n\right] \qquad (11.5)$$

where C_v is the valve coefficient, n is the valve exponent, and z is the vertical distance between the bottom of the mixing tank and the outlet of the effluent stream. We can measure *h*. Normally, we need to estimate the reaction invariants, but for this problem, we will assume they are measured. These equations are all first order and are therefore the state equations for the system. The flow rates are all multiplied by the states, meaning that their influence on the derivatives is a product of the states and the streams. The differential equations for the effluent invariants are singular when $h = 0$. This is because if the tank is empty, the flows have to be zero.

The resulting right-hand-side function, RHSpH, is shown in the following. This follows the format needed by our RungeKutta integrator (see Recipe 6.2), that is, RHS(t,x,d), with a tilde replacing the first input, as the dynamics are independent of time. Note the data structure which is defined and returned if there are no inputs. This model has quite a few parameters which are documented in the header.

RHSpH.m

```
1  %% RHSPH Dynamics of a chemical mixing process.
2  % The process consists of an acid (HNO3) stream, buffer (NaHCO3) stream
   ,
3  % and base (NaOH) stream that are mixed in a stirred tank. The mixed
      effluent
4  % exits the tank through a valve. The chemical equilibria is modeled by
5  % introducing two reaction invariants for each inlet stream. i = 1 for
      the
6  % acid, i = 2 for the buffer, i = 3 for the base, and i = 4 for the
7  % effluent.
8  %
9  %           +           -           -         =
10 %    wAi = [H ]i - [OH ]i - [HCO3 ]i - 2[CO3 ]i
11 %                            -          =
12 %    wBi = [H2CO3]i + [HCO3 ]i + 2[CO3 ]i
13 %
14 %% Forms
15 %    d    = RHSpH
16 %    xDot = RHSpH( t, x, d )
17 %
18 %% Inputs
19 %    t              (1,1) Time, unused
20 %    x              (3,1) State [wA4;wB4;h]
21 %    d              (.)   Structure
22 %                   .wA1      (1,1) Acid invariant A, (M)
23 %                   .wA2      (1,1) Buffer invariant A, (M)
24 %                   .wA3      (1,1) Base invariant A, (M)
25 %                   .wB1      (1,1) Acid invariant B, (M)
26 %                   .wB2      (1,1) Buffer invariant B, (M)
27 %                   .wB3      (1,1) Base invariant B, (M)
28 %                   .a        (1,1) Cross-sectional area of mixing
      tank (cm2)
29 %                   .cV       (1,1) Valve coefficient
30 %                   .n        (1,1) Valve exponent
```

```
31  %                         .z          (1,1) Vertical distance between bottom
        of
32  %                                     tank and outlet of effluent (cm)
33  %                         .q1         (1,1) Volumetric flow of HNO3   (ml/s)
34  %                         .q2         (1,1) Volumetric flow of NaHCO3 (ml/s)
35  %                         .q3         (1,1) Volumetric flow of NaOH   (ml/s)
36  %
37  %% Outputs
38  %   xDot          (3,1) State derivative
39  %
40  %% Reference
41  % Henson, M. A. and D. E. Seborg. (1997.) Nonlinear Process
46  % All rights reserved.
47
48  function xDot = RHSpH( ~, x, d )
49
50  if( nargin < 1 )
51    % Note: Cv was omitted in the reference; we calculated it assuming a
          constant
52    % liquid level in the tank of 14 cm.
53    d = struct('wA1',0.003,'wA2',-0.03,'wA3',-3.05e-3,...
54               'wB1',0.0,  'wB2', 0.03,'wB3', 5.0e-5,...
55               'a',207,'cV',4.5860777,'n',0.607,'z',11.5,...
56               'q1',16.6,'q2',0.55,'q3',15.6);
57    xDot = d;
58    if( nargout == 0 )
59      disp('RHSpH struct:');
60    end
61    return;
62  end
63
64  wA4   = x(1);
65  wB4   = x(2);
66  h     = x(3);
67
68  hA    = 1/(h*d.a);
69
70  xDot  = [hA*( (d.wA1 - wA4)*d.q1 + (d.wA2 - wA4)*d.q2 + (d.wA3 - wA4)*d
        .q3 );...
71          hA*( (d.wB1 - wB4)*d.q1 + (d.wB2 - wB4)*d.q2 + (d.wB3 - wB4)*d
            .q3 );...
72          d.q1 + d.q2 + d.q3 - d.cV*(h + d.z)^d.n];
```

The default values in the data structure are drawn from the data in the reference, with the exception of C_v; this was neglected by the reference so we calculated a value for an equilibrium tank level.

302

```
1  ans =
2
3       wA1:  0.003
4       wA2:  -0.03
5       wA3:  -0.00305
6       wB1:  0
7       wB2:  0.03
8       wB3:  5e-05
9         a:  207
10       cV:  4.5861
11        n:  0.607
12        z:  11.5
13       q1:  16.6
14       q2:  0.55
15       q3:  15.6
```

In this chapter, we are interested in control about an equilibrium point. We could rewrite the equations as linear equations in which each state and input is replaced with, for example

$$q_3 = q_{30} + \delta q_3 \tag{11.6}$$
$$h = h_0 + \delta h \tag{11.7}$$

where δq_3 is small. We could then formally derive the linear control system. We will leave that for the interested reader and just go ahead and implement a linear control system.

11.2 Sensing the pH of the Chemical Process
Problem
The pH sensor is modeled by a nonlinear equation that cannot be solved explicitly for pH.

Solution
Use the MATLAB `fzero` function to solve for pH.

How It Works
The equation for pH is[1]

$$0 = W_{a4} + 10^{(\text{pH}-14)} - 10^{-\text{pH}} + W_{b4}\frac{1 + 2 \times 10^{(\text{pH}-pK_2)}}{1 + 10^{(pK_1-\text{pH})} + 10^{(\text{pH}-pK_2)}} \tag{11.8}$$

Recall that W_{a4} and W_{b4} are the reaction invariants for the mixed effluent as defined in Recipe 11.1. pK_1 and pK_2 are the base-10 logarithms of the H_2CO_3 and HCO_3^- disassociation constants.

$$pK_a = -\log_{10} K_a$$

The function that generates the measurement is `PHSensor`. Its inputs are two of the states of the system, W_{A4} and W_{B4}, and the dissociation constants.

PHSensor.m

```
1   %% PHSENSOR Model pH measurement of a mixing process
2   % Compute pH as a function of wA4 and wB4 and also the slope of pH with
3   % respect to those states. Requires the use of fzero.
4   %
5   %% Forms
6   %   pH = PHSensor( x, d )
7   %   d = PHSensor('struct')
8   %
9   %% Inputs
10  %   x    (2,:)  State [wA4;wB4]
11  %   d    (.)    Data structure
12  %               .pK1    (1,1) Base 10 log of a disassociation constant
        (H2CO3)
13  %               .pK2    (1,1) Base 10 log of a disassociation constant
        (HCO3-)
14  %
15  %% Outputs
16  %   pH   (:,:) pH of the solution
17  %
18  %% Reference
19  % Henson, M. A. and D. E. Seborg. Nonlinear Process control, Prentice-
        Hall,
20  % 1997. pp. 207-210.
```

The body of `PHSensor` calls `fzero` to compute the pH. This requires an objective function that will be searched for a zero near the input point. We use a neutral pH of 7.0 as the initial condition for the optimization. We could have warm started the function by using the last value that was computed. This is a good approach if the inputs don't change quickly. The function is vectorized for multiple input states, computing a square matrix of pH with the combinations of W_{a4} and W_{b4}.

```
43
44  % Compute the pH starting from neutral
45  n   = size(x,2);
46  pH  = zeros(n,n);
47  pH0 = 7.0;
48  for k = 1:n
49    for j = 1:n
50      d.wA4   = x(1,k);
51      d.wB4   = x(2,j);
52      pH(k,j) = fzero( @Fun, pH0, [], d );
53    end
54  end
```

■ **TIP**   Use `fzero` to solve for the zero point for complex single equations.   Use `fminzero` for sets of equations with multiple values to be found that minimized the function.

Notice that as per our usual pattern, we have defined a data structure d for passing data to the sensor model. Our two parameters are pK_1 and pK_2.

```
38  % Default data structure
39  if( ischar(x) )
40    pH = struct('pK1',-log10(4.47e-7),'pK2',-log10(5.62e-11));
41    return
42  end
```

Equation 11.8 is embodied in the subfunction `Fun` which is passed to `fzero`.

```
71  function y = Fun( pH, d )
72  %%% PHSensor>Fun Function to be zeroed via fzero
73  %   y = Fun( pH, d )
74
75  y   = d.wA4 + 10^(pH-14) - 10^(-pH) ...
76          + d.wB4*(1 + 2*10^(pH-d.pK2))/(1 + 10^(d.pK1-pH) + 10^(pH-d.
              pK2));
```

We include a demo in the function as suggested in the best practices described in the style recipes. The demo specifies a range of values for the states – the invariants – based on the numbers in the reference.

```
28  % Demo
29  if( nargin < 1 )
30    disp('Demo of PHSensor...');
31    x(1,:)  = linspace(-9e-4,0);
32    x(2,:)  = linspace(0,1e-3);
33    d       = PHSensor('struct');
34    PHSensor( x, d );
35    return
36  end
```

The results are plotted at the end of the main function using `mesh`. The mesh is the default plot.

```
56  % If no outputs, plot pH
57  if( nargout == 0 )
58    h = figure;
59    set(h,'Name','PH Sensor');
60    mesh(pH)
61    xlabel('WB4');
62    ylabel('WA4');
63    zlabel('pH')
```

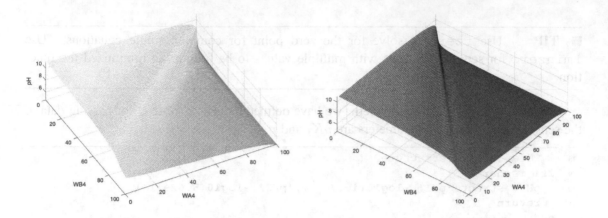

Figure 11.2: surf *and* mesh *plots of the pH sensor output.*

```
64    grid on
65    rotate3d on
66
67    clear pH
68  end
```

The plotting uses the mesh function. It is important to remember that the rows of p correspond to W_{a4} and the columns to W_{b4}. Columns are x and rows are y in the mesh plot. Figure 11.2 shows the mesh plot and also the alternative surf plot (with 'edgecolor' set to 'none'). Note the two MATLAB commands:

```
grid on
rotate3d on
```

Always use the on to be certain that the commands are executed rather than toggled. Otherwise, you can get unexpected results if you have just run another script or function with those commands.

Notice that the relationship between the pH and the reaction invariants is highly nonlinear. We would ideally like a relationship

$$pH = c \begin{bmatrix} W_{a4} \\ W_{b4} \end{bmatrix} \tag{11.9}$$

where c is a 2-by-2 matrix with constant coefficients. This might be true in the flat regions but is not true in the "waterfall" region.

When using this function in a simulation, we need it to run as fast as possible, without any diagnostics installed for fzero. During debugging, however, you may need additional information. fzero can display information on each iteration by setting the Display option. For instance, with the 'iter' setting, it will print out information for each iteration. The updated fzero call is

```
1   pH(k,j) = fzero( @Fun, pH0, optimset('Display','iter'), d );
```

and the results for a single state are

```
>> d = PHSensor('struct')
d =
    pK1: 6.3497
    pK2: 10.25

>> pH = PHSensor( [-4.32e-4;5.28e-4], d )

Search for an interval around 7 containing a sign change:
 Func-count    a          f(a)             b            f(b)
    Procedure
    1              7   -2.39821e-07             7   -2.39821e-07   initial
         interval
    3        6.80201   -4.16536e-05       7.19799    3.09792e-05   search

Search for a zero in the interval [6.80201, 7.19799]:
 Func-count    x          f(x)             Procedure
    3        7.19799    3.09792e-05        initial
    4         7.0291    4.96758e-06        interpolation
    5        7.00028   -1.88684e-07        interpolation
    6        7.00133    3.84365e-09        interpolation
    7        7.00131    2.86799e-12        interpolation
    8        7.00131              0        interpolation

Zero found in the interval [6.80201, 7.19799]

pH =

        7.0013
```

This was a very rapid solution as it is very near the starting point of 7.0. Note that fzero first found an interval containing a sign change, then searched for the zero. fzero can also output diagnostic information when complete instead of printing it during operation. For instance, if the call is

```
1   [pH(k,j),fval,exitflag,output] = fzero( @Fun, pH0, [], d );
```

then the output structure will be available, such as

```
1   output =
2
3        intervaliterations: 1
4                 iterations: 5
5                  funcCount: 8
6                  algorithm: 'bisection, interpolation'
7                    message: 'Zero found in the interval [6.80201, 7.19799]'
```

Note that the algorithm used, that is, bisection, is listed along with the total number of iterations and function evaluations. Consider a slight variation of the input state, lowering Wb4 to $4e^{-4}$ M. The number of iterations jumps significantly.

```
>> pH = PHSensor( [-4.32e-4;4e-4], d )
output =

    intervaliterations: 8
             iterations: 7
              funcCount: 24
              algorithm: 'bisection, interpolation'
                message: 'Zero found in the interval [4.76, 9.24]'
pH =
        9.022
```

Note in particular that viewing the diagnostic information for your problem can help confirm if your tolerances are suitable. Consider the final iterations of the previous case:

```
Search for a zero in the interval [4.76, 9.24]:
 Func-count      x           f(x)             Procedure
     17          9.24     2.04571e-05         initial
     18        9.04068    1.42294e-06         interpolation
     19        9.02583    2.87604e-07         interpolation
     20        9.02207    5.59178e-09         interpolation
     21        9.02199    2.25694e-11         interpolation
     22        9.02199     1.7808e-15         interpolation
     23        9.02199    5.42101e-20         interpolation
     24        9.02199    5.42101e-20         interpolation
```

The search pushed the function value all the way down to 5.4e-20, which may be more restrictive than needed. The default tolerances can be viewed by getting the default options structure using optimset.

```
>> options = optimset('fzero')

options =

        Display: 'notify'
     MaxFunEvals: []
```

308

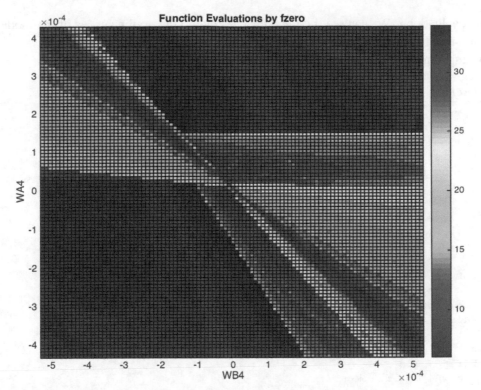

Figure 11.3: *Function evaluations for the PHSensor algorithm.*

```
      MaxIter: []
       TolFun: []
         TolX: 2.2204e-16
  FunValCheck: 'off'
    OutputFcn: []
     PlotFcns: []
```

The default tolerance on the function value, `TolX`, is 2.2204e-16. Note that we passed in the name of the selected optimization routine, `fzero`, to `optimset`. The same can be done with `fminbnd` and `fminsearch`.

■ **TIP**   Use `optimset` with the name of the optimization function to get the default options structure.

Now consider that you want to evaluate how `fzero` is performing over a range of inputs. Assume that we create a separate function for `Fun` and make a script to record extra data from `output` during a run. Figure 11.3 shows a plot using `pcolor` of the resulting recorded

309

function evaluations. We can see the rapid changes due to the nonlinearities of the model. The maximum number of function evaluations does not exceed 35.

The augmented code creating the plot is shown as follows:

SensorTest.m

```
1    %% Script for debugging PHSensor algorithm
6
7    % Nominal operating conditions from the reference
8    x10 = -4.32e-4;
9    x20 = 5.28e-4;
10
11   x        = [];
12   x(1,:)   = linspace(2*x10,0);
13   x(2,:)   = linspace(0,2*x20);
14   d        = PHSensor('struct');
15
16   % Compute the pH starting from neutral
17   n     = size(x,2);
18   pH    = zeros(n,n);
19   fEvals = zeros(n,n);
20   pH0 = 7.0;
21   for k = 1:n
22     for j = 1:n
23       d.wA4   = x(1,k);
24       d.wB4   = x(2,j);
25       % Options: TolX, Display, FunValCheck
26       % ('TolX',1e-10);
27       % ('Display','iter')
28       % ('FunValCheck','on') % no errors found for demo
29       options = optimset('FunValCheck','on');
30       [pH(k,j),fval,exitflag,output] = fzero( @Fun, pH0, options, d );
31       fEvals(k,j) = output.funcCount;
32     end
33   end
34
35   figure('Name','PH Sensor');
36   surf(pH,'edgecolor','none')
37   xlabel('WB4');
38   ylabel('WA4');
39   zlabel('pH')
40   grid on
41   rotate3d on
42
43   figure('Name','Evaluations')
44   s = pcolor(x(2,:),x(1,:),fEvals);
45   set(s,'edgecolor','none')
46   xlabel('WB4');
```

```
47   ylabel('WA4');
48   colormap jet
49   title('Function Evaluations by fzero')
```

11.3   Controlling the Effluent pH

Problem

We want to control the pH level in the mixing tank when the flow of acid or base varies.

Solution

We will vary the base stream, $i = 3$, to maintain the pH. This means changing the value of q_3 using a proportional-integral controller. This will allow us to handle step disturbances.

How It Works

A proportional-integral controller is of the form

$$u = K \left(1 + \frac{1}{\tau} \int \right) y \tag{11.10}$$

where u is the control, y is the measurement, τ is the integrator time constant, and K is the forward (proportional) gain. The control is $u = q_3$ and the measurement is $y = $ pH. This makes this a single-input-single-output process. However, the connection between q_3 and pH involves three dynamical states h, W_{a4}, and W_{b4}, and the relationship between the states and pH is nonlinear. Another issue is that q_3 cannot be negative, that is, we cannot extract the base from the tank. This should not pose a problem if the equilibrium q_3 is high enough.

Despite these potential problems, this very simple controller will work for this problem for a fairly wide range of disturbances. The equilibrium value is input with a perturbation that has a proportional and integral term. The integral term uses a simpler Euler integration. The full script is described in the next recipe; here, we will call attention to the lines implementing the control.

The control variables are defined in the following. The pH set point is neutral, that is, a pH of 7. kF is the forward gain and tau is the time constant from Equation 11.10, set to 2 and 60 seconds, respectively. q3Set is the nominal set point for the base flow rate, taken from the reference.

PHProcessSim.m

```
41   %% Control design
42   pHSet   = 7.0;
43   tau     = 60.0;  % (sec)
44   kF      = 2.0;   % forward gain
45   q3Set   = 15.6;  % (ml/s)
```

The following code snippet shows the implementation that takes place in a loop. The error is calculated as the difference between the modeled pH measurement and the pH set point. Note the Euler integration, where `intErr` is updated using simply the time step times the error. Note also that we have a flag, `controlIsOn`, which allows us to run the script in open loop, without the control being applied.

```
68    % Proportional-integral Control
69    err = pH - pHSet;
70    if controlIsOn
71      d.q3    = q3Set - kF*(err + intErr/tau);
72      intErr = intErr + dT*err;
73    else
74      d.q3 = q3Set;
75    end
```

To rigorously determine the forward gain and time constant for this problem, we would need to linearize the right-hand side for the simulation at the operating point and do a rigorous single-input-single-output control design that would involve Bode plots, root locus, and other techniques. This is beyond the scope of this book. For now, we simply select values which produce a reasonable response.

11.4   Simulating the Controlled pH Process

Problem

We want to simulate the stirred tank – mixing three streams: acid, buffer, and base – to demonstrate the control of the pH level. The base stream will be our control variable in response to perturbations in the buffer and acid streams.

Solution

We will write a script `PHProcessSim` with the controller starting at an equilibrium state. We will use the proportional-integral controller as derived in the previous recipe. We will structure the script to allow us to insert pulses in either or both the acid and buffer streams.

How It Works

Our disturbances d are deviations in q_1 and q_2:

$$d = \begin{bmatrix} q_1 \\ q_2 \end{bmatrix} \tag{11.11}$$

which are the acid and buffer streams. The base stream is q_3 and the reaction invariants for the mixed effluent are W_{a4} and W_{b4}. The system, as coded, is shown in Figure 11.4.

We specify the user inputs to the script first. We are putting it into an equilibrium state and will investigate small disturbances from steady state. There is a flag, `controlIsOn`, for turning the control system on or off. The time step and duration are determined by iterating over a few values.

PHProcessSim.m

```
12  %% User inputs
13
14  % Time (sec)
15  tEnd = 60*60;
16  dT   = 1.0;
17
18  % States
19  wA4 = -4.32e-4; % Reaction invariant A for effluent stream (M)
20  wB4 =  5.28e-4; % Reaction invariant B for effluent stream (M)
21  h   = 14.0;     % liquid level (cm)
22
23  % Closed or open loop
24  controlIsOn = true;
```

The disturbances are generated as pulses to `d.q1` and `d.q2`. The user parameters are the size and the start/stop times of the pulses. This setup will allow us to run cases similar to the reference.

```
26  % Disturbances
27  % The pulses will be applied according to the start and end times of
          tPulse:
28  %     q1 = q10 + deltaQ1;  and   q2 = q20 + deltaQ2;
29  % Small pulse: 0.65 ml/s in q2
30  % Large pulses: 2.0 ml/s
31  % Very large pulses: 8.0 ml/s
32  deltaQ1 = 8.0; % +/- 1.5
33  deltaQ2 = 0.0; % 0.65 1.45 0.45 -0.55 % values from reference
34  tPulse1 = [5 15]*60;
35  tPulse2 = [5 35]*60;
```

In the remainder of the script, we obtain the data structures defined in the previous recipes, specify the control parameters, and create the simulation loop. The measurements of the invariants are assumed to be exact; in practice, they need to be estimated. However, we should always test the controller under ideal conditions first, to understand its behavior without complications.

313

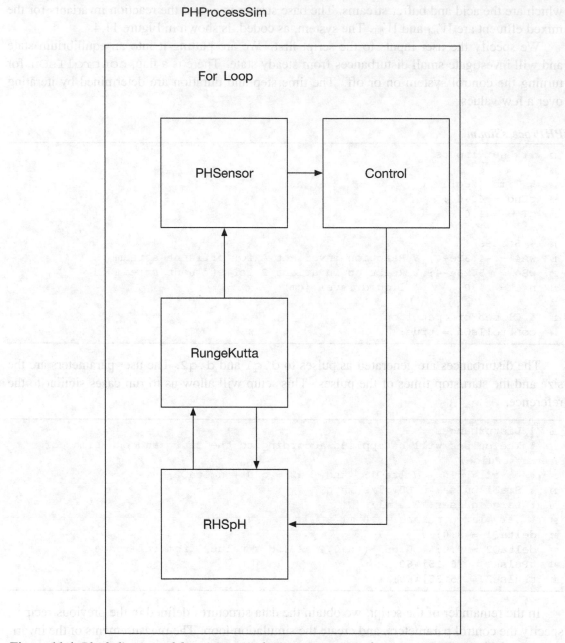

Figure 11.4: *Block diagram of the mixing simulation representing the process in Figure 11.1.*

The pH measurement is modeled using PHSensor from Recipe 11.2. The right-hand side for the process is defined in Recipe 11.1. Integration is performed using the RungeKutta defined in Chapter 7.

```
37  %% Data format
38  dSensor = PHSensor('struct');
39  d       = RHSpH;
40
41  %% Control design
42  pHSet   = 7.0;
43  tau     = 60.0; % (sec)
44  kF      = 2.0;  % forward gain
45  q3Set   = 15.6; % (ml/s)
46  q10     = d.q1;
47  q20     = d.q2;
48
49  %% Run the simulation
50
51  % Number of sim steps
52  n = ceil(tEnd/dT);
53
54  % Plotting arrays
55  xP = zeros(7,n);
56  t  = (0:n-1)*dT;
57
58  % Initial states
59  x       = [wA4;wB4;h];
60  intErr = 0;
61
62  for k = 1:n
63    % Measurement
64    dSensor.wA4 = x(1);
65    dSensor.wB4 = x(2);
66    pH          = PHSensor( x, dSensor );
67
68    % Proportional-integral Control
69    err = pH - pHSet;
70    if controlIsOn
71      d.q3    = q3Set - kF*(err + intErr/tau);
72      intErr = intErr + dT*err;
73    else
74      d.q3 = q3Set;
75    end
76
77    % Disturbance
78    if( t(k) > tPulse1(1) && t(k) < tPulse1(2) )
79      d.q1 = q10 + deltaQ1;
80    else
81      d.q1 = q10;
82    end
83
```

315

```
84    if( t(k) > tPulse2(1) && t(k) < tPulse2(2) )
85        d.q2 = q20 + deltaQ2;
86    else
87        d.q2 = q20;
88    end
89
90    % Store data for plotting
91    xP(:,k) = [x;pH;d.q1;d.q2;d.q3];
92
93    % Integrate one step
94    x = RungeKutta( @RHSpH, 0, x, dT, d );
95    end
96
97    %% Plot
98    [t,tL]   = TimeLabel(t);
99    yL       = {'W_{a4}' 'W_{b4}' 'h' 'pH' 'q_1' 'q_2' 'q_3'};
100   tTL      = 'PH Process Control';
101   if ~controlIsOn
102     tTL = [tTL ' - Open Loop'];
103   end
104   PlotSet( t, xP,'x label',tL,'y label',yL,'plot title',tTL,'figure title
          ',tTL)
105   PlotSet( t, xP([4 7],:),'x label',tL,'y label',yL([4 7]),'plot title',
          tTL,'figure title',tTL)
```

Now, we will give results for running this script with some different pulses. The nominal plot gives all three states, the measured pH, and the flow rates for the acid, base, and buffer streams. A more compact plot shows just the pH and the commanded value of q_3. We added a line in the plotting code to amend the plot title for an open loop response, so that if we run the script repeatedly, we can more easily identify the plots.

■ **TIP**   Use your control flags and string variables to customize the names of your plots.

Figure 11.5 shows the closed loop response with no disturbances at all, run for 30 simulated minutes. We can see that the values from the reference have not produced an exact equilibrium, but that the values achieved are quite close. The reaction invariant Wb4 changes by less than 0.005×10^{-4}, the liquid level h by less than 0.1 cm, and the base flow rate q3 by about 0.05 ml/s. This is the equivalent to a very small step response. Note the settling time is about 5 minutes. These results give us confidence that we have coded the problem correctly. We will see this initial response in the following simulations, before the perturbations are applied.

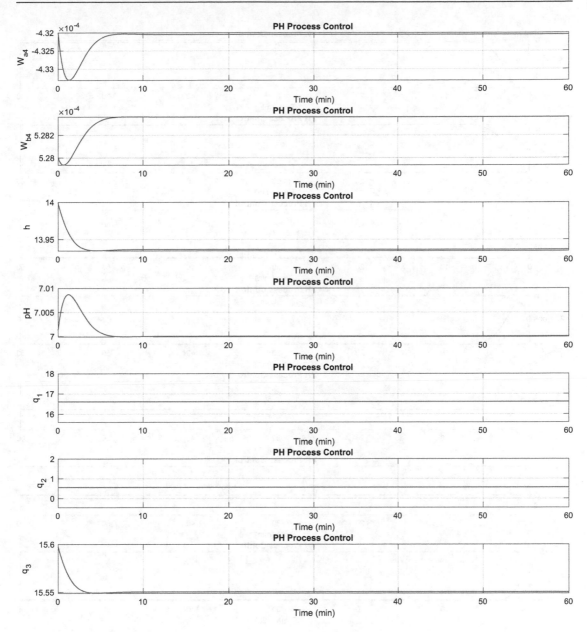

Figure 11.5: *Closed loop response with no perturbations.*

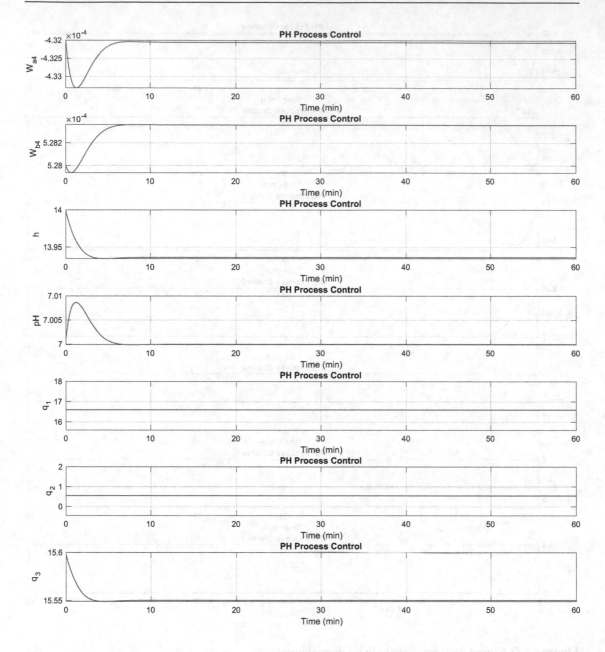

Figure 11.6: *Open loop response with a 0.65 ml/s pulse in q_2.*

Next, Figure 11.6 shows the open loop response with a pulse of 0.65 ml/s in the buffer stream starting at 20 minutes and ending at 40 minutes. Note that the pH rises to nearly 7.4, and q_3 does in fact stay constant at our set point. Figure 11.7 shows the closed loop transients in the pH and base flow q_3. The pH rise is limited to less than 7.2, and the pH and base flow

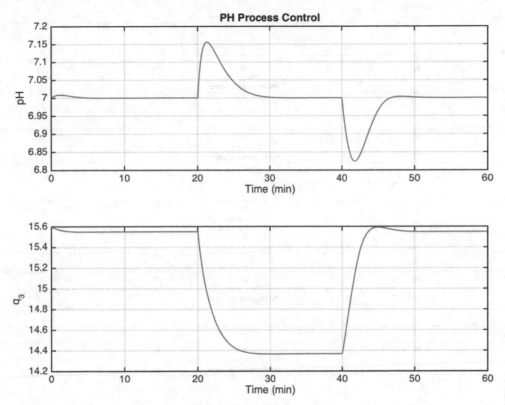

Figure 11.7: *Performance of the controller with a 0.65 ml/s pulse in q_2.*

rate reach equilibrium within about 10 minutes of the start and end of the pulse. This compares favorably with the plots in the reference, which compare adaptive and nonadaptive nonlinear control schemes.

Figure 11.8 shows the transients with larger offset perturbations of 2 ml/s in both q_1 and q_2. The pulse in q_1 is applied from 5 to 15 minutes and the pulse in q_2 from 25 to 40 minutes. Figure 11.9 has plots of just the pH and control flow q_3.

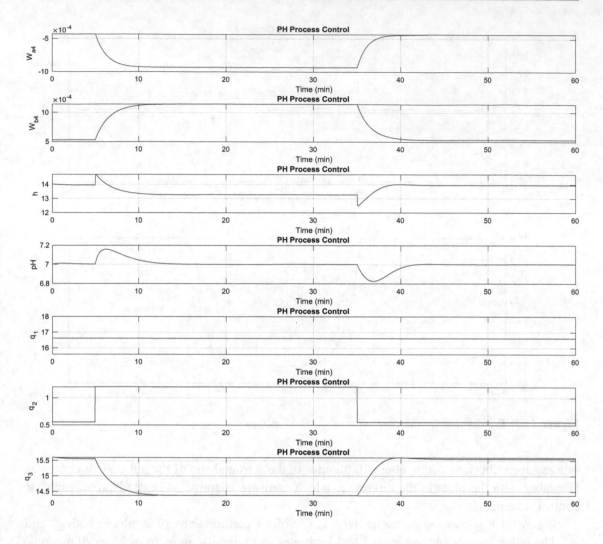

Figure 11.8: *Performance of the controller with large perturbations of 2 ml/s in q_1 and q_2.*

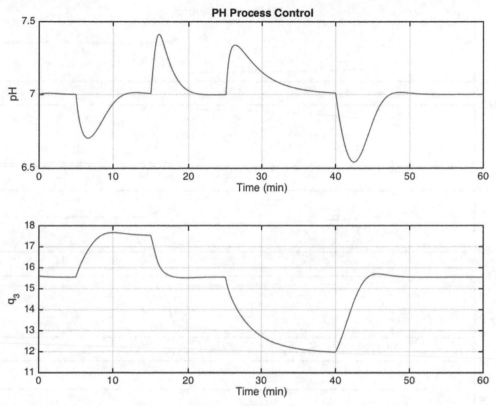

Figure 11.9: *Larger plots of the q_1 /q_2 perturbation results.*

Figure 11.10 shows the transients with a very large perturbation in q_2 of 8 ml/s from 5 to 35 minutes. The controller no longer works very well, with a much longer settling time than the previous examples and the base flow rate q_3 still dropping at the end of the pulse. A pulse value of 10 ml/s causes the simulation to "blow up" or produce imaginary values. It is always necessary to see the limits of the control performance in a nonlinear system.

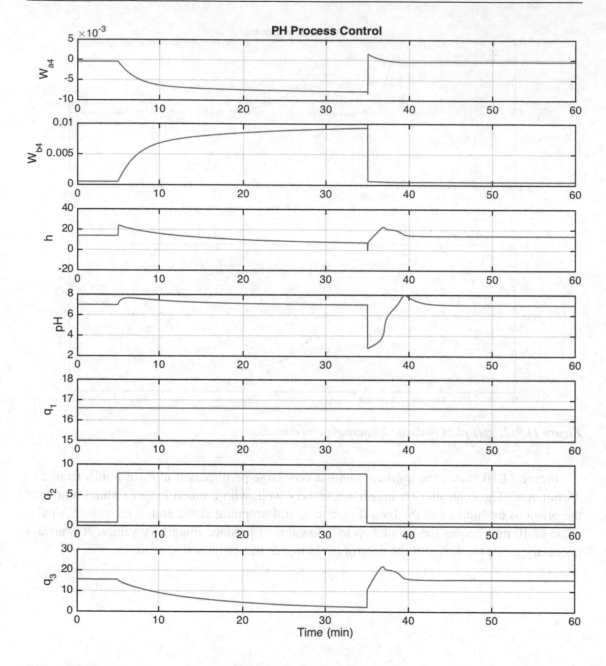

Figure 11.10: *Performance of the controller with a very large perturbation (8 ml/s) in q_2.*

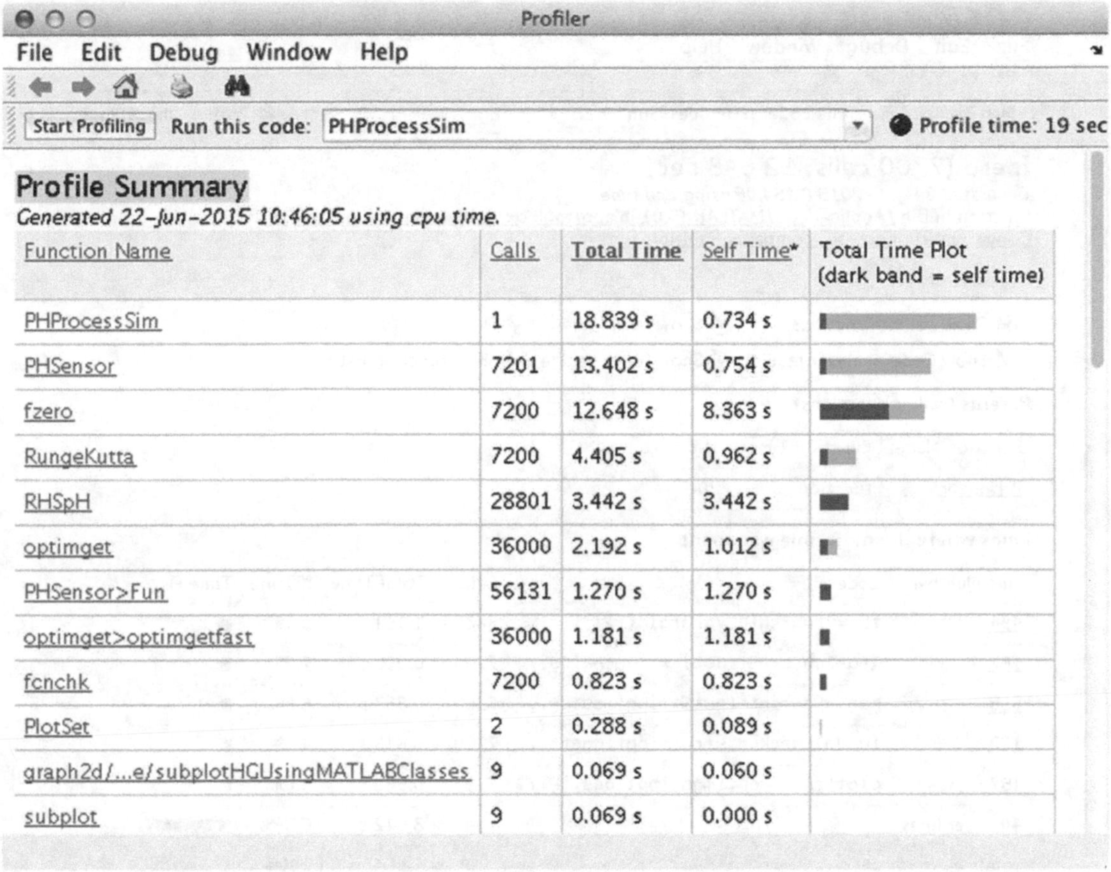

Function Name	Calls	Total Time	Self Time*	Total Time Plot (dark band = self time)
PHProcessSim	1	18.839 s	0.734 s	
PHSensor	7201	13.402 s	0.754 s	
fzero	7200	12.648 s	8.363 s	
RungeKutta	7200	4.405 s	0.962 s	
RHSpH	28801	3.442 s	3.442 s	
optimget	36000	2.192 s	1.012 s	
PHSensor>Fun	56131	1.270 s	1.270 s	
optimget>optimgetfast	36000	1.181 s	1.181 s	
fcnchk	7200	0.823 s	0.823 s	
PlotSet	2	0.288 s	0.089 s	
graph2d/...e/subplotHGUsingMATLABClasses	9	0.069 s	0.060 s	
subplot	9	0.069 s	0.000 s	

Figure 11.11: *Profiler summary from the simulation.*

Finally, since we are using a numerical optimization routine, it is instructive to profile the simulation to determine the proportion of the execution time spent on `fzero`. The Profiler can be accessed from the command window button called "Run and Time." The summary from running our pH simulation is shown in Figure 11.11. Out of nearly 19 seconds spent in the simulation, fully 12.6 seconds are spent inside `fzero` itself. Only 4.4 seconds were spent integrating of which 3.4 seconds were spent in the right-hand side. The summary has hyperlinks to the individual functions, which are timed line by line. Figure 11.12 shows the time spent inside `fzero`, with percentages calculated in addition to absolute times. Our objective function, `PHSensor>Fun`, was called 56,131 times, taking only 1.27 seconds (10% of the execution time). Significant chunks of time were spent in `sprintf` and `optimget`.

○ ○ ○   Profiler

File   Edit   Debug   Window   Help

←   →   ⌂   🖶   🔍

[Start Profiling] Run this code: | PHProcessSim   ▼ | ● Profile time: 19 sec

fzero (7200 calls, 12.648 sec)

Generated 24-Jun-2015 09:54:06 using cpu time.
function in file /Applications/MATLAB_R2014b.app/toolbox/matlab/optimfun/fzero.m
Copy to new window for comparing multiple runs

[Refresh]

☑ Show parent functions   ☑ Show busy lines   ☑ Show child functions

☑ Show Code Analyzer results ☑ Show file coverage ☑ Show function listing

Parents (calling functions)

Function Name	Function Type	Calls
PHSensor	function	7200

Lines where the most time was spent

Line Number	Code	Calls	Total Time	% Time	Time Plot
494	fb = FunFcn(b,varargin{:});	36879	1.151 s	9.1%	■
167	[FunFcn,errStruct] = fcnchk(Fu...	7200	0.923 s	7.3%	■
525	msg = sprintf(getString(messag...	7200	0.863 s	6.8%	■
132	funValCheck = strcmp(optimget(...	7200	0.615 s	4.9%	▮
157	plotfcns = optimget(options,'P...	7200	0.585 s	4.6%	▮
All other lines			8.512 s	67.3%	■■■■■■■
Totals			12.648 s	100%	

Children (called functions)

Function Name	Function Type	Calls	Total Time	% Time	Time Plot
optimget	function	36000	2.192 s	17.3%	■■
PHSensor>Fun	subfunction	56131	1.270 s	10.0%	■
fcnchk	function	7200	0.823 s	6.5%	■
Self time (built-ins, overhead, etc.)			8.363 s	66.1%	■■■■■■■
Totals			12.648 s	100%	

Figure 11.12: *Profiler results for* fzero.

■ **TIP**   Always do a run with the Profiler when you are implementing a numerical search or optimization routine. This gives you insight into the number of iterations used and any unsuspected bugs in your code.

In this case, there is not much optimization that can be done as most of the time is spent in `fzero` itself and not in our objective function, but we wouldn't have known that without running the analysis. Whenever you are using numerical tools and have a script or function taking more than a second or two to run, analysis with Profiler is merited.

Summary

This chapter has demonstrated how to write the dynamics and implement a simple control law for a chemical process. The process is highly nonlinear, but we can control the process with a simple proportional-integral controller. The pH sensor does not have a closed form solution, and we use the MATLAB `fzero` function to find the pH from the invariants. We demonstrated the use of MATLAB plotting functions `mesh` and `surf` for showing three-dimensional data. We use our simulation script to evaluate the performance of the controller for a variety of conditions and run the script in the Profiler to analyze the time spent on the numerical routines. Table 11.1 lists the code developed in the chapter.

Table 11.1: *Chapter Code Listing*

File	Description
PHSensor	Model pH measurement of a mixing process
PHProcessSim	Simulation of a pH neutralization process
RHSpH	Dynamics of a chemical mixing process
SensorTest	Script to test the sensor algorithm

CHAPTER 12

■ ■ ■

Aircraft

Our aircraft model will be a three-dimensional point mass model. This models the translational dynamics in three dimensions. Translation is motion in the x, y, and z directions. An aircraft controls its motion by changing its orientation with respect to the wind (banking and angle of attack) and by changing the thrust its engine produces. In our model, we assume that our airplane can instantaneously change its orientation and thrust for control purposes. This simplifies our model but at the same time allows us to simulate most aircraft operations, such as takeoff, level flight, and landing. We also assume that the mass of the aircraft does not vary with time.

12.1 Creating a Dynamical Model of an Aircraft

Problem

We need a numerical model to simulate the three-dimensional trajectory of an aircraft in the atmosphere. The model should allow us to demonstrate control of the aircraft from takeoff to landing.

Solution

We will build a six-state model using flight path coordinates. Our controls will be the roll angle, angle of attack, and thrust. We will not simulate the attitude dynamics of the aircraft. The attitude dynamics are necessary if we want to simulate how long it takes for the aircraft to change the angle of attack and roll angle. In our model, we will assume the aircraft can instantaneously change the angle of attack, roll angle, and thrust.

How It Works

Our aircraft will have six states, needed to simulate the velocity and position in three dimensions, and three controls. Our controls will be the roll angle, ϕ; angle of attack, α; and thrust T. We aren't going to use Cartesian (x, y, z) coordinates and their time derivatives (i.e., velocities) as states; instead, we will use flight path coordinates. Flight path coordinates are shown in two dimensions in Figure 12.1. The roll, ϕ, is about the x axis, and the heading ψ is out of the page. Drag D is opposite to the velocity vector. The angle of attack α is adjusted to change the lift

© Michael Paluszek and Stephanie Thomas 2020

M. Paluszek and S. Thomas, *MATLAB Recipes*,

https://doi.org/10.1007/978-1-4842-6124-8_12

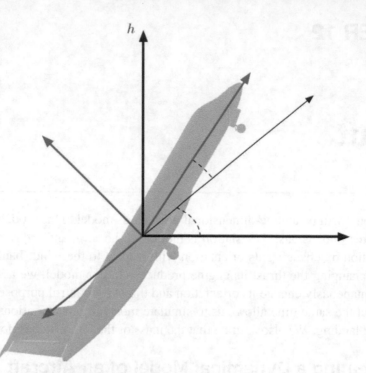

Figure 12.1: *Flight path coordinates in two dimensions.*

L and drag. The thrust vector T is aligned with the body x axis. The flight path angle γ is the angle between the x axis and the velocity vector V. The state vector, s, is

$$s = \begin{bmatrix} v \\ \gamma \\ \psi \\ x \\ y \\ h \end{bmatrix} \tag{12.1}$$

v is the velocity magnitude, γ is the flight path angle in the xz plane, ψ is the angle in the xy plane, and x, y, h are the Cartesian coordinates. h is the altitude or the z coordinate. The dynamical equations are

$$\begin{bmatrix} \dot{v} \\ \dot{\gamma} \\ \dot{\psi} \\ \dot{x} \\ \dot{y} \\ \dot{h} \end{bmatrix} = \begin{bmatrix} \frac{T\cos\alpha - D}{m} - g\cos\gamma \\ \frac{(L+T\sin\alpha)\cos\phi - mg\cos\gamma}{mv} \\ \frac{(L+T\sin\alpha)\sin\phi}{mv\cos\gamma} \\ v\cos\gamma\cos\psi \\ v\cos\gamma\sin\psi \\ v\sin\gamma \end{bmatrix} \tag{12.2}$$

g is the gravitational acceleration, m is the mass, D is the drag force, and L is the lift force. Define the dynamic pressure as

$$q = \frac{1}{2}\rho v^2 \tag{12.3}$$

where ρ is the atmospheric density. Our simple lift and drag model is

$$D = q\left(C_{D_0} + kC_L^2\right) \tag{12.4}$$

$$L = qC_L \tag{12.5}$$

$$C_L = C_{L_\alpha}\alpha \tag{12.6}$$

The first equation is called the drag polar. C_{D_0} is the drag at zero lift. k is the drag from the lift coupling coefficient. Polar comes from the C_L^2 term. This lift model is only valid for small angles of attack α as it does not account for stall, which is when the airflow becomes detached from the wing and the lift goes to zero rapidly.

Our RHS function that implements these equations is RHSAircraft. Notice that the equations are singular when $v = 0$ in Equation 12.2. We warn about this in the header. If called without arguments, the functions return the data structure. This is a handy way of making a complex function easier to use. It also gives the user an idea of what the parameters should be. All parameters must be in Meters-Kilogram-Second (MKS) units.

RHSAircraft.m

```
1   %% RHSAIRCRAFT Dynamics for a six DOF point mass aircraft model.
2   %% Form
3   %    d               = RHSAircraft;
4   %    [sDot, D, LStar] = RHSAircraft( t, s, d )
5   %% Description
6   % Computes the right hand side for a point mass aircraft. If you call
        it
7   % without any arguments, it will return the default data structure.
8   % sDot(2) and sDot(3) will be infinite when v = 0. The default
        atmosphere
9   % model is AtmosphericDensity which uses an exponential atmosphere.
10  %
11  %% Inputs
12  %    t       (1,1)    Time (unused)
13  %    s       (6,1)    State vector [v;gamma;psi;x;y;h]
14  %    d       (.)      Data structure
15  %                     .m         (1,1) Aircraft mass
16  %                     .g         (1,1) Gravitational acceleration
17  %                     .thrust    (1,1) Thrust
18  %                     .alpha     (1,1) Angle of attack
19  %                     .phi       (1,1) Roll angle
20  %                     .s         (1,1) Surface area
21  %                     .cD0       (1,1) Zero lift drag
22  %                     .k         (1,1) Lift drag coupling term
23  %                     .cLAlpha   (1,1) Lift coefficient
24  %                     .density   (1,1) Pointer to the atmospheric
```

```
25  %                                              density function
26  %
27  %% Outputs
28  %    sDot   (6,1)    State vector derivative d[v;gamma;psi;x;y;h]/dt
29  %    D      (1,1)    Drag
30  %    LStar  (1,1)    Lift/angle of attack
```

The function body is shown in the following. We assemble the state derivative, sDot, as one array since the terms are simple. Each element is on a separate line for readability. We return D and LStar as auxiliary outputs for use by the equilibrium calculation.

```
36  function [sDot, D, LStar] = RHSAircraft( ~, s, d )
37
38  % Default data structure
39  if( nargin == 0 )
40    sDot = struct('m',5000, 'g', 9.806, 'thrust',0,'alpha',0, 'phi',0,...
41                  'cLAlpha',2*pi,'cD0',0.006,'k',0.06,'s',20,'density',
                    @AtmosphericDensity);
42    if( nargout == 0 )
43      disp('RHSAircraft struct:')
44    end
45    return
46  end
47
48  % Save as local variables
49  v     = s(1);
50  gamma = s(2);
51  psi   = s(3);
52  h     = s(6);
53
54  % Trig functions
55  cG    = cos(gamma);
56  sG    = sin(gamma);
57  cPsi  = cos(psi);
58  sPsi  = sin(psi);
59  cB    = cos(d.phi);
60  sB    = sin(d.phi);
61
62  % Exponential atmosphere
63  rho   = feval(d.density,h);
64
65  % Lift and Drag
66  qS    = 0.5*rho*d.s*v^2;    % dynamic pressure
67  cL    = d.cLAlpha*d.alpha;
68  cD    = d.cD0 + d.k*cL^2;
69  LStar = qS*d.cLAlpha;
70  L     = qS*cL;
71  D     = qS*cD;
72
73  % Velocity derivative
74  % sDot is d[v;gamma;psi;x;y;h]/dt
```

```
75   lT   = L + d.thrust*sin(d.alpha);
76   sDot = [ (d.thrust*cos(d.alpha) - D)/d.m - d.g*sG;...
77          (lT*cB - d.m*d.g*cG)/(d.m*v);...
78          lT*sB/(d.m*v*cG);...
79          v*cPsi*cG;...
80          v*sPsi*cG;...
81          v*sG];
```

A more sophisticated right-hand side would pass function handles for the drag and lift calculations so that the user could use their own model. We pass a function handle for the atmospheric density calculation to allow the user to select their density function. We could have done the same for the aerodynamics model. This would make RHSAircraft more flexible.

Notice that we had to write an atmospheric density model, AtmosphericDensity, to provide as a default for the RHS function. This model uses an exponential equation for the density, which is the simplest possible representation. The function has a demo demonstrating the model which uses a log scale for the plot. This function also plots the results if no outputs are requested. This is also useful for helping users figure out what a function does.

AtmosphericDensity.m

```
1    %% ATMOSPHERICDENSITY Compute atmospheric density from an exponential
         model.
2    % Computes the atmospheric density at the given altitude using an
3    % exponential model. Produces a demo plot up to an altitude of 100 km.
4    %% Form
5    %   rho = AtmosphericDensity( h )
6    %
7    %% Inputs
8    %   h      (1,:)    Altitude (m)
9    %
10   %% Outputs
11   %   rho    (1,:)    Density (kg/m^3)
16
17   function rho = AtmosphericDensity( h )
18
19   % Demo
20   if( nargin < 1 )
21     disp('Demo of AtmosphericDensity');
22     h = linspace(0,100000);
23     AtmosphericDensity( h );
24     return
25   end
26
27   % Density
28   rho = 1.225*exp(-2.9e-05*h.^1.15);
29
30   % Plot if no outputs are requested
31   if( nargout < 1 )
32     PlotSet(h,rho,'x label','h (m)','y label','Density (kg/m^3)',...
33          'figure title','Exponential Atmosphere',...
```

331

```
34                'plot title','Exponential Atmosphere',...
35                'plot type','y log');
36    clear rho
37  end
```

12.2   Finding the Equilibrium Controls for an Aircraft Using Numerical Search

Problem

We want to find roll angles, thrusts, and angles of attack that cause the velocity, flight path angle, and bank angle state (roll angle) derivatives to be zero. This is a point of equilibrium. This is commonly called the trim condition.

Solution

We will use the Downhill Simplex algorithm, via the MATLAB function `fminsearch`, to find the equilibrium angles. `fminsearch` supports multivariable unconstrained optimization. The optimization toolbox in MATLAB provides additional functions with more options, such as handling constraints, that is, limits on the controls.

How It Works

The first step is to find the controls that produce a desired equilibrium state, known as the set point. Define the set point as the vector:

$$\begin{bmatrix} v_s \\ \gamma_s \\ \psi_s \end{bmatrix} \quad (12.7)$$

with set values for the velocity v_s, heading ψ_s, and flight path angle γ_s. We want to find controls that will have the aircraft at an equilibrium state. That means that if the controls are set just right, those quantities will not change. For example, a level flight in an aircraft is an equilibrium state. The altitude, speed, and direction are close to constant. Substitute these into the first three dynamical equations from Equation 12.2 and set the left-hand side to zero.

$$0 = \begin{bmatrix} T\cos\alpha - D(v_s, \alpha) - mg\cos\gamma_s \\ (L(v_s, \alpha) + T\sin\alpha)\cos\phi - mg\cos\gamma_s \\ (L(v_s, \alpha) + T\sin\alpha)\sin\phi \end{bmatrix} \quad (12.8)$$

The controls are the angle of attack, α; roll angle, ϕ; and thrust, T. Since we have three equations in three unknowns, we can get a single solution. An easy way to solve for the equilibrium controls is to use `fminsearch`. `fminsearch` works pretty well for this kind of problem. The result that it finds may not be the only possible solution. You might find other solutions from starting with a different initial guess since `fminsearch` is not a global optimization function. This routine will find the three controls that zero the three equations.

The function, `EquilibriumControl.m`, uses `fminsearch` in a loop to handle multiple states. Within the loop, we compute an initial guess of the control. The thrust will need to balance the drag so we compute this at zero angle of attack. The lift must balance gravity so we compute the angle of attack from that relationship. Without a reasonable initial guess, the algorithm will converge to a local minimum but not necessarily the global minimum. The cost function is nested within the control function. The function can solve for multiple sets of states, hence the n.

EquilibriumControl.m

```
26  function [u, c] = EquilibriumControl( x, d, tol )
43
44  n    = size(x,2);
45  u    = zeros(3,n);
46  c    = zeros(1,n);
47  p    = optimset('TolFun',tol);
48  % additional options during testing:
49  %'PlotFcns',{@optimplotfval,@PlotIteration},'Display','iter','MaxIter
        ',50);
50  for k = 1:n
51    [~,D,LStar] = RHSAircraft(0,x(:,k),d);
52    alpha       = d.m*d.g/LStar;
53    u0          = [D;alpha;0];
54    [umin,cval,exitflag,output] = fminsearch( @Cost, u0, p, x(:,k), d );
55    u(:,k)      = umin;
56    c(k)        = Cost( u(:,k), x(:,k), d );
57  end
```

The default output is to plot the results.

```
59  % Plot if no outputs are specified
60  if( nargout == 0 )
61    yL = {'T (N)', '\alpha (rad)', '\phi (rad)' 'Cost'};
62    s  = 'Equilibrium Control:Controls';
63    PlotSet(1:n,[u;c],'x label','set','y label',yL, ...
64            'plot title',s, 'figure title',s);
65
66    yL = {'v' '\gamma' '\psi' 'h'};
67    s  = 'Equilibrium Control:States';
68    PlotSet(1:n,x([1:3 6],:),'x label','set','y label',yL, ...
69            'plot title',s,'figure title',s);
70    clear u
71  end
```

The cost sub (subfunction) function is shown in the following. We use a quadratic cost that is the unweighted sum of the squares of the state derivatives. The cost is the quantity that `fminsearch` tries to make as small as possible.

```
74  %%% EquilibriumControl>Cost
75  % Find the cost of a given control u.
76  %
```

```
77  %   c = Cost( u, x, d )
78  function c = Cost( u, x, d )
79
80  d.thrust        = u(1);
81  d.alpha    = u(2);
82  d.phi      = u(3);
83
84  xDot       = RHSAircraft(0,x,d);
85  y          = xDot(1:3);
86  c          = sqrt(y'*y);
```

The function has a built-in demo that looks at the thrust and angle of attack at a constant velocity but increasing altitude, from 0 to 10 km. Built-in demos are always good, even if you are the only person who ever uses the function.

```
29  if( nargin < 1 )
30    disp('Demo of EquilibriumControl with variable altitude:');
31    x = [200*ones(1,101);...
32        zeros(4,101);...
33        linspace(0,10000,101)];
34    d = RHSAircraft;
35    EquilibriumControl( x, d )
36    return;
37  end
```

Figure 12.2 shows the states for which the controls are calculated in the built-in demo. Figure 12.3 shows the resulting controls. As expected, the angle of attack goes up with altitude, but the thrust goes down. The decreasing air density reduces the drag and lift, so we need to decrease the thrust but increase the angle of attack to generate more lift. The roll angle is nearly zero.

In Figure 12.3, we also plot the cost. The cost should be nearly zero if the function is working as desired.

During debugging while writing a function requiring optimization, it may be helpful to have additional insight into the numerical search process. While we only need umin, consider the additional outputs available from fminsearch in this version of the function call.

```
[umin,cval,exitflag,output] = fminsearch( @Cost, u0, p, x(:,k), d )
```

The output structure will include the number of iterations, and the exit flag will indicate the exit condition of the function: whether the tolerance was reached (1), the maximum number of allowed iterations was exceeded (0), or if a user-supplied output function terminated the search (-1). We put a breakpoint in the script to check these outputs. For a state of $v = 200$ and $h = 300$ at $k = 4$, the output will be

```
>> u0
u0 =

    2880.4
```

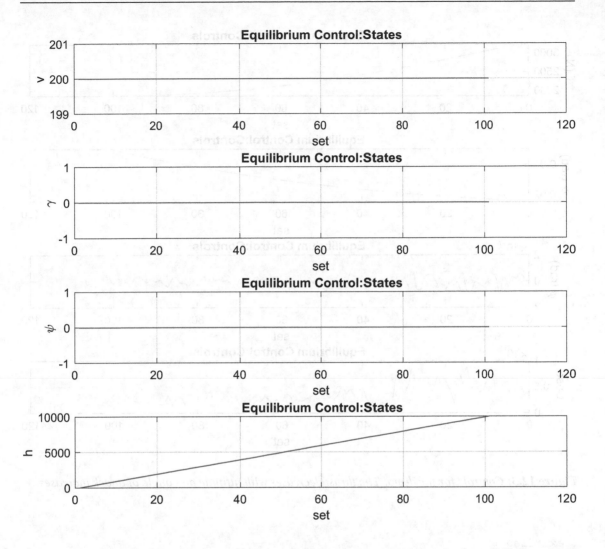

Figure 12.2: *States for the demo. Only the altitude (h) is changing, following the desired flight path.*

```
      0.016255
            0

>> [umin,cval,exitflag,output] = fminsearch( @Cost, u0, p, x(:,k), d );
umin =

       3180.7
      0.016237
     3.459e-08

cval =

   3.2473e-09
```

335

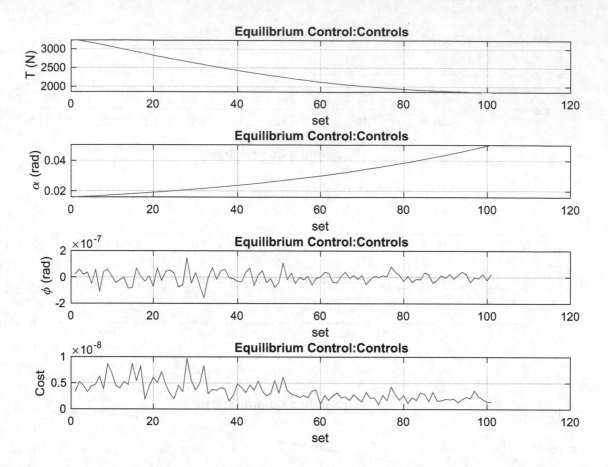

Figure 12.3: *Controls for the demo. The thrust decreases with altitude and angle of attack increases.*

```
exitflag =

    1

output =

    iterations: 141
     funcCount: 260
     algorithm: 'Nelder-Mead simplex direct search'
       message: 'Optimization terminated:
the current x satisfies the termination criteria using...'
```

So we can see that the search required 141 iterations and that the thrust increased to 3180.7 N from our initial guess of 2880.4 N. The resulting cost is 3×10^{-9}. For more information, set the Display option of fminsearch to iter or final, with the default being notify. In this case, the options look like

```
p   = optimset('TolFun',tol,'Display','iter');
```

and the following type of output will be printed to the command window:

```
>> d = RHSAircraft;
>> x = [200;0;0.5;0;0;5000];
>> EquilibriumControl( x, d )

   Iteration    Func-count      min f(x)           Procedure
       0             1          0.0991747
       1             4          0.0817065          initial simplex
       2             6          0.0630496          expand
       3             7          0.0630496          reflect
       4             9          0.0133862          expand
       5            10          0.0133862          reflect
       6            11          0.0133862          reflect
       7            13          0.0133862          contract outside
       8            15          0.0133862          contract inside

       ...
      49            87          2.1869e-05         contract inside
      50            89          2.1869e-05         contract outside

Exiting: Maximum number of iterations has been exceeded
         - increase MaxIter option.
         Current function value: 0.000022
```

For additional insight, we can add a plot function to be called at every iteration. MATLAB provides some default plot functions, for example, optimplotfval plots the cost function value at every iteration. You have to actually open optimplotfval in an editor to learn the necessary syntax. We add the function to the optimization options like this:

```
1  p    = optimset('TolFun',tol,'PlotFcns',@optimplotfval);
```

and Figure 12.4 is generated from the first iteration of the demo.

We can see that the cost value was nearly constant, on this plot, for the final 100 iterations. You can add additional plots using a cell array for PlotFcns, and each plot will be given its own subplot axis automatically by MATLAB. For tough numerical problems, you might want to generate a surface and trace the iterations of the optimization. For our problem, we add our custom plot function PlotIteration, and the results look like Figure 12.5.

```
1  p = optimset('TolFun',tol,'PlotFcns',{@optimplotfval,@PlotIteration});
```

We wrote two functions, one to generate the surface and a second to plot the iteration step. MATLAB sets the iteration value to zero during initialization, so in that case we generate the surface from the given initial state. For all other iteration values, we plot the cost on the surface using an asterisk.

```
85  y           = xDot(1:3);
86  c           = sqrt(y'*y);
88
```

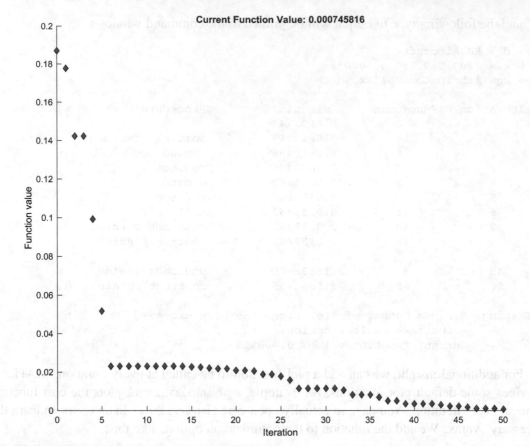

Figure 12.4: *Function value plot using* `optimplotfval`.

```
89   %%% EquilibriumControl>PlotIteration
90   % Plot an iteration of the numerical search.
91   %
92   %   stop = PlotIteration(u0,optimValues,state,varargin)
93   function stop = PlotIteration(u0,optimValues,state,varargin)
94
95   stop = false;
96   x0 = varargin{1};
97   d  = varargin{2};
98   switch state
99       case 'iter'
100          if optimValues.iteration == 0
101              a = PlotSurf( x0, u0, d );
102          end
103          plot3(u0(1),u0(2),optimValues.fval,'k*');
104  end
106
107  %%% EquilibriumControl>PlotSurf
108  % Plot a surface using the given initial state for a range of controls.
```

338

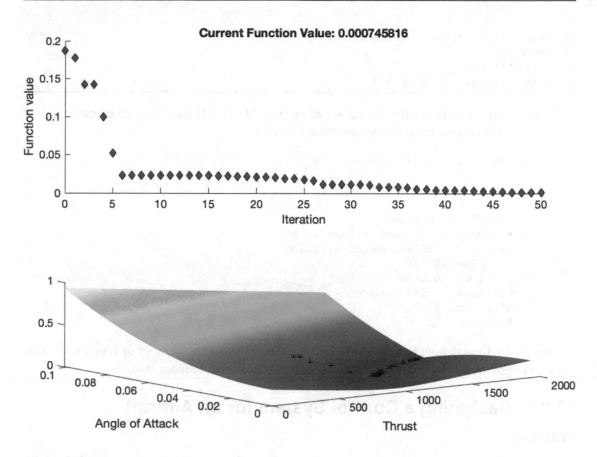

Figure 12.5: *Custom optimization plot function using a surface.*

```
109  % MATLAB will already have an empty axis available for plotting.
110  %
111  %    a = PlotSurf( x0, u0, d )
112  function a = PlotSurf( x0, u0, d )
113
114  u1 = linspace(max(u0(1)-1000,0),u0(1)+1000);
115  u2 = linspace(0,max(2*u0(2),0.1));
116  u3 = u0(3);
117  cvals = zeros(100,100);
118
119  for m = 1:100
120    for l = 1:100
121      cvals(l,m) = Cost( [u1(m);u2(l);u3], x0, d );
122    end
123  end
124
125  s = surf(u1,u2,cvals);
126  set(s,'edgecolor','none');
127  a = gca;
```

```
128  set(a,'Tag','equilibriumcontrol');
129  hold on;
130  xlabel('Thrust')
131  ylabel('Angle of Attack')
```

Note, finally, that to see the default set of options MATLAB uses for `fminsearch`, call `optimset` with the name of the optimization function.

```
>> options = optimset('fminsearch')

options =

        Display: 'notify'
     MaxFunEvals: '200*numberofvariables'
         MaxIter: '200*numberofvariables'
          TolFun: 0.0001
            TolX: 0.0001
     FunValCheck: 'off'
       OutputFcn: []
        PlotFcns: []
```

We see that the default tolerances are equal, at 0.0001, and the number of function evaluations and iterations is dependent on the number of variables in the input state x.

12.3 Designing a Control System for an Aircraft

Problem

We want to design a control system for an aircraft that will control the trajectory and allow for a three-dimensional motion.

Solution

We will use dynamic plant inversion to feedforward the desired controls for the aircraft. Proportional controllers will be used for thrust, angle of attack, and roll angle to adjust the nominal controls to account for disturbances such as wind gusts. We will not use feedback control of the roll angle to control the heading, ψ. This is left as an exercise for the reader.

How It Works

Recall from the dynamical model in Recipe 12.6 that our aircraft state is

$$\begin{bmatrix} v & \gamma & \psi & x & y & h \end{bmatrix} \tag{12.9}$$

where v is the velocity, γ is the flight path angle, ψ is the heading, x and y are the coordinates in the flight plane, and h is the altitude. The states are the values of the system that change with time. They can be dynamical quantities, such as v, γ, and psi, or kinematical quantities, x, y, and h. The derivatives of the kinematical quantities are proportional to the dynamical quantities. For example, $\dot{x}$ is proportional to v. Our control variables are the roll angle ϕ, angle of attack α, and thrust T.

340

Our controller is of the form

$$T = T_s + k_T(v_s - v) \tag{12.10}$$

$$\alpha = \alpha_s + k_\alpha(\gamma_s - \gamma) \tag{12.11}$$

If the state is at s, then the controls should be at the values T_s, α_s, ϕ_s which are the equilibrium controls. The gains push the states in the right direction. The gains are a function of the flight condition. We need to expand the first two dynamical equations from Equation 12.2.

$$\begin{bmatrix} \dot{v} \\ \dot{\gamma} \end{bmatrix} = \begin{bmatrix} \frac{T\cos\alpha - q\left(C_{D_0} + k(C_{L_\alpha}\alpha)^2\right)}{m} - g\cos\gamma \\ \frac{(qC_{L_\alpha}\alpha + T\sin\alpha)\cos\phi - mg\cos\gamma}{mv} \end{bmatrix} \tag{12.12}$$

Linearize and drop the terms not involving the controls, which are thrust T and angle of attack α.

$$\begin{bmatrix} \dot{v} \\ \dot{\gamma} \end{bmatrix} = \begin{bmatrix} \frac{T}{m} \\ \frac{q\cos\phi C_{L_\alpha}\alpha}{mv} \end{bmatrix} \tag{12.13}$$

We want the time constants τ_γ and τ_v so that our equations become

$$\begin{bmatrix} \dot{v} \\ \dot{\gamma} \end{bmatrix} = \begin{bmatrix} -\frac{v}{\tau_v} \\ -\frac{\gamma}{\tau_\gamma} \end{bmatrix} \tag{12.14}$$

Therefore

$$k_T = \frac{m}{\tau_v} \tag{12.15}$$

$$k_\alpha = \frac{mv}{q\cos\phi C_{L_\alpha}\tau_\gamma} \tag{12.16}$$

This is what happens when you perform a coordinated turn. Basically, this equation shows that in order to maintain the same level of responsiveness in longitudinal control (effective time constant of τ_γ) during a turn (when ϕ is nonzero), the control gain on α must be increased. The bank angle during a turn causes a small reduction in the lift force for a given angle of attack. The flight path angle control is achieved by modulating α to vary the lift force. To maintain the same flight path angle response during a turn, we would have to maintain the same lift force through a corresponding increase in the angle of attack. We put the control system in the function AircraftControl.

AircraftControl.m

```
29  function [T, alpha] = AircraftControl( s, d, tauGamma, tauV, vSet,
        gammaSet )
30
31  u       = EquilibriumControl( s, d );
32  v       = s(1);
33  gamma   = s(2);
```

```
34  h        = s(6);
35
36  rho      = feval(d.density,h);
37  qS       = 0.5*d.s*rho*v^2;
38  kV       = d.m/tauV;
39  kGamma   = (d.m*v)/(qS*cos(d.phi)*d.cLAlpha*tauGamma);
40  T        = u(1) + kV   *(vSet     - v);
41  alpha    = u(2) + kGamma*(gammaSet - gamma);
```

The performance of the control system will be shown in the simulation recipe. The function requires information about the flight conditions including the atmospheric density. It first uses EquilibriumControl to find the controls that are needed when we are at the set point. The aircraft data structure is required. Additional inputs are the time constants for the controllers and the set points. We compute the atmospheric density in the function using feval and the input function handle. This should be the same computation as is done in RHSAircraft.

12.4   Plotting a 3D Trajectory for an Aircraft

Problem

We want to plot the trajectory of the aircraft in three dimensions and show the aircraft axes and times along the trajectory.

Solution

We use the MATLAB plot3 function with custom code to draw the aircraft axes and times at select points on the trajectory. The resulting figure is shown in Figure 12.6.

How It Works

We use plot3 to draw the 3D display. Our function Plot3DTrajectory.m allows for argument pairs via varargin.

Plot3DTrajectory.m

```
1   %% PLOT3DTRAJECTORY Plot the trajectory of an aircraft in 3D.
2   %% Form
3   %    Plot3DTrajectory( x, varargin )
4   %% Decription
5   % Plot a 3D trajectory of an aircraft with times and local axes. Type
6   % Plot3DTrajectory for a demo.
7   %
8   %% Inputs
9   %   x           (6,:)    State vector [v;gamma;phi;x;y;h]
10  %   varargin    {:}      Parameters
11  %
12  %% Outputs
13  % None.
```

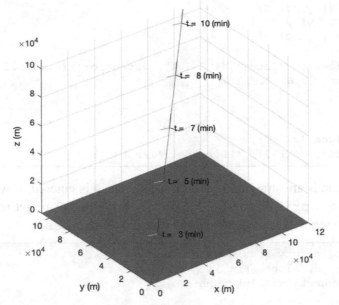

Figure 12.6: *Demo of the aircraft trajectory function.*

We skip the demo code for now and show the drawing code next. There are similarities with our 2D plotting function, `PlotSet`. We use `text` to insert time labels and a patch object to draw the ground.

```
67  % Draw the figure
68  h = figure;
69  set(h,'Name',figTitle);
70  plot3(x(4,:),x(5,:),x(6,:));
71  xlabel(xLabel);
72  ylabel(yLabel);
73  zlabel(zLabel);
74
75  % Draw time and axes
76  if( ~isempty(t) && ~isempty(tIndex) )
77    [t,~,tL] = TimeLabel(t);
78    for k = 1:length(t)
79      s = sprintf('  t = %3.0f (%s)',t(k), tL);
80      i = tIndex(k);
81      text(x(4,i),x(5,i),x(6,i),s);
82      DrawAxes(x(:,i),alpha(1,i),phi(1,i));
83    end
84  end
85
86  % Add the ground
87  xL    = get(gca,'xlim');
88  yL    = get(gca,'ylim');
89  v     = [xL(1) yL(1) 0;...
```

343

```
90              xL(2)  yL(1)  0;...
91              xL(2)  yL(2)  0;...
92              xL(1)  yL(2)  0];
93
94   patch('vertices',v,'faces',[1 2 3 4],'facecolor',[0.65 0.5 0.0],'
            edgecolor',[0.65 0.5 0.0]);
95   grid on
96   rotate3d on
97   axis image
98   zL    = get(gca,'zlim');
99   set(gca,'zlim',[0 zL(2)],'ZLimMode','manual');
```

The plot commands are straightforward. If a time array is entered, it will draw the times along the track using `sprintf` and `text`. We use `TimeLabel` to get reasonable units. It will also draw the aircraft axes using the nested function `DrawAxes`.

```
101  %%% Plot3DTrajectory>DrawAxes subfunction
102  %    DrawAxes( x, alpha, phi )
103  function DrawAxes( x, alpha, phi )
104
105  gamma = x(2);
106  psi   = x(3);
107
108  % Aircraft frame is x forward, y out the right wing and z down
109  u0    = [1 0 0;0 1 0;0 0 -1];
110
111  cG    = cos(gamma+alpha);
112  sG    = sin(gamma+alpha);
113  cP    = cos(psi);
114  sP    = sin(psi);
115  cR    = cos(phi);
116  sR    = sin(phi);
117
118  u     =  [cP -sP   0;sP cP  0;  0    0   1]...
119          *[cG    0 -sG; 0  1  0;sG    0  cG]...
120          *[1    0   0;0 cR sR; 0 -sR cR]*u0;
121
122  % Find a length for scaling of the axes
123  xL    = get(gca,'xlim');
124  yL    = get(gca,'ylim');
125  zL    = get(gca,'zlim');
126
127  l     = sqrt((xL(2)-xL(1))^2 + (yL(2)-yL(1))^2 + (zL(2)-zL(1))^2)/20;
128
129  x0    = x(4:6);
130  for k = 1:3
131    x1    = x0 + u(:,k)*l;
132    c     = [0 0 0];
133    c(k)  = 1;
134    line([x0(1);x1(1)],[x0(2);x1(2)],[x0(3);x1(3)],'color',c);
135  end
```

344

This function draws an axis system for the aircraft, x out the nose, y out the right wing, and z down. It uses the state vector so it needs to convert from γ and ψ to rotation matrices. The axis system is in wind axes.

The function takes parameter pairs to allow the user to customize the plot. The parameter pairs are processed here just after the defaults are set:

```
34  % Defaults
35  xLabel    = 'x (m)';
36  yLabel    = 'y (m)';
37  zLabel    = 'z (m)';
38  figTitle       = 'Trajectory';
39  t         = [];
40  tIndex    = [];
41  alpha     = 0.02*ones(1,size(x,2));
42  phi       = 0.25*pi*ones(1,size(x,2));
43
44  for k = 1:2:length(varargin)
45    switch lower(varargin{k})
46      case 'x label'
47        xLabel     = varargin{k+1};
48      case 'y label'
49        yLabel     = varargin{k+1};
50      case 'z label'
51        zLabel     = varargin{k+1};
52      case 'figure title'
53        figTitle   = varargin{k+1};
54      case 'time'
55        t          = varargin{k+1};
56      case 'time index'
57        tIndex     = varargin{k+1};
58      case 'alpha'
59        alpha      = varargin{k+1};
60      case 'phi'
61        phi        = varargin{k+1};
62      otherwise
63        error('%s is not a valid parameter',varargin{k});
64    end
65  end
```

We use `lower` in the `switch` statement to allow the user to input capital letters and not have to worry about case issues. Most of the parameters are straightforward. The time input could have been done in many ways. We chose to allow the user to enter specific times for the time labels. As part of this, the user must enter the indices to the state vector.

The function includes a demo. You can type `Plot3DTrajectory` and get the example trajectory shown in Figure 12.6. In the case of a graphics function, the demo literally shows the user what the graphics should look like and provides examples about how to use the function.

```
21  % Demo
22  if( nargin  < 1 )
23    disp('Demo of Plot3DTrajectory:');
24    l = linspace(0,1e5);
25    x = [200*ones(1,100);...
26          (pi/4)*ones(1,100);...
27          (pi/4)*ones(1,100);1;1;1];
28    t = [200 300 400 500 600];
29    k = [20   40   60   80 100];
30    Plot3DTrajectory( x, 'time', t, 'time index', k, 'alpha',0.01*ones
          (1,100) );
31    return;
32  end
```

12.5   Simulating the Controlled Aircraft

Problem

We want to simulate the motion of the aircraft with the trajectory controls.

Solution

We will create a script with the control system and flight dynamics. The dynamics will be propagated by `RungeKutta`. This is a fourth-order method, meaning the truncation errors go as the fourth power of the time step. Given the typical sample time for a flight control system, the fourth order is sufficiently accurate for flight simulations. We will display the results using our 3D plotting function `Plot3DTrajectory` described in the previous recipe.

How It Works

The simulation script reads the data structure from `RHSAircraft` and changes values to match an F-35 fighter. The model only involves the thrust and drag, and even these are very simple models. The initial flight path angle and velocity are set. We turn on the control and establish the set points and time constants for the velocity and flight path angle states. For the output, we plot the states, control, and a 3D trajectory.

346

AircraftSim.m

```
1   %% A trajectory control simulation of an F-35 aircraft.
2   % The dynamics of a point mass aircraft is simulated.
3   %% See also
4   % RungeKutta, RHSAircraft, PDControl, EquilibriumControl
```

The script begins with obtaining our default data structure from the RHS function.

```
10  %% Data structure for the right hand side
11  d            = RHSAircraft;
12
13  %% User initialization
14  d.m          = 13300.00; % kg
15  d.s          = 204.00; % m^2
16  v            = 200; % m/sec
17  fPA          = pi/6;  % rad
18
19  % Initialize duration and delta time
20  tEnd         = 40;
21  dT           = 0.1;
22
23  % Controller
24  controlIsOn = true;
25  tauV         = 1;
26  tauGamma     = 1;
27  d.phi        = 0;
28  vSet         = 220;
29  gammaSet     = pi/8;
30
31  %% Simulation
32  % State vector
33  x            = [v;fPA;0;0;0;0];
34
35  % Plotting and number of steps
36  n            = ceil(tEnd/dT);
37  xP           = zeros(length(x)+2,n);
38
39  % Find non-feedback settings
40  [~,D,LStar] = RHSAircraft(0,x,d);
41  thrust0      = D;
42  alpha0       = d.m*d.g/LStar;
43
44  % Run the simulation
45  for k = 1:n
46
47    if( controlIsOn )
```

```
48        [d.thrust, d.alpha] = AircraftControl( x, d, tauGamma, tauV, vSet,
            gammaSet );
49     else
50       d.thrust    = thrust0;
51       d.alpha     = alpha0;
52     end
53
54     % Plot storage
55     xP(:,k)         = [x;d.thrust;d.alpha];
56
57     % Right hand side
58     x           = RungeKutta(@RHSAircraft,0,x,dT,d);
59
60   end
61
62   %% Plotting
63   [t,tL] = TimeLabel((0:(n-1))*dT);
64
65   yL = {'T (N)', '\alpha (rad)'};
66   s  = 'Aircraft Sim:Controls';
67   PlotSet(t,xP(7:8,:),'x label',tL,'y label',yL,'plot title',s, 'figure
          title',s);
68
69   yL = {'v' '\gamma' '\psi' 'x' 'y' 'h'};
70   s  = 'Aircraft Sim:States';
71   PlotSet(t,xP(1:6,:),'x label',tL,'y label',yL,'plot title',s,'figure
          title',s);
72
73   k = floor(linspace(2,n,8));
74   t = t(k);
75   Plot3DTrajectory( xP, 'time', t, 'time index', k, 'alpha', xP(8,:) );
```

If the control is off, we set the thrust and angle of attack to constant values to balance the drag and gravity. The set points for velocity and flight path angle are slightly different than the initial conditions. This will allow us to demonstrate the transient response of the controller.

The states are shown in Figure 12.7. The velocity and flight path angle converge to their set points.

The controls are shown in Figure 12.8. The controls reach their steady values. The thrust and angle of attack change as the plane climbs. The thrust drops because the drag drops and the angle of attack increases to maintain the lift/gravity balance.

The 3D trajectory is shown in Figure 12.9. As expected, it climbs at a nearly constant angle at a constant velocity.

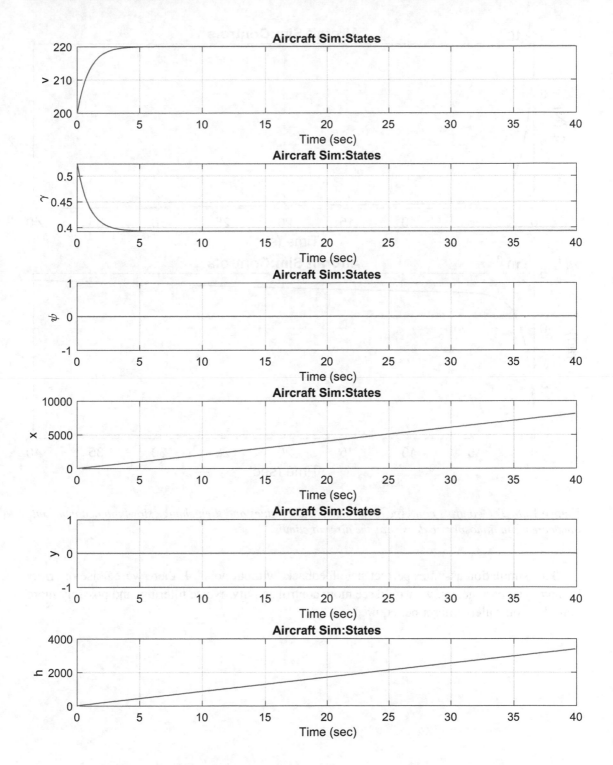

Figure 12.7: *Velocity and flight path angle converge to their desired values.*

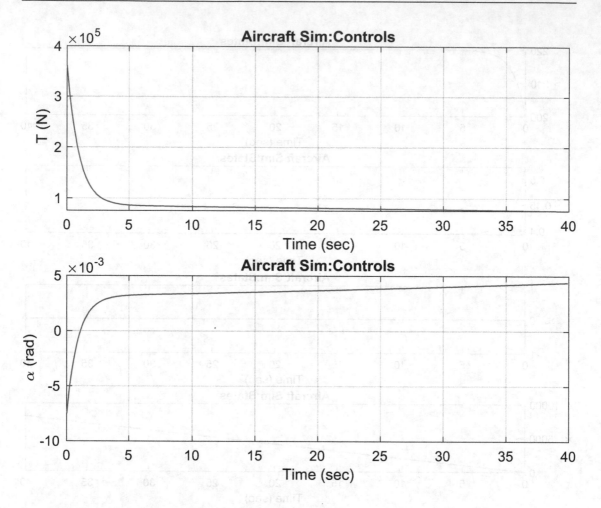

Figure 12.8: *The controls converge to the steady-state values and then change slowly to accommodate the decrease in atmospheric density as the aircraft climbs.*

This simulation assumes perfect state feedback, without noise. If there were noise or errors in the model parameters, we would see more control activity. Noise filtering, and possibly more complex controllers, might be required.

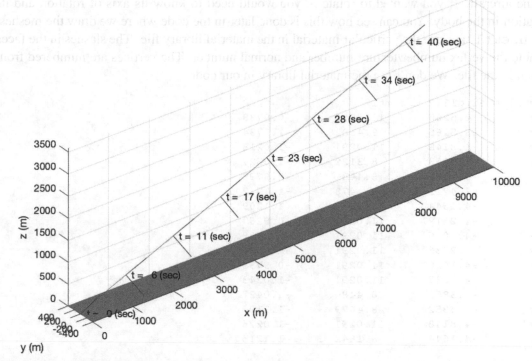

Figure 12.9: *The aircraft trajectory.*

12.6   Draw an Aircraft

Problem

We want to visualize the orientation of an aircraft in three dimensions.

Solution

We will write a MATLAB visualization tool that will show an aircraft in three dimensions. The model will be from a Wavefront OBJ file.

How It Works

There are many visualization formats. One of the easiest to use is the Wavefront OBJ format. This is a text file with lines representing vertices, faces, surface normals, and texture coordinates. We will only use the faces and vertices. Some lines from the `Gulfstream.obj` file are shown in the following. `v` means vertex; `f` means face. There are only three vertices per face. OBJ can handle faces with any number of vertices, but graphics engines are more efficient when all of the polygons are triangles. `mtllib` gives the name of the material library that will be used to get textures and colors for the surfaces. `g` and `o` break the file into components. No hierarchy information is given in the file. If you want to articulate the object, you need to get the information from another source. For example, for the vertices, give its absolute position

351

in the aircraft. If you wanted to rotate it, you would need to know its axis of rotation and its location in the body. You can see how this is done later in the code where we draw the meshes.

usemtl says use the particular material in the material library file. The slashes in the faces denote the vertex number/texture number and normal number. The vertices are numbered from 1 to n in the file. We don't use the material library in our code.

```
mtllib airgl312_mac.mtl
v        -4.5265      11.0291      -0.9738
v        -3.5165       8.4082      -1.1734
v        -4.8148      11.0291      -1.0218
v        -3.7923       8.3129      -1.0527
v        -3.4301       8.4481      -0.9798
v        -3.4212       8.4478      -1.1732
v        -4.1644       8.1842      -1.1209
v        -4.2112      11.0291      -0.9267
v        -4.2016      11.0291      -1.0708
v        -4.2836      11.0291      -1.0696
v        -4.1638      11.0291      -0.9735
v        -4.1637      11.0291      -1.0243
v        -3.3853       8.4480      -1.0427
v        -3.3852       8.4479      -1.1100
v        -4.8148      11.0291      -1.0225
v        -4.1644       8.1842      -1.1215

usemtl gulf351

o AileronL
g AileronL

s off
f 9/13/21 6/4/22 2/3/20
f 2/3/20 10/10/19 9/13/21
f 2/3/20 16/1/18 15/11/17
```

The function LoadOBJ loads an OBJ file and breaks it into components using the g lines. From the preceding example, you can see that one component will be AileronL.

The function DrawAircraft creates a 3D view of the aircraft and animates it.

The main loop is a switch function. Actions are "initialize," "update," "movie," and "close." The "initialize" action sets up the axes.

DrawAircraft.m

```
33  function [h,mV] = DrawAircraft( action, g, h, x, t, tU )
34
35  if( nargin < 1 )
36    Demo
37    return
38  end
39
40  switch( lower(action) )
```

```
41    case 'initialize'
42
43        h.fig = NewFigure( g.name );
44        axes('DataAspectRatio',[1 1 1],'PlotBoxAspectRatio',[1 1 1] );
45
46        xlabel('X (m)')
47        ylabel('Y (m)')
48        zlabel('Z (m)')
49
50        grid on
51        view(3)
52        rotate3d on
53        hold off
54        drawnow
55
56        n = length(g.component);
57        h.h = zeros(1,n);
58
59        for k = 1:n
60          h.h(k) = DrawMesh(g.component(k));
61        end
62
63    case 'update'
64        UpdateMesh(h,g.component,x, t, tU);
65
66    case 'movie'
67        mV = UpdateMesh(h,g.component,x,  t, tU, 1 );
68
69    case 'close'
70        close(h.fig);
71
72    otherwise
73        warning('%s not available',action);
74  end
```

The code in `DrawMesh` draws the vertices and faces for the component m. The only parts of the components used are the vertices and faces. The Phong lighting is a type of OpenGL lighting that approximates real lighting. Notice that the edge lighting can be different from the face lighting.

```
76  %% DrawAircraft>DrawMesh
77  function h = DrawMesh( m )
79
80  h = patch(   'Vertices', m.v, 'Faces',   m.f, 'FaceColor', [0.8 0.8
         0.8],...
81               'EdgeColor','none','EdgeLighting', 'phong',...
82               'FaceLighting', 'phong');
```

The following code updates the patches for the aircraft. It also updates the time `uicontrol`. It gets the vertices for each component, rotates the vertices, and then translates them. The GUI automatically changes the limits so that the plane remains centered. You will see, when you run the demo, the axis numbers change as the aircraft moves. We also call the function `DrawAlphaBeta` to draw axes and annotate them. `getframe` gets the information in the frame for use in a movie.

```
85   %% DrawAircraft>UpdateMesh
86   function mV = UpdateMesh( h, c, x, t, tU, ~ )
87
88   hL = [];
89   hAB = [];
90
91   if( nargin > 3 )
92     mV(1:size(x,2)) = getframe(h.fig);
93   else
94     mV = [];
95   end
96
97   s = sprintf('%6.2f %s',t(1),tU);
98   hT = uicontrol( h.fig,'style','edit','string',s,'position',[10 10 100
         20]);
99
100  qY = Mat2Q([1 0 0;0 -1 0;0 0 -1]);
101
102  for j = 1:size(x,2)
103    m = Q2Mat(x(7:10,j));
104    r = x(1:3,j);
105    for k = 1:length(c)
106      v       = (m*c(k).v')';
107      v(:,1)  = v(:,1) + r(1);
108      v(:,2)  = v(:,2) + r(2);
109      v(:,3)  = v(:,3) + r(3);
110      set(h.h(k),'vertices',v);
111    end
112    xL = get(gca,'xlim');
113    yL = get(gca,'ylim');
114    zL = get(gca,'zlim');
115
116    if(~isempty(hL) )
117      delete(hL);
118    end
119    hL = light('position',10*[xL(2) yL(2) zL(2)]);
120
121    r = x(1:3,j);
122    v = x(4:6,j);
123    q = QMult(qY,x(7:10,j));
124
125    if( ~isempty(hAB) )
126      delete(hAB);
```

```
127    end
128
129    set(hT,'string',sprintf('%6.2f %s',t(j),tU));
130    hAB = DrawAlphaBeta( r, q, Unit(v), 14 );
131    if( nargin > 3 )
132      mV(j) = getframe(h.fig);
133    end
134    pause(0.1);
135    drawnow
136  end
```

The following code runs the demo. We need to generate the position, velocity, and orientation. The orientation is defined by an attitude or orientation quaternion. Euler angles are another possibility, but they are computationally slower than quaternions. A quaternion can be thought of as an axis of rotation and an angle about that axis. The four elements are not independent since the sum of the squares of the elements of a quaternion equals 1. Quaternions are used in computer graphics and for spacecraft. A quaternion can be replaced by a transformation matrix. However, the latter is harder to handle since it is 3-by-3, while the quaternion is a 4-by-1 array. It is simpler to store a set of "n" 4-by-1 quaternions (e.g., as a 4-by-n matrix) than it is to store a set of "n" 3-by-3 rotation matrices.

The demo loads the aircraft and then calls the movie action in `DrawAircraft`. This returns a pointer to a movie frame that can later be saved. If you don't want a movie, call `DrawAircraft` with "update." The rotation of the aircraft is the product of two matrices. One is a constant pitch (rotation about Y) matrix and the other a time-varying roll (rotate about X) matrix. You can create arbitrary rotation matrices by multiplying multiple rotation matrices together.

We create an array consisting of the column matrix [1;0;-0.2] by dot multiplication with a row array of ones. This is a relatively new MATLAB feature. MATLAB figures out that you want 100 copies of the column array.

```
139  function Demo
140
141  g          = LoadOBJ('Gulfstream.obj');
142
143  h     = DrawAircraft( 'initialize', g );
144
145  dToR = pi/180;
146  n    = 100;
147  z    = linspace(100,400,n);
148  x    = linspace(0,40000,n);
149  a    = linspace(0,pi/4,n);
150  c    = cos(a);
151  s    = sin(a);
152  q    = zeros(4,n);
```

355

```
153  cY   = cos(15*dToR);
154  sY   = sin(15*dToR);
155  mY   = [cY 0 -sY;0 1 0;sY 0 cY];
156
157  for k = 1:n
158    q(:,k) = Mat2Q([1 0 0;0 c(k) s(k);0 -s(k) c(k)]*mY);
159  end
160  v    = 100*[1;0;-0.2].*ones(1,100);
161  s    = [x;zeros(1,n);z;v;q];
162  t    = linspace(0,1000,n);
163  [~,mV] = DrawAircraft( 'movie', g, h, s, t, 'sec' );
164
165  SaveMovie( mV, 'Gulfstream' )
```

The image at the end of the demo is shown in Figure 12.10.

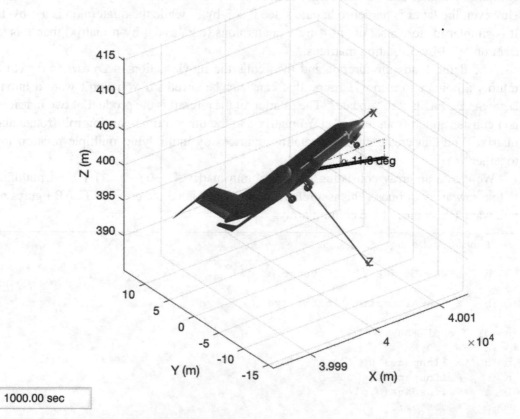

Figure 12.10: *The aircraft at the end of the demo.*

356

Summary

This chapter has demonstrated how to write the dynamics for a point mass aircraft. We learned how to find the equilibrium control state using a search algorithm. This includes utilizing the debug output available from MATLAB for its optimization algorithms and adding custom plotting for each search iteration. We learned how to design a control system to maintain a desired velocity, bank angle, and flight path angle. We learned how to make 3D plots with annotations of both text and other drawing objects. We also learned how to pass function handles to other functions to make functions more versatile. Table 12.1 lists the code developed in the chapter.

Table 12.1: *Chapter Code Listing*

File	Description
DrawAircraft	Draw a 3D model of an aircraft
RHSAircraft	Six degrees of freedom RHS for a point mass aircraft
AtmosphericDensity	Atmospheric density as a function of altitude from an exponential model
EquilibriumControl	Find the equilibrium control for a point mass aircraft
AircraftControl	Compute the angle of attack and thrust for a 3D point mass aircraft
Plot3DTrajectory	Plot a 3D trajectory of an aircraft with times and local axes
AircraftSim	A trajectory control simulation of an aircraft

CHAPTER 13

■ ■ ■

Spacecraft Attitude Control

Spacecraft pointing control is an essential technology for all robotic and manned spacecraft. A control system consists of sensing, actuation, and the dynamics of the spacecraft itself. Spacecraft control systems are of many types, but in this chapter, we will be concerned only with three axis pointing. We will use reaction wheels for actuation.

Reaction wheels are used for control through the conservation of angular momentum. The torque on the reaction wheel causes it to spin one way and the spacecraft to spin in the opposite direction. Momentum removed from the spacecraft is absorbed in the wheel. Reaction wheels are classified as momentum exchange devices because they exchange momentum with the rest of the spacecraft. You can reorient the spacecraft using reaction wheels without any external torques. Before reaction wheels were introduced, thrusters were often used for orientation control. This would consume the propellant which is undesirable since when you run out of propellant, the spacecraft can no longer be used.

The spacecraft is modeled as a rigid body except for the presence of three reaction wheels that rotate about orthogonal (perpendicular) axes. One rotates about the x axis, one rotates about the y axis, and the third about the z axis. The shaft of the motor is attached to the rotor of the wheel which is attached to the spacecraft. The torque applied between the wheel and spacecraft causes the wheel and spacecraft to move in opposite angular directions. We will assume that we have attitude sensors that measure the orientation of the spacecraft. We will also assume that our wheels are ideal with just viscous damping friction.

13.1 Creating a Dynamical Model of the Spacecraft

Problem

The spacecraft is a rigid body with three wheels. Each wheel is connected to the spacecraft as shown in Figure 13.1.

© Michael Paluszek and Stephanie Thomas 2020
M. Paluszek and S. Thomas, *MATLAB Recipes*,
https://doi.org/10.1007/978-1-4842-6124-8_13

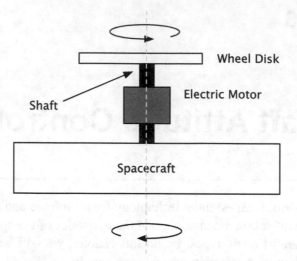

Figure 13.1: *A reaction wheel. The reaction wheel platter spins in one direction, and the spacecraft spins in the opposite direction.*

Solution

The equations of motion are written using angular momentum conservation. This produces a dynamical model known as the Euler equations with the addition of the spinning wheels. This is sometimes known as a gyrostat model.

How It Works

The spacecraft model can be partitioned into dynamics, including the dynamics of the reaction wheels, and the kinematics of the spacecraft. If we assume that the wheels are perfectly symmetric, are aligned with the three body axes, and have a diagonal inertia matrix, we can model the spacecraft dynamics with the following coupled first-order differential equations:

$$I\dot{\omega} + \omega^{\times}\left(I\omega + I_w(\omega_w + \omega)\right) + I_w(\dot{\omega}_w + \dot{\omega}) = T \tag{13.1}$$

$$I_w\left(\dot{\omega}_w + \dot{\omega}\right) = T_w \tag{13.2}$$

I is the 3-by-3 inertia matrix of the spacecraft and does not include the inertia of the wheels.

$$I = \begin{bmatrix} I_{xx} & I_{xy} & I_{xz} \\ I_{yx} & I_{yy} & I_{yz} \\ I_{zx} & I_{zy} & I_{zz} \end{bmatrix} \tag{13.3}$$

The inertia matrix is symmetric so $I_{xy} = I_{yx}, I_{xz} = I_{zx}, I_{zy} = I_{yz}$. ω is the angular rate vector for the spacecraft seen in the spacecraft frame.

$$\omega = \begin{bmatrix} \omega_x \\ \omega_y \\ \omega_z \end{bmatrix} \tag{13.4}$$

ω_w is the angular rate of the reaction wheels:

$$\omega_w = \begin{bmatrix} \omega_1 \\ \omega_2 \\ \omega_3 \end{bmatrix} \tag{13.5}$$

for wheels 1, 2, and 3. 1 is aligned with x, 2 with y, and 3 with z. In this way, the reaction wheels form an orthogonal set and can be used for three-axis control. T is the external torque on the spacecraft which can include external disturbances such as solar pressure or aerodynamic drag and thruster or magnetic torquer coil torques. T_w is the internal torque on the wheels; I_w is the scalar polar inertia of the wheels (we assume that they all have the same polar inertia). We can substitute the second equation into the first to simplify the equations.

$$I\dot{\omega} + \omega^\times \left(I\omega + I_w(\omega_w + \omega)\right) + T_w = T \tag{13.6}$$
$$I_w\left(\dot{\omega}_w + \dot{\omega}\right) = T_w \tag{13.7}$$

This term

$$T_e = \omega^\times \left(I\omega + I_w(\omega_w + \omega)\right) \equiv \omega \times h \tag{13.8}$$

is known as the Euler torque. If the angular rates are small, we can set this term to zero and the equations simplify to

$$I\dot{\omega} + T_w = T \tag{13.9}$$
$$I_w\left(\dot{\omega}_w + \dot{\omega}\right) = T_w \tag{13.10}$$

For kinematics, we will use quaternions. A quaternion is a four-parameter representation of the orientation of the spacecraft with respect to the inertial frame. We could use angles since we really only need three states (dynamical quantities) to specify the orientation. The problem with Euler angles is that they introduce singularities, that is, certain orientations where an angle is undefined, and therefore they are not suitable for simulations. The derivative of the quaternion from the inertial frame to the body frame is

$$\dot{q} = \frac{1}{2}\begin{bmatrix} 0 & \omega^T \\ -\omega & \omega^\times \end{bmatrix} \tag{13.11}$$

The term $\omega^\times$ is the skew symmetric matrix that is the equivalent of the cross product and is

$$\omega^\times = \begin{bmatrix} 0 & -\omega_z & \omega_y \\ \omega_z & 0 & -\omega_x \\ -\omega_y & \omega_x & 0 \end{bmatrix} \tag{13.12}$$

The skew matrix always has zeros on the diagonal, and the matrix is equal to the negative of its transpose. The wheel torque is a combination of friction torque and control torque. Reaction wheels are usually driven by brushless direct current (DC) motors that have the back electromotive force canceled by current feedback within the motor electronics. The total reaction wheel torque is therefore

$$T_w = T_c + T_f \tag{13.13}$$

where T_c is the commanded reaction wheel torque and T_f is the friction torque. A simple friction model is

$$T_f = -k_d \omega_k \tag{13.14}$$

k_d is the damping coefficient. If k_d is large, we can compensate for it proactively by feeding the expected friction torque forward into the controller. This requires careful calibration of the wheel to determine the damping coefficient.

First, we will define the data structure for the model that is returned by the dynamics right-hand-side function if there are no inputs. The name of the function is RHSSpacecraftWith RWA.m. We use RWA to mean "Reaction Wheel Assembly." We say "assembly" because the reaction wheel is assembled from bearings, wheel, shaft, support structure, and power electronics. Spacecraft are built up of assemblies.

The default unit vectors for the wheel are along orthogonal axes, that is, x, y, and z. The default inertia matrix is the identity matrix, making the spacecraft a sphere. The default reaction wheel inertias are 0.001. All of the nonspinning parts of the wheels are lumped in with the inertia matrix.

RHSSpacecraftWithRWA.m

```
1   %% RHSSPACECRAFTWITHRWA Compute the dynamics for a spacecraft with
       reaction wheels.
33  function [xDot, hECI] = RHSSpacecraftWithRWA( ~, x, d )
34
35  % Default data structure
36  if( nargin == 0 )
37    xDot = struct('inr',eye(3), 'torque',[0;0;0],'inrRWA',
         0.001*[1;1;1],...
38                  'torqueRWA',[0;0;0],'uRWA',eye(3), 'damping',[0;0;0]);
39    if( nargout == 0 )
40      disp('RHSSpacecraftWithRWA struct:')
41    end
42    return
43  end
```

The dynamical equations for the spacecraft are given in the following lines of code. We need to compute the total wheel torque because it is applied both to the spacecraft and the wheels. We use the backslash operator to multiply the equations by the inverse of the inertia matrix. The inertia matrix is positive definite symmetric so specialized routines can be used to speed computation of this inverse. It is a good idea to avoid computing inverses as they can be ill-conditioned, meaning that small errors in the matrix can result in large errors in the inverse.

We save the elements of the state vector as local variables with meaningful names to make reading the code easier. This also eliminates unnecessary multiple extraction of submatrices.

You will notice that the omegaRWA variable reads from element 8 to the end of the vector using the end keyword. This allows the code to handle any number of reaction wheels. You might just want to control one axis with a wheel or have more than three wheels for redundancy. Be sure that the inputs in d match the number of wheels. We also input unit vectors for each

362

wheel. The unit vector is the axis or rotation for each. As a consequence, the wheels do not have to be aligned with x, y, and z, that is, do not have to be orthogonal.

```
45   % Save as local variables
46   q        = x(1:4);
47   omega    = x(5:7);
48   omegaRWA = x(8:end);
49
50   % Total body fixed angular momentum
51   h = d.inr*omega + d.uRWA*(d.inrRWA.*(omegaRWA + d.uRWA'*omega));
52
53   % Total wheel torque
54   tRWA = d.torqueRWA - d.damping.*omegaRWA;
```

Note that uRWA is an array of the reaction wheel unit vectors, that is, the spin vectors. In computing h, we have to transform ω into the wheel frame using the transpose of uRWA and then transform back before adding the wheel component to the core component, $I\omega$. The wheel dynamics are given next, note the use of the backslash operator to solve the set of linear equations for $\dot{\omega}$, omegaDotCore:

```
56   % Core angular acceleration
57   omegaDotCore = d.inr\(d.torque - d.uRWA*tRWA - cross(omega,h));
```

The total state derivative is in these lines:

```
59   % Wheel angular acceleration
60   omegaDotWheel   = tRWA./d.inrRWA - d.uRWA'*omegaDotCore;
61
62   % State derivative
63   sW    = [        0 -omega(3) omega(2);...
64            omega(3)         0 -omega(1);...
65           -omega(2) omega(1)         0]; % skew symmetric matrix
66   qD    = 0.5*[0, omega';-omega,-sW];
67   xDot  = [qD*q;omegaDotCore;omegaDotWheel];
```

The total inertial angular momentum is an auxiliary output. In the absence of external torques, it should be conserved so it is a good test of the dynamics. A simple way to test angular momentum conservation is to run a simulation with anger rates for all the states and then rerun it with a smaller time step. The change in angular momentum should decrease as the time step is decreased.

```
69   % Output the inertial angular momentum
70   if( nargout > 1 )
71     hECI = QTForm( q, h );
72   end
```

13.2   Computing Angle Errors from Quaternions

Problem

We want to point the spacecraft to a new target attitude (orientation) with the three reaction wheels or maintain the attitude given an external torque on the spacecraft.

Solution

We will make three proportional-derivative (PD) controllers, one for each axis. We need a function to take two quaternions and compute the small angles between them as input to these controllers.

How It Works

If we are pointing at an inertial target and wish to control about that orientation, we can simplify the rate equations by approximating ω as $\dot{\theta}$ which is valid for small angles when the order of rotation doesn't matter and the Euler angles can be treated as a vector.

$$\dot{\theta} = \omega \tag{13.15}$$

We will also multiply both sides of the Euler equation, Equation 13.9, by I^{-1}, to solve for the derivatives. Note that T_w, the torque from the wheels, is equivalent to Ia_w, where a is the acceleration. Our system equations now become

$$\ddot{\theta} + a_w = a \tag{13.16}$$
$$I_w \left(\dot{\omega}_w + \dot{\omega} \right) = -T_w \tag{13.17}$$

The first equation is now three decoupled second-order equations, just as in our Chapter 7. We can stabilize this system with our standard PD controller.

We need attitude angles as input to the PD controllers to compute our control torques. Our examples will only be for small angular displacements from the nominal attitude. We will pass the control code a target quaternion, and it will compute Δ angles, or we will impose a small disturbance torque.

In these cases, the attitude can be treated as a vector where the order of the rotations doesn't matter. A quaternion derived from small angles is

$$q_\Delta \approx \begin{bmatrix} 1 \\ -\theta_1/2 \\ -\theta_2/2 \\ -\theta_3/2 \end{bmatrix} \tag{13.18}$$

We find the required error quaternion q_Δ by multiplying the target quaternion, q_T, with the transpose of the current quaternion:

$$q_\Delta = q^T q_T \tag{13.19}$$

This algorithm to compute the angles is implemented in the following code. The quaternion multiplication is made a subfunction. This makes the code cleaner and easier to see how it relates to the algorithm. `QMult` is written to handle multiple quaternions at once so the function is easy to vectorize. `QPose` finds the transpose of the quaternion. Both of these functions would normally be separate functions, but in this chapter, they are only associated with the error computation code so we put them in the same file.

ErrorFromQuaternion.m

```
1   %% ERRORFROMQUATERNION Compute small angle error between two
        quaternions.
19  function deltaAngle = ErrorFromQuaternion( q, qTarget )
20
21  deltaQ      = QMult( QPose(q), qTarget );
22  deltaAngle  = -2.0*deltaQ(2:4);
24
25  %% ErrorFromQuaternion>QMult Multiply two quaternions
26  % Q2 transforms from A to B and Q1 transforms from B to C
27  % so Q3 transforms from A to C.
28  %
29  %    Q3 = QMult( Q2 ,Q1 )
30  function Q3 = QMult( Q2 ,Q1 )
31
32  Q3 = [Q1(1,:).*Q2(1,:) - Q1(2,:).*Q2(2,:) - Q1(3,:).*Q2(3,:) - Q1(4,:)
        .*Q2(4,:);...
33      Q1(2,:).*Q2(1,:) + Q1(1,:).*Q2(2,:) - Q1(4,:).*Q2(3,:) + Q1(3,:)
            .*Q2(4,:);...
34      Q1(3,:).*Q2(1,:) + Q1(4,:).*Q2(2,:) + Q1(1,:).*Q2(3,:) - Q1(2,:)
            .*Q2(4,:);...
35      Q1(4,:).*Q2(1,:) - Q1(3,:).*Q2(2,:) + Q1(2,:).*Q2(3,:) + Q1(1,:)
            .*Q2(4,:)];
37
38  %% ErrorFromQuaternion>QPose Transpose of a quaternion
39  % The transpose requires changing the sign of the angle terms.
40  %
41  %    q = QPose(q)
42  function q = QPose(q)
43
44  q(2:4,:) = -q(2:4,:);
```

The control system is implemented in the simulation loop (in the next recipe) with the following code:

SpacecraftSim.m

```
57  % Find the angle error
58  angleError = ErrorFromQuaternion( x(1:4), qTarget );
59  if( controlIsOn )
60    u = [0;0;0];
61    for j = 1:3
62      [u(j), dC(j)] = PDControl('update',angleError(j),dC(j));
63    end
64  else
```

```
65        u   = [0;0;0];
66    end
67
68    % Wheel torque is on the left hand side
69    d.torqueRWA = d.inr*u;
```

13.3   Simulating the Controlled Spacecraft

Problem

We want to test our attitude controller and see how it performs.

Solution

The solution is to build a MATLAB script in which we design the PD controller matrices and then simulate the controller in a loop, applying the calculated torques until the desired quaternion is attained or until the disturbance torque is canceled.

How It Works

We build a simulation script for the controller, SpacecraftSim. The first thing we do with the script is to set parameters at the top of the file to check the angular momentum conservation by running the simulation for 300 seconds at time steps of 0.1 and 1 second and comparing the magnitude of the angular momentum in the two test cases. The control is turned off by setting the controlIsOn flag to false. In the absence of external torques, if our equations are programmed correctly, the momentum should be constant. We will however see the growth in the momentum due to error in the numerical integration. The growth should be much lower in the first case than the second case as the smaller time step makes the integration more exact. Remember that for fourth-order Runge-Kutta, the error goes as the fourth power of the time step. Note that we give the spacecraft random initial rates in both omega and omegaRWA and a nonspherical inertia to help catch any bugs in the dynamics code.

```
1    tEnd        = 300;
2    dT          = 0.1;
3    controlIsOn = false;
4    qECIToBody  = [1;0;0;0];
5    omega       = [0.01;0.02;-0.03]; % rad/sec
6    omegaRWA    = [5;-3;2];  % rad/sec
7    d.inr       = [3 0 0;0 10 0;0 0 5];  % kg-m^2
```

Figure 13.2 shows the results of the tests using the above initialization code. The momentum growth is four orders of magnitude lower in the test with a 0.1 second time step indicating that the dynamical equations conserve angular momentum as they should. The shape of the growth does not change and will depend on the relative magnitudes of the various angular rates.

We initialize the script by using our data structure feature of the RHS function. This is shown in the following with parameters for a run with the control system on. The rates are

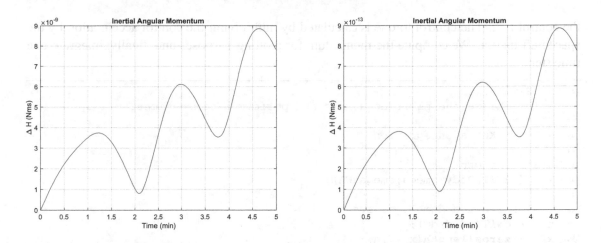

Figure 13.2: *Angular momentum conservation for 1 second and 0.1 second time steps. The growth is four orders of magnitude lower in the 0.1 second test, to $1e^{-13}$ from $1e^{-9}$.*

now initialized to zero, and we use the time step of 1 second, which showed sufficiently small momentum growth in our previous test.

```
1   %% Spacecraft reaction wheel simulation script
2   % An attitude control simulation using reaction wheels.
13  %% Data structure for the right hand side
14  d          = RHSSpacecraftWithRWA;
```

The control system is designed here. Note the small value of wN and the unit damping ratio. The frequency of the disturbances on a spacecraft is quite low, and the wheels have torque limits, leading to a wN much smaller than the robotics recipe. All three controllers are identical.

```
28  %% Control design
29  % Design a PD controller. The same controller is used for all 3 axes.
30  dC            = PDControl( 'struct' );
31  dC(1).zeta    = 1;
32  dC(1).wN      = 0.02;
33  dC(1).wD      = 5*dC(1).wN;
34  dC(1).tSamp   = dT;
35  dC(1)         = PDControl( 'initialize', dC(1) );
36
37  % Make all 3 axis controllers identical
38  dC(2)         = dC(1);
39  dC(3)         = dC(1);
```

The simulation loop follows. As always, we initialize the plotting array with zeros. By allocating memory for the array, we speed up the code as memory allocation is usually slow. The first step in the loop is finding the angular error between the current state and the target attitude. Next, the control acceleration is calculated or set to zero, depending on the value of the control flag. The control torque is calculated by multiplying the control acceleration by the spacecraft inertia. We compute the momentum for plotting purposes and, finally, integrate one time step.

```
41   %% Simulation
42   % Initialize the plotting arrays and perform a fixed timestep loop
         using
43   % Runge-Kutta integration.
44
45   % State vector
46   x = [qECIToBody;omega;omegaRWA];
47
48   % Plotting and number of steps
49   n    = ceil(tEnd/dT);
50   xP   = zeros(length(x)+7,n);
51
52   % Find the initial angular momentum
53   [~,hECI0] = RHSSpacecraftWithRWA(0,x,d);
54
55   % Run the simulation
56   for k = 1:n
57     % Find the angle error
58     angleError = ErrorFromQuaternion( x(1:4), qTarget );
59     if( controlIsOn )
60       u = [0;0;0];
61       for j = 1:3
62         [u(j), dC(j)] = PDControl('update',angleError(j),dC(j));
63       end
64     else
65       u   = [0;0;0];
66     end
67
68     % Wheel torque is on the left hand side
69     d.torqueRWA = d.inr*u;
70
71     % Get the delta angular momentum
72     [~,hECI]   = RHSSpacecraftWithRWA(0,x,d);
73     dHECI      = hECI - hECI0;
74     hMag       = sqrt(dHECI'*dHECI);
75
76     % Plot storage
77     xP(:,k)         = [x;d.torqueRWA;hMag;angleError];
78
79     % Right hand side
80     x   = RungeKutta(@RHSSpacecraftWithRWA,0,x,dT,d);
81   end
```

Our output is entirely two-dimensional plots. We break them up into pages with one to three plots per page. This makes them easily readable on most computer displays.

```
83  %% Plotting
84  % Generate plots of the attitude, body and wheel rates, control torque,
        angular
85  % momentum, and anglular error. If there is no external disturbance
        torque than
86  % angular momentum should be conserved.
87  [t,tL] = TimeLabel((0:(n-1))*dT);
88
89  yL    = {'q_s', 'q_x', 'q_y', 'q_z'};
90  PlotSet( t, xP(1:4,:), 'x label', tL, 'y label', yL,...
91    'plot title', 'Attitude', 'figure title', 'Attitude');
92
93  yL    = {'\omega_x', '\omega_y', '\omega_z'};
94  PlotSet(t, xP(5:7,:), 'x label', tL, 'y label', yL,...
95    'plot title', 'Body Rates', 'figure title', 'Body Rates');
96
97  yL    = {'\omega_1', '\omega_2', '\omega_3'};
98  PlotSet( t, xP(8:10,:), 'x label', tL, 'y label', yL,...
99    'plot title', 'RWA Rates', 'figure title', 'RWA Rates');
100
101 yL    = {'T_x (Nm)', 'T_y (Nm)', 'T_z (Nm)'};
102 PlotSet( t, xP(11:13,:), 'x label', tL, 'y label', yL,...
103    'plot title', 'Control Torque', 'figure title', 'Control Torque');
104
105 yL    = {'\Delta H (Nms)'};
106 PlotSet( t, xP(14,:), 'x label', tL, 'y label', yL,...
107    'plot title', 'Inertial Angular Momentum', 'figure title', 'Inertial
        Angular Momentum');
108
109 yL    = {'\theta_x (rad)', '\theta_y (rad)', '\theta_z (rad)'};
110 PlotSet( t, xP(15:17,:), 'x label', tL, 'y label', yL,...
```

Note how PlotSet makes plotting much easier to set up and read code than if we use MATLAB's built-in plot and supporting functions. You do lose some flexibility. The y axis labels use LaTeX notation. LaTeX is a technical publications package. This provides limited LaTeX syntax, such as Greek letters, subscripts, and superscripts. You can set the plotting to full LaTeX mode to get access to all LaTeX commands and formatting.

Note that we compute the angle error directly from the target and true quaternion. This represents our attitude sensor. In real spacecraft, attitude estimation is quite complicated. Multiple sensors, such as combinations of magnetometers, GPS, and earth and sun sensors, are used, and often rate-integrating gyros are employed to smooth the measurements. Star cameras or trackers are popular for three-axis sensing and require converting images in a camera to attitude estimates. You can't use gyros by themselves because they do not provide an initial orientation with respect to the inertial frame.

We will run two tests. The first shows that our controllers can compensate for a body-fixed disturbance torque. The second is to show that the controller can reorient the spacecraft.

The following is the initialization code for the disturbance torque test. The initial and target attitudes are the same, a unit quaternion, but there is a small disturbance torque.

```
1  % Initialize duration, delta time states and inertia
2  tEnd       = 600;
3  dT         = 1;
4  controlIsOn = true;
5  qECIToBody = [1;0;0;0];
6  omega      = [0;0;0]; % rad/sec
7  omegaRWA   = [0;0;0];  % rad/sec
8  d.inr      = [3 0 0;0 10 0;0 0 5];  % kg-m^2
9  qTarget    = QUnit([1;0;0.0;0]);
10 d.torque   = [0;0.0001;0]; % Disturbance torque (N)
```

We are running the simulations to 600 seconds to see the transients settle out. The distur-bance torque is very small, which is typical for spacecraft. We make the torque single axis to make the responses clearer. Figure 13.3 shows the complete set of output plots.

The disturbance causes a change in attitude around the y axis. This offset is expected with a PD controller. The control torque eventually matches the disturbance, and the angular error reaches its maximum. The PD control method will have steady-state error for constant (or more generally, nonzero-mean) disturbances. The integral control would be required to compensate for such disturbances.

The y wheel rate grows linearly as it has to absorb all the momentum produced by the torque. We don't limit the maximum wheel rate. In a real spacecraft, the wheel would soon saturate, reaching its maximum allowed speed. Our control system would need to have other actuators to desaturate the wheel. The inertial angular momentum also grows linearly as is expected with a constant external torque.

We now do an attitude correction around the x axis. The following is the initialization code:

```
1  % Initialize duration, delta time states and inertia
2  tEnd       = 600;
3  dT         = 1;
4  controlIsOn = true;
5  qECIToBody = [1;0;0;0];
6  omega      = [0;0;0]; % rad/sec
7  omegaRWA   = [0;0;0];  % rad/sec
8  d.inr      = [3 0 0;0 10 0;0 0 5];  % kg-m^2
9  qTarget    = QUnit([1;0.004;0.0;0]); % Normalize
10 d.torque   = [0;0;0]; % Disturbance torque
```

We command a small attitude offset around the x axis, which is done by changing the second element in the quaternion. We unitize the quaternion to prevent numerical issues. Figure 13.4 shows the output plots.

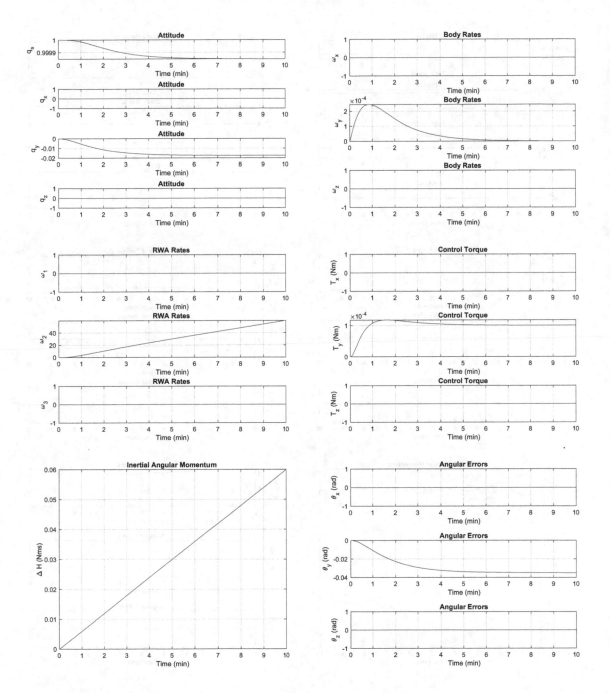

Figure 13.3: *Controlling a suddenly applied external torque.*

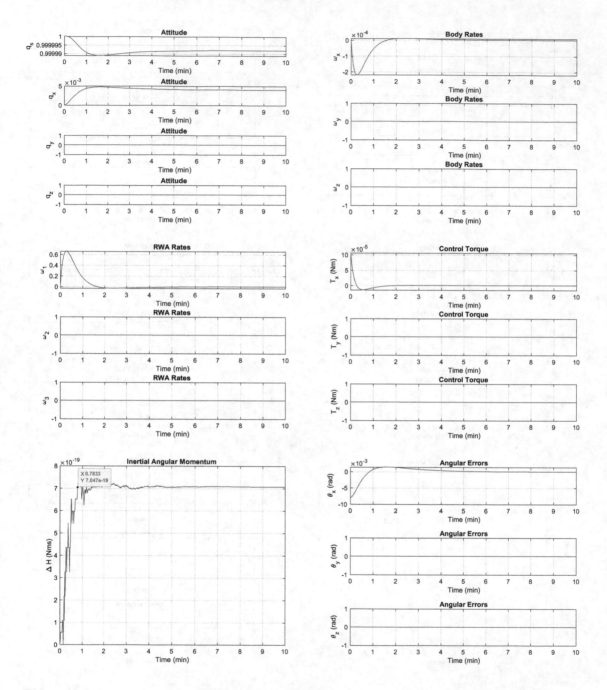

Figure 13.4: *Response to a small change in attitude.*

In this case, the angular error around the x axis is reduced to zero. The inertial angular momentum remains "constant" although it jumps around a bit due to truncation error in the numerical integration. This is expected and it is good to keep checking the angular momentum with the control system running. If it doesn't remain nearly constant, it means that the simulation probably has an error in the dynamical equations. Internal torques do not change the inertial angular momentum. This is why reaction wheels are called "momentum exchange devices." They exchange momentum with the spacecraft body but don't change the total inertial angular momentum.

The attitude rates remain small in both cases so that the Euler coupling torques are small. This justifies our earlier decision to treat the spacecraft as three double integrators. It also justifies our quaternion error to small angle approximation.

13.4   Performing Batch Runs of a Simulation

Problem

We've used our simulation script to verify momentum conservation and test our controller, but note how we have to change lines at the top by hand for each case. This is fine for development but can make it very difficult to reproduce results; we don't know the initial conditions that generated any particular plot. We may want to run our simulation for a whole set of inputs and do a Monte Carlo analysis.

Solution

We'll create a new function based on our script with inputs for our critical parameters. A new data structure will store both our inputs and the outputs so we can save individual runs to matfiles. This will make it possible to replot the results of any run in the future, or redo runs from the stored inputs, for example, if you find and fix a bug in the controller.

How It Works

Start from the simulation script copied into a new file. Add a function signature. Replace the initialization variables with an input structure. Perform the simulation, then save the input structure along with your generated output. The resulting function header is shown in the following listing. The `input` structure includes our RHS data, controller data, and simulation timing data.

SpacecraftSimFunction.m

```
1  %% SPACECRAFTSIMFUNCTION Spacecraft reaction wheel simulation function
2  % Perform a simulation of a spacecraft with reaction wheels given a
3  % particular initial state. If there are no inputs it will perform a
     demo for an
4  % open loop system. If there are no outputs it will create plots via
5  % PlotSpacecraftSim.
6  %% Form
7  %   d = SpacecraftSimFunction( x0, qTarget, input )
```

```
8   %% Inputs
9   %  x0      (7+n,1) Initial state
10  %  qTarget (4,1) Target quaternion
11  %  input     (.)   Data structure
12  %                     .rhs    (.)  RHS data
13  %                     .pd     (:)  Controllers
14  %                     .dT     (1,1) Timestep
15  %                     .tEnd (1,1) Duration
16  %                     .controlIsOn Flag
17  %% Outputs
18  %  d         (.)   Data structure
19  %                     .input    (.)    Input structure
20  %                     .x0      (7+n,1) Initial state
21  %                     .qTarget (4,1) Target quaternion
22  %                     .xPlot (7+n,:) State data
23  %                     .dPlot (4+n,:) Torque and angle error data
24  %                     .tPlot (1,:)  Time data
25  %                     .yLabel  {}    State labels
26  %                     .dLabel  {}    Data labels
27  %                     .tLabel  ''    Time label string
28  %% See also
29  % RHSSpacecraftWithRWA, ErrorFromQuaternion, PDControl, RungeKutta,
        TimeLabel,
30  % PlotSpacecraftSim
```

Now, we can write a script that calls our simulation function in a loop. The possibilities are endless – you can test different targets, vary the initial conditions for a Monte Carlo simulation, and apply different disturbance torques. You can perform statistical analysis on your results or identify and plot individual runs based on some criteria. In this example, we will find the maximum control torque applied in each run.

BatchSimRuns.m

```
1   %% Script performing multiple runs of spacecraft simulation
2   % Perform runs of SpacecraftSimFunction in a loop with varying initial
3   % conditions. Find the max control torque applied for each case.
4   %% See also
5   % SpacecraftSimFunction
10
11  %% Initialization
12  sim = struct;
13  % Initialize duration, delta time states and inertia
14  sim.tEnd       = 600;
15  sim.dT         = 1;
16  sim.controlIsOn = true;
17
18  % Spacecraft state
19  qECIToBody = [1;0;0;0];
20  omega      = [0;0;0]; % rad/sec
21  omegaRWA   = [0;0;0];  % rad/sec
22  x0 = [qECIToBody;omega;omegaRWA];
```

```
23
24    % Target quaternions
25    qTarget = QUnit([1;0.004;0.0;0]); % Normalize
26
27    %% Control design
28    % Design a PD controller
29    dC              = PDControl( 'struct' );
30    dC(1).zeta      = 1;
31    dC(1).wN        = 0.02;
32    dC(1).wD        = 5*dC(1).wN;
33    dC(1).tSamp     = sim.dT;
34    dC(1)           = PDControl( 'initialize', dC(1) );
35
36    % Make all 3 axis controllers identical
37    dC(2)           = dC(1);
38    dC(3)           = dC(1);
39
40    sim.pd = dC;
41
42    %% Spacecraft model
43    % Make the spacecraft nonspherical; no disturbances
44    rhs         = RHSSpacecraftWithRWA;
45    rhs.inr     = [3 0 0;0 10 0;0 0 5];  % kg-m^2
46    rhs.torque = [0;0;0]; % Disturbance torque
47    sim.rhs     = rhs;
49
50    %% Simulation loop
51    clear d;
52    for k = 1:10;
53      % change something in your initial conditions and simulate
54      x0(5) = 1e-3*k;
55      thisD = SpacecraftSimFunction( x0, qTarget, sim );
56
57      % save the run results as a mat-file
58      thisDir = fileparts(mfilename('fullpath'));
59      fileName = fullfile(thisDir,'Output',sprintf('Run%d',k));
60      save(fileName,'-struct','thisD');
61
62      % store the run output
63      d(k) = thisD;
64    end
65
66    %% Perform statistical analysis on results
67    % ... as you wish
68    for k = 1:length(d)
69      tMax(k) = max(max(d(k).dPlot(2:4,:)));
70    end
71    figure;
72    plot(1:length(d),tMax);
73    xlabel('Run')
74    ylabel('Torque (Nm)')
75    title('Maximum Control Torque');
```

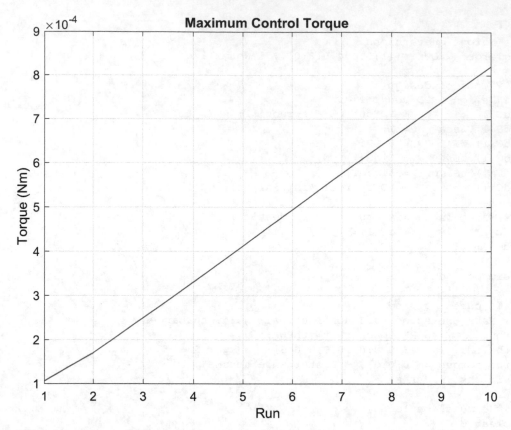

Figure 13.5: *Maximum control torque over ten simulation runs.*

```
76
77  % Plot a single case
78  kPlot = 4;
79  PlotSpacecraftSim( d(4) );
```

Figure 13.5 shows the maximum torque results. Each run has a larger initial angular velocity. We expect to see this trend, because the torque control is proportional to the angular rate.

An individual run's output is shown as follows:

```
>> d(1)

ans =

       input: [1x1 struct]
          x0: [10x1 double]
     qTarget: [4x1 double]
       xPlot: [10x600 double]
       dPlot: [7x600 double]
       tPlot: [1x600 double]
```

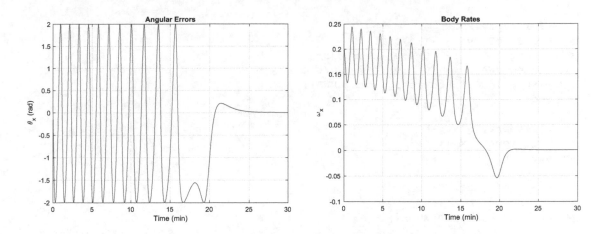

Figure 13.6: *Control response to a large rate in x. The rate does damp out, eventually!*

```
tLabel: 'Time (min)'
yLabel: {1x10 cell}
dLabel: {1x7 cell}
```

As another interesting example, we can give the spacecraft a higher initial rate and see how the controller responds. From the command line, we change the initial rate around the x axis to be 0.2 rad/sec and call the simulation function with no outputs, so that it will generate the full suite of plots. We see that the response takes a long time, over 20 minutes, but the rate does eventually damp out. Figure 13.6 shows the damping response.

The full simulation function is shown in the following. The built-in demo performs an open loop simulation of the default spacecraft model with no control, as with the momentum conservation test performed in the previous recipe (Figure 13.2).

```
36  function d = SpacecraftSimFunction( x0, qTarget, input )
37
38  % Handle inputs
39  if nargin == 0
40    % perform an open loop simulation
41    disp('SpacecraftSimFunction: Open loop simulation for 10 minutes.
          Initial rates are random.')
42    input = struct;
43    input.rhs  = RHSSpacecraftWithRWA;
44    input.pd   = [];
45    input.dT   = 1; % sec
46    input.tEnd = 600; % sec
47    input.controlIsOn = false;
48    x0 = [1;0;0;0;1e-3*randn(6,1)];
49    SpacecraftSimFunction( x0, [], input );
50    return;
51  end
52
```

```
53   if isempty(x0)
54     qECIToBody   = [1;0;0;0];
55     omega        = [0;0;0]; % rad/sec
56     omegaRWA     = [0;0;0];  % rad/sec
57     x0 = [qECIToBody;omega;omegaRWA];
58   end
59
60   if isempty(qTarget)
61     qTarget = x0(1:4);
62   end
63
64   % State vector
65   x = x0;
66   nWheels = length(x0)-7;
67
68   % Plotting and number of steps
69   n    = ceil(input.tEnd/input.dT);
70   xP   = zeros(length(x),n);
71   dP   = zeros(7,n);
72
73   % Find the initial angular momentum
74   d = input.rhs;
75   [~,hECI0] = RHSSpacecraftWithRWA(0,x,d);
76
77   % Run the simulation
78   for k = 1:n
79     % Control
80     u = [0;0;0];
81     angleError = [0;0;0];
82     if( input.controlIsOn )
83       % Find the angle error
84       angleError = ErrorFromQuaternion( x(1:4), qTarget );
85       % Update the controllers individually
86       for j = 1:nWheels
87         [u(j), input.pd(j)] = PDControl('update',angleError(j),input.pd(j
                  ));
88       end
89     end
90
91     % Wheel torque
92     d.torqueRWA = d.inr*u;
93
94     % Get the delta angular momentum
95     [~,hECI]  = RHSSpacecraftWithRWA(0,x,d);
96     dHECI     = hECI - hECI0;
97     hMag      = sqrt(dHECI'*dHECI);
98
99     % Plot storage
100    xP(:,k)       = x;
101    dP(:,k)       = [hMag;d.torqueRWA;angleError];
102
103    % Right hand side
```

378

```
104    x           = RungeKutta(@RHSSpacecraftWithRWA,0,x,input.dT,d);
105  end
106
107  [t,tL] = TimeLabel((0:(n-1))*input.dT);
108
109  % Record initial conditions and results
110  d = struct;
111  d.input    = input;
112  d.x0       = x0;
113  d.qTarget = qTarget;
114  d.xPlot = xP;
115  d.dPlot = dP;
116  d.tPlot = t;
117  d.tLabel = tL;
118
119  y = cell(1,nWheels);
120  for k = 1:nWheels
121    y{k} = sprintf('\\omega_%d',k);
122  end
123  d.yLabel = [{'q_s','q_x','q_y','q_z','\omega_x','\omega_y','\omega_z'}
         y];
124  d.dLabel = {'\Delta H (Nms)','T_x (Nm)', 'T_y (Nm)', 'T_z (Nm)', ...
125    '\theta_x (rad)', '\theta_y (rad)', '\theta_z (rad)'};
126
127  if nargout == 0
128    PlotSpacecraftSim( d );
129  end
```

The plotting code is put into a separate function that accepts the output data structure. We create and save the plot labels in the simulation function. This allows us to replot any saved output. We add a statement to check for nonzero angle errors before creating the control and angle error plots, since they are not needed for open loop simulations.

■ **TIP**   Use the fields in your structure for plotting without renaming the variables locally, so you can copy/paste individual plots to the command line after doing a run of your simulation.

PlotSpacecraftSim.m

```
1  %% PLOTSPACECRAFTSIM Plot the spacecraft simulation output
2  %% Form
3  %   PlotSpacecraftSim( d )
4  %% Inputs
5  %   d  (.)    Simulation data structure
6  %% Outputs
7  %   None.
12
```

```
13   function PlotSpacecraftSim( d )
14
15   t   = d.tPlot;
16
17   yL = d.yLabel(1:4);
18   PlotSet( d.tPlot, d.xPlot(1:4,:), 'x label', d.tLabel, 'y label', yL
        ,...
19     'plot title', 'Attitude', 'figure title', 'Attitude');
20
21   yL = d.yLabel(5:7);
22   PlotSet(d.tPlot, d.xPlot(5:7,:), 'x label', d.tLabel, 'y label', yL,...
23     'plot title', 'Body Rates', 'figure title', 'Body Rates');
24
25   yL = d.yLabel(8:end);
26   PlotSet( t, d.xPlot(8:end,:), 'x label', d.tLabel, 'y label', yL,...
27     'plot title', 'RWA Rates', 'figure title', 'RWA Rates');
28
29   yL = d.dLabel(1);
30   PlotSet( d.tPlot, d.dPlot(1,:), 'x label', d.tLabel, 'y label', yL,...
31     'plot title', 'Inertial Angular Momentum',...
32     'figure title', 'Inertial Angular Momentum');
33
34   if any(d.dPlot(5:end,:)~=0)
35     yL = d.dLabel(2:4);
36     PlotSet( d.tPlot, d.dPlot(2:4,:), 'x label', d.tLabel, 'y label', yL
          ,...
37       'plot title', 'Control Torque', 'figure title', 'Control Torque');
38     yL = d.dLabel(5:end);
39     PlotSet( d.tPlot, d.dPlot(5:end,:), 'x label', d.tLabel, 'y label',
          yL,...
40       'plot title', 'Angular Errors', 'figure title', 'Angular Errors');
41   end
```

An interesting exercise for the reader would be to replace the fixed disturbance input, d.torque, with a function handle that calls a disturbance function. This forms the basis of spacecraft simulation in our Spacecraft Control Toolbox, where the disturbances are calculated from the spacecraft geometry and space environment as it rotates and moves along its orbit.

Summary

This chapter has demonstrated how to write the dynamics and implement a simple control law for a spacecraft with reaction wheels. Our control system is only valid for small angle changes and will not work well if the angular rates on the spacecraft get large. In addition, we do not consider the torque or momentum limits on the reaction wheels. We also learned about quaternions and how to implement kinematics of rigid body with quaternions. We showed how to get angle errors from two quaternions. Table 13.1 lists the code developed in the chapter.

Table 13.1: *Chapter Code Listing*

File	Description
RHSSpacecraftWithRWA	RHS for spacecraft with reaction wheels
ErrorFromQuaternion	Spacecraft simulation script
SpacecraftSim	Spacecraft simulation script
SpacecraftSimBatch	Spacecraft simulation function
BatchSimRuns	Multiple runs of the spacecraft simulation
PlotSpacecraftSim	Plot the simulation results
QTForm	Transform a vector opposite the direction of the quaternion
QUnit	Normalize a quaternion

CHAPTER 14

■ ■ ■

Automobiles

Automobile technology has gone from the mundane to the cutting edge over the past decade. New technologies such as electric cars and autonomous driving are making automotive engineering one of the most exciting areas for engineers.

In this chapter, we will give recipes covering a wide range of automotive topics, including dynamics and autonomous driving.

14.1 Automobile Dynamics

Problem

We need to model the car dynamics. We will limit this to a planar model in two dimensions. We are modeling the location of the car in x/y and the angle of the wheels which allows the car to change direction.

Solution

Write a right-hand-side function that can be called by the RungeKutta integration function.

How It Works

We will need two functions for the dynamics of the automobile. RHSAutomobile is used by the simulation. RHSAutomobile has the full dynamical model including the engine and steering model. Aerodynamic drag, rolling resistance, and side force resistance (the car doesn't slide sideways without resistance) are modeled. RHSAutomobile handles multiple automobiles. An alternative would be to have one automobile function and call RungeKutta once for each automobile. The latter approach works in all cases, except when you want to model collisions. In many types of collisions, two cars collide and then stick, effectively becoming a single car. Each vehicle has six states. They are

1. x position

2. y position

3. x velocity

© Michael Paluszek and Stephanie Thomas 2020
M. Paluszek and S. Thomas, *MATLAB Recipes*,
https://doi.org/10.1007/978-1-4842-6124-8_14

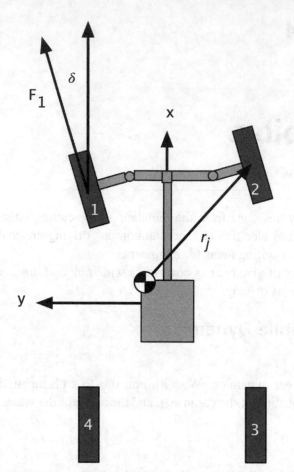

Figure 14.1: *Planar automobile dynamical model. The contact and rolling friction are included in F_k where k is one of the four wheels.*

4. y velocity

5. Heading

6. Angular rate about vertical

The velocity derivatives are driven by the forces and the angular rate derivative by the torques. The planar dynamical model is illustrated in Figure 14.1 [2]. Unlike the reference, we constrain the rear wheels to be fixed and the angles for the front wheels to be the same.

The dynamical equations are written in the rotating frame:

$$m(\dot{v}_x - 2\omega v_y) = \sum_{k=1}^{4} F_{k_x} - qC_{D_x}A_x u_x \tag{14.1}$$

$$m(\dot{v}_y + 2\omega v_x) = \sum_{k=1}^{4} F_{k_y} - qC_{D_y}A_y u_y \tag{14.2}$$

$$I\dot{\omega} = \sum_{k=1}^{4} r_k^\times F_k \tag{14.3}$$

where m is the total mass of the car, v is the translational velocity, ω is the angular rate about vertical, I is the inertia about the vertical axis, and F_k is the kth component of force. C_D is the drag coefficient, and A_x and A_y are the areas in the x and y directions used to compute the drag. This is treating the car as two flat plates normal to the flow. The dynamic aerodynamic pressure is

$$q = \frac{1}{2}\rho\sqrt{v_x^2 + v_y^2} \tag{14.4}$$

and

$$v = \begin{bmatrix} v_x \\ v_y \end{bmatrix} \tag{14.5}$$

The unit vector is

$$u = \frac{\begin{bmatrix} v_x \\ v_y \end{bmatrix}}{\sqrt{v_x^2 + v_y^2}} \tag{14.6}$$

The gravitational force is mg where g is the acceleration of gravity. The force at the tire contact point, where the tire touches the road, for tire k is

$$F_{t_k} = \begin{bmatrix} T/\rho - F_r \\ -F_c \end{bmatrix} \tag{14.7}$$

where T is the torque and ρ is the radius of the tire. F_r is the rolling friction and is

$$F_r = f_0 + K_1 v_{t_x}^2 \tag{14.8}$$

where v_{t_x} is the velocity in the tire frame in the rolling direction. f_0 is the velocity-independent force, and K_1 is the velocity coefficient. For front-wheel drive cars, the torque, T, is zero for the rear wheels. The contact friction is

$$F_c = \frac{1}{4}\mu_c mg\frac{v_{t_y}}{|v_t|} \tag{14.9}$$

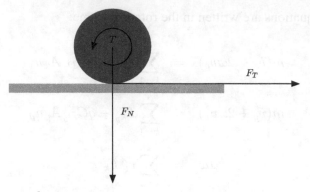

Figure 14.2: *Wheel force and torque.*

This is the force perpendicular to the normal rolling direction of the wheel that is into or out of the paper in Figure 14.2. The velocity term ensures that the friction force does not cause limit cycling. That is, when the y velocity is zero, the force is zero. μ_c is a constant for the tires. This model assumes that lateral friction is only applied when slipping.

The transformation from the tire to body frame is

$$c = \begin{bmatrix} \cos\delta & -\sin\delta \\ \sin\delta & \cos\delta \end{bmatrix} \tag{14.10}$$

where δ is the steering angle so that

$$F_k = cF_{t_k} \tag{14.11}$$

$$v_t = c^T \begin{bmatrix} v_x \\ v_y \end{bmatrix} \tag{14.12}$$

The kinematical equation that related the yaw angle and yaw angular rate is

$$\dot{\theta} = \omega \tag{14.13}$$

and the inertial velocity V, the velocity needed to tell you where the car is going, is

$$V = \begin{bmatrix} \cos\theta & -\sin\theta \\ \sin\theta & \cos\theta \end{bmatrix} v \tag{14.14}$$

The dynamics simulation right-hand side computes the dynamics of the automobile. RHSAutomobile can simulate multiple cars.

RHSAutomobile.m

```
1   %% RHSAUTOMOBILE Right hand side for a 2D automobile.
2   % Use AutomobileInitialize to set up d
3   %% Form
4   %   xDot = RHSAutomobile( t, x, d )
```

```
5   %
6   %% Inputs
7   %   t             Time, unused
8   %   x     (6*n,1) State, [x;y;vX;vY;theta;omega]
9   %   d     (1,1)   Data structure
10  %                     .car    (n,1)   Car data structure
11  %                        .mass  (1,1) Mass (kg)
12  %                        .delta (1,1) Steering angle (rad)
13  %                        .r     (2,4) Position of wheels (m)
14  %                        .cDF   (1,1) Frontal drag coefficient
15  %                        .cDS   (1,1) Side drag coefficient
16  %                        .cF    (1,1) Friction coefficient
17  %                        .fT    (1,1) Traction force (N)
18  %                        .areaF (1,1) Frontal area for drag (m^2)
19  %                        .areaS (1,1) Side area for drag (m^2)
20  %                        .fRR   (1,2) [f0 K]
21  %
22  %% Outputs
23  %   x     (6*n,1)  d[x;y;vX;vY;theta;omega]/dt
24  %
```

This is a designer's choice. It allows for it to simulate automobile interactions more easily. If the cars are always separate, we could have one right-hand side per car.

The beginning just initializes the arrays and constants.

```
26  function xDot = RHSAutomobile( ~, x, d )
27
28  % Constants
29  g       = 9.806; % Acceleration of gravity (m/s^2)
30  n       = length(x);
31  nS      = 6; % Number of states
32  xDot    = zeros(n,1);
33  nAuto   = n/nS;
```

The for loop cycles through all of the cars. The first part is the kinematics.

```
35  j = 1;
36  % State [j j+1 j+2 j+3  j+4   j+5]
37  %        x  y   vX  vY  theta omega
38  for k = 1:nAuto
39      vX    = x(j+2,1);
40      vY    = x(j+3,1);
41      theta = x(j+4,1);
42      omega = x(j+5,1);
43
44      % Car angle
45      c = cos(theta);
```

```
46      s = sin(theta);
47
48      % Inertial frame
49      v       = [c -s;s c]*[vX;vY];
50
51      delta = d.car(k).delta;
52      c       = cos(delta);
53      s       = sin(delta);
54      mCToT = [c s;-s c];
```

The next part computes forces and torques.

```
56      % Find the rolling resistance of the tires
57      vTire     = mCToT*[vX;vY];
58      f0        = d.car(k).fRR(1);
59      K1        = d.car(k).fRR(2);
60
61      fRollingF = f0 + K1*vTire(1)^2;
62      fRollingR = f0 + K1*vX^2;
63
64      % This is the side force friction
65      fFriction = d.car(k).cF*d.car(k).mass*g;
66      fT        = d.car(k).radiusTire*d.car(k).torque;
67
68      fF        = [fT - fRollingF;-vTire(2)*fFriction];
69      fR        = [   - fRollingR;-vY        *fFriction];
70
71      % Tire forces
72      f1 = mCToT'*fF;
73      f2 = f1;
74      f3 = fR;
75      f4 = f3;
76
77      % Aerodynamic drag
78      vSq    = vX^2 + vY^2;
79      vMag   = sqrt(vSq);
80      q      = 0.5*1.225*vSq;
81      fDrag = q*[d.car(k).cDF*d.car(k).areaF*vX;...
82                     d.car(k).cDS*d.car(k).areaS*vY]/vMag;
83
84      % Force summations
85      f = f1 + f2 + f3 + f4 - fDrag;
86
87      % Torque
88      T = Cross2D( d.car(k).r(:,1), f1 ) + Cross2D( d.car(k).r(:,2), f2 )
            + ...
89          Cross2D( d.car(k).r(:,3), f3 ) + Cross2D( d.car(k).r(:,4), f4 )
            ;
```

Finally, we assemble the derivative array that is returned.

```
91    % Right hand side
92    xDot(j,   1) = v(1);
93    xDot(j+1,1) = v(2);
94    xDot(j+2,1) = f(1)/d.car(k).mass + omega*vY;
95    xDot(j+3,1) = f(2)/d.car(k).mass - omega*vX;
96    xDot(j+4,1) = omega;
97    xDot(j+5,1) = T/d.car(k).inr;
98
99    j = j + nS;
100   end
```

14.2   Modeling the Automobile Radar

Problem

The sensor utilized for this example will be the automobile radar. The radar measures azimuth, range, and range rate.

Solution

Build a radar model in a MATLAB function. The function will use analytical derivations of range and range rate.

How It Works

The radar model is extremely simple. It assumes the radar measures the line-of-site range, range rate, and azimuth, the angle from the forward axis of the car. The model skips all the details of radar signal processing and outputs those three quantities. This type of simple model is always the best when you start a project. Later on, you will need to add a very detailed model that has been verified against test data to demonstrate that your system works as expected.

The position and velocity of the radar are entered through the data structure. This does not model the signal-to-noise ratio of a radar. The power received by a radar goes as $\frac{1}{r^4}$. In this model, the signal goes to zero at the maximum range that is specified in the function. The range is found from the difference in position between the radar and the target. If δ is the difference, we write

$$\delta = \begin{bmatrix} x - x_r \\ y - y_r \\ z - z_r \end{bmatrix} \tag{14.15}$$

The range is then

$$\rho = \sqrt{\delta_x^2 + \delta_y^2 + \delta_z^2} \tag{14.16}$$

The delta velocity is

$$\nu = \begin{bmatrix} v_x - v_{x_r} \\ v_y - v_{y_r} \\ v_z - v_{z_r} \end{bmatrix} \tag{14.17}$$

In both equations, the subscript r denotes the radar. The range rate is

$$\dot{\rho} = \frac{\nu^T \delta}{\rho} \tag{14.18}$$

The AutoRadar function handles multiple targets and can generate radar measurements for an entire trajectory. This is really convenient because you can give it your trajectory and see what it returns. This gives you a physical feel for the problem without running a simulation. It also allows you to be sure the sensor model is doing what you expect. This is important because all models have assumptions and limitations. It may be that the model really isn't suitable for your application. For example, this model is two-dimensional. If you are concerned about your system getting confused about a car driving across a bridge above your automobile, this model will not be useful in testing that scenario.

AutoRadar.m

```
57  function [y, v] = AutoRadar( x, d )
58
59  % Demo
60  if( nargin < 1 )
61    if( nargout == 0 )
62      Demo;
63    else
64      y = DataStructure;
65    end
66        return
67  end
68
69  m    = size(d.kR,2);
70  n    = size(x,2);
71  y    = zeros(3*m,n);
72  v    = ones(m,n);
73  cFOV = cos(d.fOV);
74
75  % Build an array of random numbers for speed
76  ran = randn(3*m,n);
77
78  % Loop through the time steps
79  for j = 1:n
80    i     = 1;
81    s     = sin(d.theta(j));
```

```
82    c       = cos(d.theta(j));
83    cIToC = [c s;-s c];
84
85    % Loop through the targets
86    for k = 1:m
87      xT      = x(d.kR(:,k),j);
88      vT      = x(d.kV(:,k),j);
89      th      = x(d.kT(1,k),j);
90      s       = sin(th);
91      c       = cos(th);
92      cTToIT  = [c -s;s c];
93      dR      = cIToC*(xT - d.xR(:,j));
94      dV      = cIToC*(cTToIT*vT - cIToC'*d.vR(:,j));
95      rng     = sqrt(dR'*dR);
96      uD      = dR/rng;
97
98      % Apply limits
99      if( d.noLimits || (uD(1) > cFOV && rng < d.maxRange) )
100       y(i  ,j)  = rng               + d.noise(1)*ran(i  ,j);
101       y(i+1,j)  = dR'*dV/y(i,j)     + d.noise(2)*ran(i+1,j);
102       y(i+2,j)  = atan(dR(2)/dR(1)) + d.noise(3)*ran(i+2,j);
103     else
104       v(k,j)      = 0;
105     end
106     i   = i + 3;
107   end
108 end
```

Built-in plotting is provided. This makes it easier for a user to understand the function.

```
110 % Plot if no outputs are requested
111 if( nargout < 1 )
112   [t, tL]       = TimeLabel( d.t );
113
114   % Every 3rd y is azimuth
115   i       = 3:3:3*m;
116   y(i,:)        = y(i,:)*180/pi;
117
118   yL      = {'Range (m)' 'Range Rate (m/s)', 'Azimuth (deg)' 'Valid
          Data'};
119   PlotSet(t,[y;v],'x label',tL','y label',yL,'figure title','Auto Radar
          ',...
120           'plot title','Auto Radar');
121
122   clear y
123 end
```

The function returns a default data structure to get the user started.

```
125  %% AutoRadar>DataStructure
126  function d = DataStructure
127  %% Default data structure
128  d.kR          = [1;2];
129  d.kV          = [3;4];
130  d.kT          = 5;
131  d.theta       = [];
132  d.xR          = [];
133  d.vR          = [];
134  d.noise       = [0.02;0.0002;0.01];
135  d.fOV         = 0.95*pi/16;
136  d.maxRange      = 60;
137  d.noLimits    = 1;
138  d.t           = [];
```

Notice that the function has a built-in demo and, if there are no outputs, will plot the results.

```
140  %% AutoRadar>Demo
141  function Demo
142  % Shows radar performance as range changes
143
144  omega         = 0.02;
145  d             = DataStructure;
146  n             = 1000;
147  d.xR          = [linspace( 0,1000,n);zeros(1,n)];
148  d.vR          = [ones(1,n);zeros(1,n)];
149  t             = linspace(0,1000,n);
150  a             = omega*t;
151  x             = [linspace(10,10+1.05*1000,n);2*sin(a);...
152                    1.05*ones(1,n); 2*omega*cos(a);zeros(1,n)];
153  d.theta       = zeros(1,n);
154  d.t           = t;
155
156  AutoRadar( x, d );
```

The radar returns the range, the range rate, and the azimuth angle of the target. Even though we are using radar as our sensor, there is no reason why you couldn't use a camera, laser rangefinder, or sonar instead. The limitation in the algorithms and software provided in this book is that it will only handle one sensor. You can get software from Princeton Satellite Systems that expands this to multiple sensors. For example, cars carry radar, cameras, and LIDAR (laser or light radar). You might want to integrate all of their measurements together. Figure 14.3 shows the internal radar demo. The target car is weaving in front of the radar. It is receding at a steady velocity, but the weave introduces a time-varying range rate.

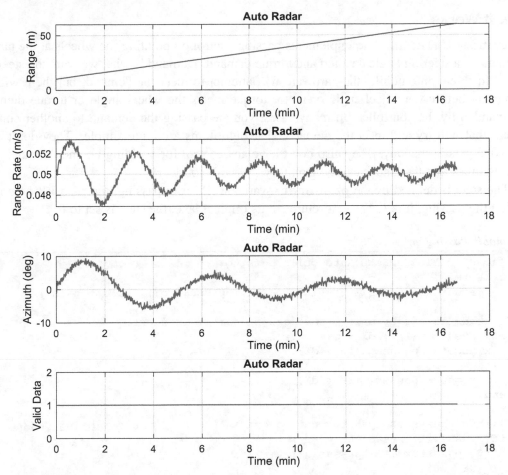

Figure 14.3: *Built-in radar demo. The target is weaving in front of the radar.*

14.3   Automobile Autonomous Passing Control

Problem

To have something interesting for our radar to measure, we need our cars to perform some maneuvers. We will develop an algorithm for a car to change lanes.

Solution

The cars are driven by steering controllers that execute basic automobile maneuvers. The throttle (accelerator pedal) and steering angle can be controlled. Multiple maneuvers can be chained together. The first function is for autonomous passing, and the second performs the lane change.

How It Works

The `AutomobilePassing` implements passing control by pointing the wheels at the target. It generates a steering angle demand and torque demand. Demand is what we want the steering to do. In a real automobile, the hardware will attempt to meet the demand, but there will be a time lag before the wheel angle or motor torque meets the wheel angle or torque demand commanded by the controller. In many cases, you are passing the demand to another control system that will try and meet the demand. The algorithms are quite simple. They don't care if anyone gets in the way. They also don't have any control for avoiding another vehicle. The code assumes that the lane is empty. Don't try this with your car!

The state is defined by the `passState` variable. Prior to passing, the `passState` is 0. During the passing, it is 1. When it returns to its original lane, the state is set to 0.

AutomobilePassing.m

```
1  %% AUTOMOBILEPASSING Automobile passing control
2  %% Form
3  %   passer = AutomobilePassing( passer, passee, dY, dV, dX, gain )
43
44 % Lead the target unless the passing car is in front
45 if( passee.x(1) + dX > passer.x(1) )
46   xTarget = passee.x(1) + dX;
47 else
48   xTarget = passer.x(1) + dX;
49 end
50
51 % This causes the passing car to cut in front of the car being passed
52 if( passer(1).passState == 0 )
53   if( passer.x(1) > passee.x(1) + 2*dX )
54     dY = 0;
55     passer(1).passState = 1;
56   end
57 else
58   dY = 0;
59 end
60
61 % Control calculation
62 target          = [xTarget;passee.x(2) + dY];
63 theta           = passer.x(5);
64 dR              = target - passer.x(1:2);
65 angle           = atan2(dR(2),dR(1));
66 err             = angle - theta;
67 passer.delta    = gain(1)*(err + gain(3)*(err - passer.errOld));
68 passer.errOld   = err;
69 passer.torque   = gain(2)*(passee.x(3) + dV - passer.x(3));
```

The second function, `AutomobileLaneChange`, performs a lane change. It implements lane change control by pointing the wheels at the target. The function generates a steering angle demand and a torque demand. The default gains work reasonably well. You should always supply defaults that make sense.

AutomobileLaneChange.m

```
1   %% AUTOMOBILELANECHANGE Automobile lane change control
2   %% Form
3   %    passer = AutomobileLaneChange( passer, dY, v, gain )
35  % Default gains
36  if( nargin < 5 )
37    gain = [0.05 80 120];
38  end
39
40  % Lead the target unless the passing car is in front
41  xTarget        = passer.x(1) + dX;
42
43  % Control calculation
44  target         = [xTarget;y];
45  theta          = passer.x(5);
46  dR             = target - passer.x(1:2);
47  angle          = atan2(dR(2),dR(1));
48  err            = angle - theta;
49  passer.delta   = gain(1)*(err + gain(3)*(err - passer.errOld));
50  passer.errOld  = err;
51  passer.torque  = gain(2)*(v - passer.x(3));
```

14.4   Automobile Animation

Problem

We want to visualize the cars as they maneuver.

Solution

Read in a file in `.obj` format. Display it using MATLAB's `patch` function. Pass the orientation and position of the automobile to the animation function and create a movie.

How It Works

We create a function to read in `.obj` files. We then write a function to draw and animate the model.

The first step is to find an automobile model. A good resource is TurboSquid: www.turbosquid.com. You will find thousands of models. We need the `.obj` format and prefer a low polygon count. Ideally, we want models with triangles. In the case of the model found for this chapter, it had rectangles so we converted them to triangles using a Macintosh application, Cheetah3D: www.cheetah3d.com. An OBJ model comes with an `.obj` file, an `.mtl` file (material file), and images for textures. We will only use the OBJ file.

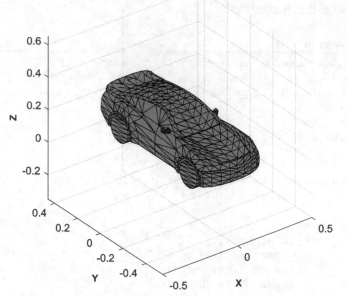

Figure 14.4: *Automobile 3D model.*

LoadOBJ loads the file and puts it into a data structure. The data structure uses the g field of the OBJ file to break the file into components. In this case, the components are the four tires and the rest of the car. The demo is just:

```
LoadOBJ( 'MyCar.obj' )
```

You do need the extension, .obj. The car is shown in Figure 14.4. The image is generated with one call to patch per component.

The first part of DrawComponents initializes the model. We save, and return, pointers to the patches so that we only have to update the vectors with each call.

DrawComponents.m

```
1  %% DRAWCOMPONENTS Draws a multi-component object
2  %% Form
3  %    h = DrawComponents( 'initialize', g )
4  %    DrawComponents( 'update', g, h, x )
25
26 function h = DrawComponents( action, g, h, x )
27
28 if( nargin < 1 )
29   Demo
30   return
31 end
32
33 switch( lower(action) )
```

396

```
34    case 'initialize'
35
36      n = length(g.component);
37      h = zeros(1,n);
38
39      for k = 1:n
40        h(k) = DrawMesh(g.component(k) );
41      end
42
43    case 'update'
44      UpdateMesh(h,g.component,x);
45
46    otherwise
47      warning('%s not available',action);
48  end
```

The mesh is drawn with a call to patch. We use the minimal set of properties. We make the edges black to make the model easier to see. The Phong reflection model is an empirical lighting model. It includes diffuse and specular lighting.

```
51  function h = DrawMesh( m )
52
53  h = patch(  'Vertices', m.v, 'Faces',   m.f, 'FaceColor', m.color,...
54              'EdgeColor',[0 0 0],'EdgeLighting', 'phong',...
55              'FaceLighting', 'phong');
```

Updating is done by rotating the vertices around the z axis and then adding the x and y positional offsets. The input array is [x;y;yaw]. We then set the new vertices. The function can handle an array of positions, velocities, and yaw angles.

```
59  function UpdateMesh( h, c, x )
60
61  for j = 1:size(x,2)
62    for k = 1:length(c)
63      cs        = cos(x(3,j));
64      sn        = sin(x(3,j));
65      b         = [cs -sn 0 ;sn cs 0;0 0 1];
66      v         = (b*c(k).v')';
67      v(:,1)    = v(:,1) + x(1,j);
68      v(:,2)    = v(:,2) + x(2,j);
69      set(h(k),'vertices',v);
70    end
71  end
```

The graphics demo AutomobileDemo implements passing control. Automobile Initialize reads in the OBJ file. The following code sets up the graphics window:

AutomobileDemo.m

```
33  % Set up the figure
34  NewFigure( 'Car Passing' )
35  axes('DataAspectRatio',[1 1 1],'PlotBoxAspectRatio',[1 1 1] );
36
37  h = [];
38  h(1,:) = DrawComponents( 'initialize', d.car(1).g );
39  h(2,:) = DrawComponents( 'initialize', d.car(2).g );
40
41  xlabel('X (m)')
42  ylabel('Y (m)')
43  zlabel('Z (m)')
44
45  set(gca,'ylim',[-4 4],'zlim',[0 2]);
46
47  grid on
48  view(3)
49  rotate3d on
50  hold off
```

During each pass through the simulation loop, we update the graphics. We call `DrawCompo`
`nents` once per car along with the stored `patch` handles for each car's components. We adjust
the limits so that we maintain a tight focus on the two cars. We could have used the camera
fields in the axes data structure for this too. We call `drawnow` after setting the new `xlim` for
smooth animation. Animation is also discussed in Chapter 4.

```
69  for k = 1:n
70    % Draw the cars
71    pos1 = x([1 2]);
72    pos2 = x([7 8]);
73    DrawComponents( 'update', d.car(1).g, h(1,:), [pos1;pi/2 + x( 5)] );
74    DrawComponents( 'update', d.car(2).g, h(2,:), [pos2;pi/2 + x(11)] );
75
76    xlim = [min(x([1 7]))-10 max(x([1 7]))+10];
77    set(gca,'xlim',xlim);
78    drawnow
79
80    for i = 1:nAuto
81      p          = 6*i-5;
82      d.car(i).x = x(p:p+5);
83    end
84
85    % Implement Control
86
87    % For all but the passing car control the velocity
88    d.car(1).torque = -10*(d.car(1).x(3) - vSet(1));
89
90    % The active car
91    if( t(k) < tEndPassing )
92      d.car(2)    = AutomobilePassing( d.car(2), d.car(1), 3, 1.3, 10 );
```

```
93    else
94        d.car(2).torque = -10*(d.car(2).x(3) - vSet(2));
95    end
96
97    % Integrate
98    x    = RungeKutta(@RHSAutomobile, 0, x, dT, d );
99  end
```

Figure 14.5 shows four points in the passing sequence.

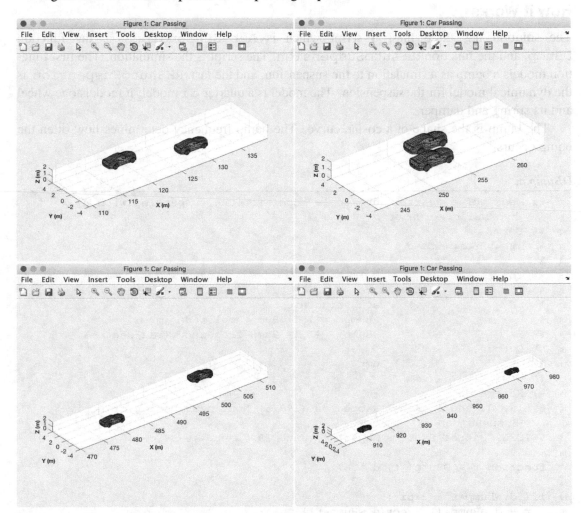

Figure 14.5: *Automobile simulation snapshots showing passing.*

14.5   Modeling an Automobile Suspension

Problem

The dynamics of an automobile suspension are important to the performance of an automobile.

Solution

Build a dynamical model of a suspension.

How It Works

The solution has three files, a script which is AutoSuspensionSim, the function DBump, and the functionRHSAutoSuspension. The script is the simulation. The first function models a bump as a simulation to the suspension, and the last, RHSAutoSuspension, is the dynamical model for the suspension. The model is a quarter car model; it models one wheel and its spring and damper.

The bump is the shape of a cosine curve. The bump frequency determines how often the bump occurs.

DBump.m

```
1  %% Model of a quarter automobile model for the suspension with a
        hydraulic actuator
2  %% Form:
3  %    r = DBump( t, d )
4  %
5  %% Inputs
6  %    t              (1,1)  Time
7  %    d              (.)    data structure
8  %                          .aBump    (1,1)  Bump amplitude (m)
9  %                          .wBump    (1,1)  Bump frequency (rad/sec)
10 %% Outputs
11 %    r              (1,1)  Bump
12 %
13 %% Reference:
14 % Lin, J. and I. Kanellakopoulos (1997.) Nonlinear Design of Active
        Suspensions.
15 % IEEE Control Systems Magazine, June 1997. pp. 45-59.
16
17 function r = DBump( t, d )
18
19 if( d.wBump*t < 2*pi )
20   r = d.aBump*(1 - cos(d.wBump*t));
21 end
```

The suspension is shown in Figure 14.6.

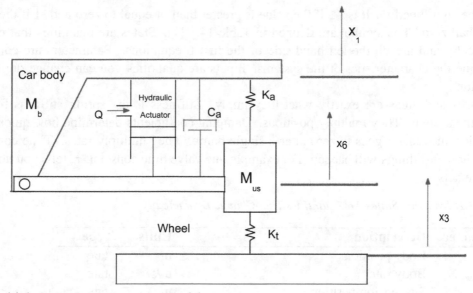

Figure 14.6: *Quarter car suspension model.*

The equations for the suspension are as follows:

$$\dot{x}_1 = x_2 \tag{14.19}$$

$$\dot{x}_2 = -\frac{1}{M_b}\left(K_a(x_1 - x_3) + C_a(x_2 - x_4) - AX_5\right) \tag{14.20}$$

$$\dot{x}_3 = x_4 \tag{14.21}$$

$$\dot{x}_4 = -\frac{1}{M_{us}}\left(K_a(x_1 - x_3) + C_a(x_2 - x_4) - AX_5 - K(x_3 - r)\right) \tag{14.22}$$

$$\dot{x}_5 = -\beta x_5 - \alpha A(x_2 - x_4) + \gamma x_6 w_3 \tag{14.23}$$

$$\dot{x}_6 = \frac{1}{\tau}(u - x_6) \tag{14.24}$$

$$w_3 = \operatorname{sgn}(P_s - \operatorname{sgn}(x_6)x_5)\sqrt{|P_s - \operatorname{sgn}(x_6)x_5|} \tag{14.25}$$

$$\alpha = \frac{4\beta_e}{V} \tag{14.26}$$

$$\beta = \alpha C_{tp} \tag{14.27}$$

$$\gamma = \alpha C_d w\sqrt{\frac{1}{\rho}} \tag{14.28}$$

$$Q = C_d w x_6\sqrt{\frac{1}{\rho}\left(P_s - \operatorname{sgn}(x_6)x_5\right)} \tag{14.29}$$

sgn is the sign function. It is +1 if the value is greater than or equal to zero and -1 if the value is less than zero. The symbol are defined in Table 14.1 [1]. States are quantities that change dynamically and are on the left-hand-side of the first 6 equations. Parameters are constants that define the characteristics of the systems. Inputs are quantities you can change during the simulation.

Masses and areas are exactly what they imply. Stiffness is the spring stiffness for that part of the system. They multiply positions. Damping coefficients determine how quickly the suspension movement goes to zero after a single bump. They multiply rates. Time constants tell you how fast things will happen. For example, the valve time constant, τ, tells you how fast the valve acts.

Table 14.1: *Symbols. States Are Quantities That Change Dynamically*

Parameter	Description	Units	Type
x_1	Body position	m	state
x_2	Body rate	m/s	state
x_3	Suspension position	m	state
x_4	Suspension rate	m/s	state
x_5	Pressure drop across the hydraulic piston	Pa	state
x_6	Valve displacement	m	state
M_b	Body mass	kg	constant
K_a	Suspension stiffness	N/m	constant
C_a	Suspension damping	N/m/s	constant
K_t	Tire stiffness	N/m	constant
τ	Valve time constant	s	constant
M_{us}	Suspension mass	kg	constant
β	Time constant	s	constant
A	Piston area	m^2	constant
w	Spool valve area	m^2	constant
P_s	Supply pressure	m^2	constant
Q	Hydraulic load flow	m^3/s	constant
C_{tp}	Total leakage coefficient	N/m^5/s	constant
V_t	Total actuator volume	m^3	constant
C_d	Discharge coefficient	kg	constant
β_s	Effective bulk modulus	N/m^2	constant
u	Control	N/m^2	input

The right-hand side returns a default data structure if it is not given inputs. This is a convenient way for the user to get a reasonable set of parameters for the model. The bump function is called from within the function. If you wanted a different disturbance, you would change this code:

RHSAutoSuspension.m

```
44   function xDot = RHSAutoSuspension( x, t, d )
45
46   if( nargin < 1 )
47     xDot = DefaultDataStructure;
48     return
49   end
50
51   f     = d.pS - sign(x(6))*x(5);
52   w3    = sign(f)*sqrt(abs(f));
53   r     = DBump( t, d );
54   mu    = 1e7;
55
56   x24   = x(2) - x(4);
57   x13   = x(1) - x(3);
58   aX5   = d.a*mu*x(5);
59
60   xDot = [ x(2);...
61          -(d.kA*x13 + d.cA*x24 - aX5)/d.mB;...
62            x(4);...
63           (d.kA*x13 + d.cA*x24 - d.kT*(x(3) - r) - aX5)/d.mUS;...
64           -d.beta*x(5) - d.alpha*d.a*x24/mu + d.gamma*x(6)*w3/mu;...
65          (d.u - x(6))/d.tau];
66
67   function d = DefaultDataStructure
68
69     %% Automobile parameters
70   d.mB    =    290; % Body mass (kg)
71   d.mUS   =     59; % Wheel mass (kg)
72   d.kA    =  16812; % Spring constant (N/m)
73   d.cA    =   1000; % Damping constant (N/(m/sec))
74   d.kT    = 190000; % Tire spring constant (N/m)
75
76   %% Hydraulic actuator parameters
77   d.alpha = 4.515e13; % N/m^5
78   d.beta  = 1; % alpha times piston leakage coefficient (1/s)
79   d.gamma = 1.545e9; % N/(m^5/2 kg^1/2)
80   d.tau   = 1/30; % Spool valve time constant (s)
81   d.pS    = 10342500; % Supply pressure (Pa)
82   d.a     = 3.35e-4; % Piston area (m^2)
83   d.u     = 0;
84
```

```
85   %% Bump disturbance
86   d.aBump = 0.025; % Bump amplitude (m)
87   d.wBump = 8*pi;   % Bump frequency (rad/sec)
88
89   d.states = { 'body disp' 'body rate' 'wheel disp' 'wheel rate' ...
90                'pressure drop' 'spool valve'};
```

An alternative would be to add a function pointer to your disturbance function like d.fun = @DBump. You would need to predefine the arguments. That is

```
r    = feval{d.fun, t, d )
```

The simulation gets the default data structure and runs a simulation. It simulates a quarter automobile model for the suspension with a hydraulic actuator. The automobile parameters and actuator parameters are defined, and a bump disturbance is considered. The natural motion is simulated for three seconds (no control), and the results for the autosuspension and hydraulic states are displayed.

AutoSuspensionSim.m

```
1    %% Simulation of an automobile suspension
2    % Simulates a quarter automobile model for the suspension with a
        hydraulic
3    % actuator. The automobile parameters and actuator parameters are
        defined
4    % and a bump disturbance is considered. The natural motion is simulated
5    % for three seconds (no control) and the results for the auto
        suspension
6    % and hydraulic states are displayed.
7    %% Reference
8    % Lin, J. and I. Kanellakopoulos (1997.) Nonlinear Design of
9    %       Active Suspensions. IEEE Control Systems Magazine, June 1997.
        pp. 45-59
10
11   %% Automobile parameters
12   d = RHSAutoSuspension;
13
14   %% State
15   % Form: [car body displacement; car body rate; wheel displacement;...
16   %        wheel rate; pressure drop across the piston;...
17   %        spool valve displacement]
18   x    = [0;0;0;0;0;0];
19   t    = 0;
20
21   %% Number of sim steps
22   tEnd    = 3;
23   nSim    = 2000;
```

```
24  dT      = tEnd/(nSim-1);
25
26  %% Plotting arrays
27  xP      = zeros(7,nSim);
28
29  %% Run the simulation
30  for k = 1:nSim
31    x                 = RK4( 'RHSAutoSuspension', x, dT, t, d );
32    t                 = t + dT;
33    xP(:,k)           = [x;DBump( t, d )];
34  end
35
36  %% Plot results
37  xP      = xP(:,1:k);
38  [t,tL]  = TimeLabel((0:k-1)*dT);
39  k1      = [1:4 7];
40  k2      = 5:6;
41  yL      = [d.states(:)' {'bump'}];
42
43  PlotSet( t, xP(k1,:),'x label',tL,'y label',yL(k1),'figure title','
        Suspension States');
44  PlotSet( t, xP(k2,:),'x label',tL,'y label',yL(k2),'figure title','
        Hydraulic States');
```

Figure 14.7 shows the suspension response.

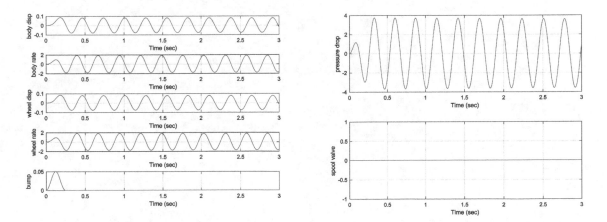

Figure 14.7: *Automobile suspension response. The left plot is the states. The right is a plot of the hydraulic states.*

405

Summary

This chapter has demonstrated how to use MATLAB to simulate automobile dynamics in 2D, automobile maneuvers, and suspensions. We also show how to animate an automobile. Table 14.2 lists the code developed in the chapter.

Table 14.2: *Chapter Code Listing*

File	Description
AutomobileDemo	Script to demonstrate automobile control
AutomobilePassing	Automobile passing controller
AutomobileLaneChange	Automobile land changing controller
AutoRadar	Automobile radar model
AutoSuspensionSim	Simulation of an automobile suspension
DBump	Bump model for the suspension simulation
RHSAutoSuspension	Dynamical model of an automobile suspension

Bibliography

[1] Lin, J. and I. Kanellakopoulos. Nonlinear Design of Active Suspensions. *IEEE Control Systems Magazine*, pages 45–59, June 1997.

[2] Matthew G. Villella. *Nonlinear Modeling and Control of Automobiles with Dynamic Wheel-Road Friction and Wheel Torque Inputs*. PhD thesis, Georgia Institute of Technology, April 2004.

© Michael Paluszek and Stephanie Thomas 2020
M. Paluszek and S. Thomas, *MATLAB Recipes*,
https://doi.org/10.1007/978-1-4842-6124-8

Index

A

Aircraft
    design control system, 340–342
    dynamical model
        atmospheric density model, 331
        dynamic pressure, 329
        flight path, 328
        MKS units, 329
        RHSAircraft, 329
    equilibrium controls
        cost function, 333
        display option, 336
        fminsearch, 332–334, 340
        optimplotfval function, 337, 338
        plot function, 339
        PlotIteration function, 337
    Gulfstream.obj file, 351
    3D trajectory plot, 342–346
    three dimensions, 351–356
    trajectory controls, 346–351
    velocity and flight path angle, 348
    Wavefront OBJ format, 351
Air turbine model, fault detection
    dynamical model, 278
    failed regulator, 286
    linear model, 277, 280
    tachometer failure, 287
AnnotatePlot, 107–108
ArgCheckFun, 80
assert function, 177
Automobiles
    animation, 395–399
    autonomous passing control, 393–395

AutoSuspensionSim
    bump frequency, 400
    bump function, 403
    DBump function, 400
    hydraulic actuator, 404
    quarter car model, 400, 401
    RHSAutoSuspension function,
        400
    simulation snapshots, 399
    stiffness, 402
    suspension response, 405
car dynamics model
    aerodynamic pressure, 385
    derivative array, 389
    dynamical equations, 385
    gravitational force, 385
    kinematical equation, 386
    for loop cycles, 387
    planar dynamical model, 384
    RHSAutomobile, 383, 386
    RungeKutta, 383
    six states, 383, 384
Phong reflection model, 397
radar model
    AutoRadar function, 390
    default data structure, 392
    LIDAR, 392
    signal-to-noise ratio, 389
    target weaving, 392, 393

B

Bluetooth devices, 168
BoxPatch, 124

© Michael Paluszek and Stephanie Thomas 2020
M. Paluszek and S. Thomas, *MATLAB Recipes*,
https://doi.org/10.1007/978-1-4842-6124-8

C

C/ C++, 4
calibrate, 172
Callbacks tab, 160–161
Car dynamics model
   aerodynamic pressure, 385
   derivative array, 389
   dynamical equations, 385
   gravitational force, 385
   kinematical equation, 386
   for loop cycles, 387
   planar dynamical model, 384
   RHSAutomobile, 383, 386
   RungeKutta, 383
   six states, 383, 384
catColorMapper, 30
Categorical arrays, 23, 24
Cell arrays, 11
Chemical process
   effluent pH control, 311–312
   mixing process
      default values, 302
      equilibrium point, 303
      reaction invariants, 300
      right-hand-side function, 299, 301
      RungeKutta integrator, 301
      volumetric flow rates, 300
   PHProcessSim
      acid, base and buffer streams, 313, 316
      block diagram, 314
      closed loop response, 316, 317
      controlIsOn, 313
      controller performance, 319, 320
      numerical optimization, 323
      open loop response, 318
      perturbation results, 321, 322
      profiler results, 324
      profiler summary, 323
      RungeKutta integrator, 315
   pH sensor
      display option, 307
      fminbnd and fminsearch, 309

function evaluations, 310
fzero, 304, 307
MATLAB fzero function, 303
mesh function, 305, 306
optimset, 308, 309
surf and mesh plots, 306
proportional-integral controller, 311–312
Classes, 25
   constructor, 204
   continuous subclass of StateSpace,
      207–209
   DensityModel, 209, 210
   discrete state space systems, 206–207
   double integrator, 205–206
   mocking framework, 209–211
   object-oriented programming, 199–201
   state space systems, 201–206
   stubDensity, 210
   unittest framework, 210
   using LAPACK, 199
Colon operator, 14–16
ColorDistribution, 118–120
CompleteTriangle, 177, 178
Custom GUI
   App Designer interface, 159
   blank GUI, 160
   box animation app, 165
   Callbacks tab, 160–161
   Code View, 161
   complete code, 164
   "Design App" button, 158
   start button code, 162–163
   stopAnimation, 161–164
   3D commands, 161

D

Data acquisition GUI, 166
   Bluetooth devices, 168
   calibrate, 172
   calibration button, 169
   callback, 166
   Clear, 172

GUIPlots, 167–168
modal dialog, 172
questdlg, 171, 172
SaveFile, 173
uicontrol, 170–171
variable set, 169–170
Datastores, 20–21
datetick, 117
2D/3D grids, 120–123
DebugLog, 189–191
Deep Learning algorithm, 173
Dot indexing, 82
Double integrator
closed loop response, 228, 229
control parameters, 225
CToDZOH function, 223, 224
DefaultStruct function, 223
digital control, 225–229
DoubleIntegratorSim, 225, 226
figures creation, 230–233
fourth-order Runge-Kutta integrator, 217–218
numerical integration, 216–219
open loop response, 226, 227
PDControl design, 221–222, 226
pole placement, 220
proportional-derivative control, 219–225
RHS function, 216
second-order system, 215–216
time axes creation, 229
Draftmark, 141
Drag test, 209

E
Euler equations, 360, 364
Explicit expansion, 55

F
Fast Fourier Transform (FFT), 31–33
Fault detection
air turbine model, 277–280
dynamical model, 278
failed regulator, 286

linear model, 277, 280
tachometer failure, 287
filter detection, 280–283
GUI
appdesigner interface, 287–289
app details, 292
blank template, 290
debugger, 295
DFGUI.mlapp, 288
execution, 296
failed regulator, 286
tachometer failure, 287
pressure regulator/tachometerfailure, 284–287
varargin function, 281
Field-oriented control (FOC), 258–261
Figures creation, multiple subplots, 230–233
FORTRAN, 4
Function handles, 17

G
getnext functions, 30
GNU Linear Programming Kit (GLPK), 47–50
Graphical user interface (GUI), fault detection
appdesigner interface, 287–289
app details, 292
blank template, 290
debugger, 295
DFGUI.mlapp, 288
execution, 296
failed regulator, 286
tachometer failure, 287
GUIPlots, 167–168

H
hasnext functions, 30
heatmap function, 112–113

I
if statements, 70–71
Images, 18–20
Integrating toolbox documentation

custom toolbox, 96
Display Custom Documentation, 93
Display Custom Examples, 93
features.html, 95
function categories.html, 95
helptoc.xml, 94–95
info.xml file, 93–94
MagneticControlDemo, 96–97
search database, 97
Supplemental Software, 93
writing HTML documentation, 93
Interactive graphics
  animate line objects
    quad plot, 150–152
    QuadPlot.m function, 149
    uimenu, 148
    video player, 149
  create custom GUI
    App Designer interface, 159
    blank GUI, 160
    box animation app, 160
    Callbacks tab, 160–161
    Code View, 160
    complete code, 164
    "Design App" button, 158
    start button code, 162–163
    stopAnimation, 161–164
    3D commands, 161
  creating simple animation
    PatchAnimation, 147
    pause command, 145
    rotating box, 144–145
    Run Section toolbar button, 146
    updating patch objects, 143
  data acquisition GUI, 166
    Bluetooth devices, 168
    calibrate, 172
    calibration button, 169
    callback, 166
    Clear, 172
    GUIPlots, 167–168
    modal dialog, 172

    questdlg, 171
    SaveFile, 173
    uicontrol, 170–171
    variable set, 169–170
  display status of running simulation,
    155–158
  playing back, 147–148
  uicontrol button, 152–155

J, K
Java, 4
JSON-formatted strings, 39–40

L
Linear plot, 103
Logical arrays, 12

M
Map containers, 25
MAT-file, 23–25
MathWorks, 3
MATLAB code coverage tools, 176
MATLAB Command-Line Help, 85–88
MATLAB debugger, 160
MATLAB errors and warnings, 191–192
MATLAB Language Primer
  application development window, 4
  cell arrays, 11
  character arrays, 8
  colon operator, 14–16
  Command History, 4
  command-line file interaction, 45–47
  Command Window, 4, 5
  creating documents, 61–63
  creating function help, 40–42
  Current Folder, 4
  data storage, locating directories for,
    42–43
  datastores, 20–21
  data types, 25
  explicit expansion, 55
  FFT function, 29–31
  File details, 4

function handles, 17
help strings, 9
images, 18–20
initializing data structure, 25–29
JSON-formatted strings, 39, 40
key functions, 8
linear algebra libraries, 3
loading binary data, 43–45
logical arrays, 12
mapreduce, 29–31
MAT-file, 23–25
MATLAB Online, 64–67
memoize function, 58–59
MEX file link to external library, 47–50
multiple inputs and outputs, 16–17
non-english strings, 38
numerics, 18
persistent and global scope, 12–14
PLOTS and APPS tabs, 4
Profiler, 4
protect IP with parsed files, 50–51
PUBLISH and VIEW tabs, 5
script subfunction, 56–58
searchable help documentation, 4, 7
single index, 6
Sparse matrices, 22, 23
spreadsheet-like graphical editor, 4, 6
strict data structures, 9–10
string concatenation, 37
strings arrays, 37, 38
substrings, 39
table data processing, 33–36
tables and categoricals, 23
Tall arrays, 22
third-party toolboxes, 5
tool for timing code, 4
using Java, 59–60
Workspace display, 4
writing to text file, 51–54
MATLAB "Mortgage" demonstration app., 158
MATLAB Online, 64–67
MATLAB style

adding built-in inputs and outputs, 78–80
adding dot indexing, 82
"built-in" inputs and outputs, 69
guidelines
    align blocks of code, 71
    if statements, 70–71
    MyFancyDataStructure, 70
    naming conventions, 72–74
    tab preferences, 72
    tab sizes, 70, 72
    variable value, 70
integrating toolbox documentation (*see*
    Integrating toolbox documentation)
MATLAB Command-Line Help, 85–88
overloading functions, 77–78
publishing code, 88–92
smart structuring of scripts, 82–85
structuring toolbox, 98–99
writing function help, 74–76
MATLAB-style string, 43
matlab.unittest package, 177
MATLAB variables, 175
memoize function, 58–59
Mocking framework, 209–211

N, O
Nelder-Mead minimizer, 243
Non-english strings, 38

P, Q
PatchAnimation, 147, 155
PatchAnimationStorage.m., 147
patch function, 123
PLOTS and APPS tabs, 4, 102
Proportional-integral differential (PID)
    controller, 227
PUBLISH and VIEW tabs, 5
Publishing code, 88–92
Python, 4

R
Radar model
    AutoRadar function, 390

default data structure, 392
 LIDAR, 392
 signal-to-noise ratio, 389
 target weaving, 392
runtests function, 176

S

SCARA robot
 arm moves., 251
 box function, 243
 controlled robot, 248–251
 control system, 246–248
 dynamical model creation, 235–238
 equations of motion, 236
 fminsearch function, 244
 initalize function, 241
 Nelder-Mead minimizer, 243
 numerical search, 243–245
 patch function, 239, 241
 PD controller, 246
 RHS function, 237, 238
 torque control, 246
 transient response, 250
 update function, 242
 VideoWriter object, 245
 visualization function, 238–243
Script subfunction, 56–58
selectIf, 186
setupOnce function, 182–183
Spacecraft attitude control
 dynamical model creation
  damping coefficient, 362
  Euler equations, 360
  inertial angular momentum, 363
  inertia matrix, 360
  for kinematics, 361
  omegaDotCore, 363
  omegaRWA variable, 362
  reaction wheel, 360
  skew matrix, 361
 quaternions, angle errors
  control system, 365

Euler equation, 364
 proportional-derivative controllers, 364
 QMult, 365
 QPose, 365
simulation script
 control torque, 376
 momentum conservation test, 377
 Monte Carlo simulation, 373, 374
 nonzero angle errors, 379
SpacecraftSim
 angular momentum, 367
 applied external torque, 371
 control acceleration, 367
 controlIsOn flag to false, 366
 fourth-order Runge-Kutta, 366
 momentum exchange, 373
 PD control method, 370
 PlotSet, 369
 test results., 366
Space vector modulation, 270–275
Sparse matrices, 22, 23
StateSpace class, 206–207
String concatenation, 37
Strings arrays, 37–38
stubDensity class, 210
Substrings, 39–40

T

table class, 180–181
Tables, 23
tabularTextDatastore, 34, 35
Tall arrays, 22
TestCase, 177
TestCase class, 181–182
Testing
 add custom warning, 176
 custom errors and warnings, 192–196
 generation of figures, 196–197
 inverse cosine, 176
 logging function, 189–191
 running test suite, 184–186
 setting verbosity levels, 186–188

unit test
    assert function, 177
    CompleteTriangle, 177
    Coverage Report, 184
    equilateral triangle test, 179
    matlab.unittest package, 177
    runtests function, 176
    setupOnce function, 182–183
    table class, 180–181
    TestCase, 177
    TestCase class, 181–182
    TriangleTest, 178
    triangle tests, 183
    trigonometric functions, 180
  verbosity level, 188
  warning and error, 191–192
TestSuite class, 185
Three-phase permanent magnet motor
  electrical torque, 257, 260
  field-oriented control, 258–261
  first-order differential equations, 255
  motor controller, 254
  mutual inductance, 255
  PI torque control, 273
  pulsewidth modulation, 262–269
  space vector modulation, 270–275
  three-phase coils, 254
  three-phase driver circuitry, 256
  voltage pulsewidths, 274
Time axes creation, 229–230
TimeDisplayGUI, 155
Time series, 25
TriangleTest, 178
Triangle tests, 183

U
uicontrol button, 152–155
UIControlDemo.m., 153

V
varargin and varargout, 77–78
verifyWarning function, 192

Visualization
  adding watermark
    company watermark, 139–140
    draft watermark, 140–142
    low-level graphics functions, 138
  additional interactive buttons, 107
  AnnotatePlot, 107–108
  camera properties, 129–132
  Catalog button, 103
  color distribution, 118–120
  command-line help, 101
  create heat map, 112–113
  custom plot page, 109–112
  custom-sized axes, 113–115
  display image, 132–134
  Generate Code command, 104
  generate 3D objects, 123–125
  graph and digraph, 135–138
  light objects
    command line get, 127
    light function, 125
    PatchWithLighting, 125–126
    Shiny box, 127–128
    SphereLighting, 128–129
  Linear plot, 103
  parametric area plot, 104
  photo-realistic rendering, 101
  Plot Edit toolbar, 104
  plot of trigonometric data, 104–106
  PLOTS tab, 102
  plotting with dates, 115–118
  subplot function, 106
  3D plots, 101
  three-dimensional visualization, 101
  trigonometric functions, 102
  2D/3D grids, 120–123
  2D plots, 101

W
WeatherFFT, 35